WITHDRAWN

D0772551

FOREIGN BODIES

Also by Simon Schama

HISTORY AND ART HISTORY

Patriots and Liberators: Revolution in the Netherlands

Two Rothschilds and the Land of Israel

The Embarrassment of Riches: An Interpretation of Dutch Culture in the Golden Age

Citizens: A Chronicle of the French Revolution

Landscape and Memory

Rembrandt's Eyes

A History of Britain, Volume I: At the Edge of the World?

A History of Britain, Volume II: The British Wars

A History of Britain, Volume III: The Fate of Empire

Rough Crossings: Britain, the Slaves and the American Revolution

The Power of Art

The American Future: A History

The Story of the Jews, Volume I: Finding the Words

The Story of the Jews, Volume II: Belonging

The Face of Britain

ESSAYS

Hang-Ups: Essays on Painting (Mostly)

Scribble, Scribble, Scribble: Writing on Politics, Ice Cream, Churchill and My Mother

Wordy: Sounding Off on High Art, Low Appetite and the Power of Memory

FICTION

Dead Certainties: Unwarranted Speculations

THEATRE

Rough Crossings (with Caryl Phillips)

Decade

FOREIGN BODIES

PANDEMICS, VACCINES, AND THE HEALTH OF NATIONS

SIMON SCHAMA

An Imprint of HarperCollins*Publishers*

HarperCollins books may be purchased for educational, business, or sales promotional use. For information, please email the Special Markets Department at SPsales@harpercollins.com.

Originally published in Great Britain in 2023 by Simon & Schuster UK Ltd.

FIRST U.S. EDITION

Library of Congress Cataloging-in-Publication Data has been applied for.

ISBN 978-1-328-97483-9

23 24 25 26 27 LBC 5 4 3 2 1

For Ginny, without whom
this could never have been written

‘A human being is a part of the whole, called by us “Universe”, a part limited in time and space. He experiences himself, his thoughts and feelings as something separated from the rest – a kind of optical delusion of his consciousness. The striving to free oneself from this delusion is the one issue of true religion. Not to nourish the delusion but to try to overcome it is the way to reach the attainable measure of peace of mind’

Albert Einstein to Rabbi Dr Robert S. Marcus,
12 February 1950

Contents

Prologue

In the end, all history is natural history.

Old Pliny knew as much, well before his last posting to the Bay of Naples, beneath the volcano that would kill him. His nephew managed to compile thirty six volumes of *The Natural History* from his uncle's encyclopaedic notes and it was still not enough. But Pliny's data-glut was more than sufficient to make the point that biology and ecology, and the play between them, are the ultimate shapers of human destiny. For the ancients, mischief of the gods notwithstanding, this was a truism. Aristotle – zoologist as well as philosopher – would not have disagreed.

At this late point in the flash in the pan that is the paltry ten millennia of human civilisation, we have returned to this chastening truth: that the matter filling million upon millions of pages of recorded history – wars and revolutions, the rise and fall of cities and empires, fevers of faith, the heaping up and the emptying out of wealth – has been circumscribed by what we have done to nature and what it has done to us.

It would, of course, be absurd to discount the transformative power of human ingenuity. But the acme of its achievement, the *natural* sciences as they have long been called, have revealed almost unimaginable powers of genetically amending the terms of human life while at the same time making startlingly apparent the sobering limits of human agency. Bio-ecological imperatives, rather than the emperors of construction and destruction, are our true rulers. And science rather than military hardware is our best defence. Perhaps it is chagrin at the realisation that the best-laid plans of mice and monsters are so many vanity projects compared to the entropy of the habitable planet, or the eruption of pandemics, that makes for reluctance to describe those existential crises in anything but the stale vocabulary of political and military history. Diseases are invaders; measures to deal with them a plot; bacteriologists and epidemiologists an alien elite, the microbe and the scientist in cahoots against homespun wisdom. The health of the world contracts into the health of nations, even when the latter cannot be sustained without ensuring the former.

Much madness has arisen from this ongoing drama of false consciousness; and many are the perils arising from its obstinate perpetuation. What follows are scenes from this late-period episode of the human comedy. And like much comedy, they could not be more serious.

i

Et in suburbia ego

> Thus it came about that oxen, asses, sheep, goats, pigs, chickens and even dogs (for all their fidelity to man) were driven away and allowed to roam freely through the fields where the crops lay abandoned and had not even been reaped let alone gathered in. And after a whole day's feasting many of these animals, as though possessing the power of reason, would return glutted in the evening to their own quarters, without any shepherd to guide them . . .
>
> Giovanni Boccaccio, *Decameron*, First Day

In March 2021, the thirteenth month of the COVID confinement, the peepers, in their vast multitudes, sang out again. Down in the swampy wetlands below our house in the Hudson valley, millions of *Pseudacris crucifer* ('Cross-bearing false locusts' but actually minute frogs) puffed up their air sacs and warbled for a mate. That's spring for you. The peepers are so tiny – an inch or so long – that you'll never see one, no matter how carefully you creep up on them. Their blown-out song bags are nearly as big as the rest of them; it's all they are: innocently inflated peeps of expectation.

They are not alone. In recent years the soprano peepers have been accompanied by a bass rhythm section – wood frogs,

Lithobates sylvaticus, a tattoo of deep quacking, punctuated by raspy burps. They and the peepers survive bitter winters by means of anti-freeze cryoprotectants stored within their bodies. When ice crystals begin to form on their skins, their livers flood the bloodstream with glucose, sending vital organs like the heart, its beating paused, into a dormant but protected state. Seventy per cent of the frogs' body water can then freeze without compromising the organs that will magically reawaken in the spring.[1] To help matters, wood frogs can recycle urea through their urine. So if you were to come across a wood frog in deep winter, or expose a tiny peeper beneath the leaf litter, their sparkling, gelid rigidity would lead you to assume they were dead. A twist from your fingers could snap a leg. So don't do that, for as the Hudson valley light goes pearly and the afternoons stretch out, the superficial body ice of the frogs melts away and, along with that decrystallising resurrection, wild singing begins: at first a mere tea-time tuning up by scattered vocalists, but by sunset building into a massive chorus, an entire Albert Hall-ful of peepers. There is always mating business to be getting on with and only a month or so to get it done. Quick, quack, peep. The teeming amphibians ecstatically multiplied, even as much of humanity sank into another engulfing wave of infection.

It's a commonplace (but no less true for being so) that the empty desolation of cities, the grim, still, silence of locked-down streets and squares, was offset by the irrepressible burgeoning of nature. We saw it – the budding and blossoming, the buzzing and butterfly-fluttering on our walks in parks and on heaths, in our gardens and on windowsills. The cheek of nature, impervious to our fearful distress, flaunting the sociability of its flocks, gaggles and mysteriously choreographed starling murmurations: the avian corps de ballet. For a while, back then, the grounding of jets and the stilling of traffic scoured away the rusty scrim of pollution from the skies. Avians replaced aviation. Ruby-throated hummingbirds were seen in unprecedented numbers around Kennedy airport, their tiny wings, beating fifty times a second, more miraculously engineered than anything attached to aircraft. Children in vast conurbations –

Beijing, Mumbai, São Paulo, Los Angeles – few of whom had ever seen a truly blue sky, now craned their necks to a clarified space hitherto glimpsed only in storybooks and video cartoons. At night, with ambient light dimmed and road traffic muted, stars glinted, needle sharp. The Milky Way sloshed across our field of vision. While we hunkered and cowered, and ordered home delivery, flora rioted; fauna trespassed. Parliaments of legislators were reduced to socially distanced barking from the hollow shell of their chambers, while parliaments of birds flocked and chattered. We tweeted with our fingers; they tweeted with their lungs. Those with the sweetest song showed off, none more liquidly around here, in the Hudson valley, than the Carolina wren nesting under our barbecue. The more we retreated into digitally numbed companionship, the more brazenly the company of animals advanced towards us. Coyotes raved in the midnight backyard. A weedy ditch by the road became the habitat of a family of water voles, pups slickly backflipping at each end of the conduit. One early summer morning, I chased a chipmunk round the house before dislodging the speedy trespasser from behind the television. Reckless possums crossed the roads at night, though their habit of playing dead before the odd oncoming car was often a prelude to being actually dead. Morning roadkill was evidence of nocturnal roamings and ramblings by hitherto seldom-seen critters. On the path leading to a local arts centre, weasels and milk snakes lay side by side, cartoonishly flattened, as if mutually KO'd in a small-hours brawl. At the entrance to our local woodland trail, a sign advised walkers not to make nice with the black bears. Everything, except us, seemed to be emboldened. Reporting on record fox sightings in her north London neighbourhood, a friend chuckled, 'It's laughing at us, nature is.' So it was: the low chuckle of gallows humour.

But the joke's on us. Things are amiss. Species are out of place, or incautiously testing human presumptions about where their place actually lies, and what its boundaries might be. Lockdowns or not, migrants, two legged and four, are on the move to wherever subsistence beckons. The domain of the wild now includes urban gardens, parks and alleyways. Even before the pandemic shut the doors on residents, wild boar from the woody slopes of Mount Carmel were seen on the streets of Haifa. But since the pandemic, the pigs – off the menu for both Muslims and Jews and thus unconcerned by hunting predators – have been regularly grazing central reservations, trotting over road crossings with a pause to drop dung, and waking sleepers with their snotty snuffling as they tear and munch on weedy backyards. Regular fights with pet dogs often end with Lulu the Pomeranian or Yossi the Labradoodle the worse for wear. For once, a Haifa schoolboy's claim that a pig had eaten his homework (along with his sweatshirt and a final slice of Margherita pizza) turned out to be true.

In north Wales, mountain goats from the Great Orme munching potted petunias off Llandudno windowsills supplied much-needed online entertainment. But the crashing of barriers between wild and domestic spaces has an ominous side. Displacement is a symptom of ecosystems under stress. The capybara roaming through the upscale gardens of houses in Nordelta on the outskirts of Buenos Aires would not be there had not the suburb been built by draining extensive areas of the Lujan River delta, robbing the three-foot-long rodents of their natural habitat. The relentless growth of Mumbai – a million new residents a year – has pushed its eastern and western suburbs into areas normally reserved for leopards, specifically the hundred square kilometres of the Sanjay Gandhi Sanctuary. Deprived of prey, the big cats have strayed beyond the preserve. At least fifty of them have taken up residence within the city, sustaining themselves from the enormous population of feral dogs, occasionally sampling an amuse-bouche of a dachshund or a Siamese cat.

When a small herd of elephants broke out of their reserve in south-west China in April 2020, and were video-tracked lumbering through a car dealership and scouting kitchens by probing through open windows with their trunks, they became an online sensation, described by one transfixed fan as 'magical'. But this was not circus time; it was a symptom of something gone awry. The causes and consequences of this ecological disruption are complicated. On the one hand, it's not good for the leopards to become Mumbai street creatures; on the other, they are doing the bloated metropolis a favour by culling the feral dog packs, which often include rabid animals. But then again there would not be so many of those wild dogs were it not for the introduction, a decade ago, of diclofenac, an anti-inflammatory drug commonly used for livestock in the 1990s, which ended up driving the third player in this urban drama – white-rumped vultures – to near extinction as a result of scavenging drugged cattle. A south Asian vulture population of 40 million in the 1980s now numbers around 19,000 forty years later. This is more than a catastrophic species loss, bad enough though that is. The dramatic depletion of vultures has unpicked the ecological threads that have tied human and animal culture together in India for centuries. The reverent freedom given to sacred cows by Hinduism, so that they might wander the streets until their bodies lie down in peaceful death, depended on the working assumption that carcasses would be cleaned by scavenging vulture flocks. Parsi sky burials, with the human deceased set on stone slabs for vulture cleansing, have likewise been affected, to the point at which that community are raising birds specifically to service their ritual. Without the vultures in Hindu cities, decomposing cattle have attracted rats and feral dogs, whose numbers have increased exponentially as the birds have disappeared. A collateral result is the steeply rising incidence of rabid attacks on humans, many of them fatal.

Mutuality between humans and animals has been dangerously disrupted. Temple monkeys, long conditioned to exist symbiotically with humans and largely dependent on pilgrims and tourists for their food, turned combative as a result of the abrupt withdrawal of their customary diet. In March 2021, the Thai temple city of Lopburi saw gangs of macaques, in their thousands, engage in violent street battles over scraps of discarded food while residents barricaded themselves in their houses against the rampaging primates. There is good reason for their fear. Macaques are reservoir carriers of herpes B – McHV1 – often lethal for humans.

Et in suburbia ego. Disruption-born contagions are happening in domestic as well as exotic places. A serious malady generated by ecological displacement arrived almost fifty years ago in the United States and parked itself on the vegetation of the American dream: the suburban lawn.[2] During my first year in New York State in 1994, it found me, and was no fun at all: three months of piercing headaches, spells of dizziness and sharp, arthritic muscle pain, before an antibiotic got the better of it. The infecting agent of Lyme disease (named after Old Lyme, Connecticut, where it was first diagnosed and analysed) is a corkscrew-shaped spirochaete found in white-footed mice and sometimes in other small mammals like chipmunks. Not only do those mice survive the excavation and shredding of the woodland habitat for house construction, they positively thrive on the alteration, overwintering in the suburban estates which have displaced their native habitat. The rodents function as reservoirs for the dormant but immanent spirochaete. Enter black-legged ticks, needing blood meals at each change in their life cycle, from larva to nymph to adult. A feed on the mice absorbs the spirochaete, which is then transferred to white-tailed deer, upon which the ticks lodge in huge numbers, especially on the ears and around the nose. The deer have themselves multiplied abundantly on the borderland between old forests and the herbicide-saturated, brilliantly lurid carpet lawns of 'colonial' McMansions. Suburbanites are accustomed to watching white-tailed deer emerge from their woodland cover to graze their shrubs or settle on lawn pasture. To compound all this, the spread of one disease, Lyme disease, has been accelerated by the social effects of another, COVID-19. During the pandemic, fear of urban contagiousness led to departures from cities by those who could afford to do so. But clearcutting for suburban construction to meet the quickened demand has brought new residents ever closer to those ubiquitous reservoirs of disease – white-footed mice. Even as house-dwellers awaited their next delivery of online-ordered groceries, black-legged ticks hung on the blades of those hyperfertilised lawns, primed for their next blood meal. Just how plentiful these biological hitchhikers are depends on the presence or absence of the natural predators of the mice, which in turn depends on the extent of woodland cover. Our house sits on a ridge above second-growth forest covering the drystone wall remains of early nineteenth-century livestock farms. That relatively undisturbed native woodland of tulip, hickory, beech, chestnut and red oak is dense enough to support red-tailed hawks and barred and great horned owls, which prey on mice and other small rodents hosting the spirochaete. But we owe the preservation of these woodlands and wetlands to the self-interested benevolence of a plutocratic dynasty whose colossal fortune was built in the nineteenth century from the intensive extraction and production of oil. The creation of a broad *cordon sanitaire* of protected country was not, of course, purely a matter of public-spirited philanthropy. Protecting Hudson River landscapes served to insulate the bucolic manorial estate of the Rockefellers from the displeasing encroachments of the motorised modern world they had done a fair bit to enable. But closing off land to all but walkers and horse-riders had the effect (and still does) of coopting the public in the conservation of a green barrier.

Even at the height of the 2020 lockdown, local families rambled the Rockefeller State Park Preserve, children hooting at the flocks of decorative sheep and glossy herds of pedigree cattle. The domestic animals are impeccably maintained by the Stone Barns Center for Food and Agriculture, committed to educating the public on the reconciliation of farming with sustainable environmental practices. Meat from the humanely raised, pasture-fed animals supplies the menu of the famous high-end restaurant located on the same land. Diners are fed edifying environmental sermons along with their exquisitely cooked meals before being handed the eye-watering bill. But this exemplary suburban pastoral with its deer-roamed glades, meadow openings, and teacher-farmers, gardeners and cooks is a far cry from the perilous hellscape burning up much of the planet from the Amazon basin to Indonesia. In the Hudson valley we have feel-good farming; elsewhere, the eco-apocalypse is most certainly now. In 2020, 12 per cent more of the Amazon rain forest was cleared by burning than in the previous year, the pall of smoke easily visible in satellite photography.

Global hamburger syndrome drives much of this devastation. Millions of acres of tropical rain forest are cleared to make way for cattle pasture or the production of crops like soy and rape processed for the needs of industrial feedlot fattening, so that burger chains will never run short of inexpensive ground meat, American breakfasts never want for crispy bacon. McDonald's alone purchases annually nearly 2 billion pounds of beef, packed from the carcasses of 7 million cattle.[3] This is why American Big Meat – companies like Tyson Foods and lobbyists like the National Cattlemen's Beef Association, on the defensive for methane emissions sent into the atmosphere by livestock and water polluted by the oceanic slurry of pig excreta – has spent millions of dollars over the past three decades attacking environmental reform. The last few intact temperate rain forests, such as the Tongass National Forest in south-east Alaska, are under pressure from mining companies to be opened for fossil fuel extraction. This is not an exclusively US problem. Between 17,400 and 26,400 hectares of Amazon rain forest are cut down every year to supply the Chinese market with Brazilian beef.[4] Many more carbon-absorbing forests across the globe have already been irrecoverably lost to the same industrial pressure. The well-known result is a catastrophic destruction of biodiversity which at current rates will send at least a million species into extinction by the end of the century.

Even more ominously, the industrial production of cheap meat has seen the massing of livestock in foodlots so crowded and engorged that mass infections can only be prevented by the pre-emptive application of antibiotics. This routine practice in turn engenders two terrifying perils. First, it invites the evolution of antibiotic-resistant strains of disease among the cattle, pigs and poultry; and second, it increases the likelihood that those diseases will make the jump from animal to human populations that will no longer be able to depend on antibiotics to fight off the infection. This bleak syndrome is no longer a possibility but a certainty.

All these dislocations have already reordered the relationship between the animal and human worlds, with dire consequences for both. Deprived of native habitat and the complex web of biological relationships required to sustain and reproduce themselves, wild animals have moved into the human worlds of tourist and shanty-town waste at the same time as swollen populations of impoverished cities have built halfway to meet them. The sight of feral animals – foxes, bears, wild cats, raccoons – rooting through discarded take-away rubbish has become an urban commonplace. Smart city rats now have big competition.

The shrinking of distance between wild and human habitats has also generated opportunities for another source of desperately needed income: the long-distance traffic in wild animals. In 2005, it was estimated that each year of the previous decade had seen the live trafficking of 40,000 primates, 640,000 reptiles, 4 million birds and 350 million fish, numbers which have almost certainly increased in the years since. In 2016, China's National Key Research and Development Programme estimated the value of wildlife trades for medical sales and food consumption at 520 billion yuan. While most proposals seeking to break the connection between the wildlife trade and zoonotic diseases argue for strict bans on the former, such measures will not be easy since a large proportion of the planet's poor, especially in tropical Africa, depends on 'bush meats' made available to those for whom Big Whoppers are well out of reach.[5] Some animal meats are also in demand at the other end of the market, both as exotic luxury food and as traditional medicine. Pangolins – scaly anteaters found in both sub-Saharan Africa and south-east Asia – are, since the enforcement of restrictions against ivory, the most commonly trafficked mammals of all. Malayan pangolins are served up in high-end restaurants in south-east Asia, especially in Vietnam where they are both the most popular wild delicacy on the menu and, at $150 a pound, the most expensive. Assuming you have remembered to order your pangolin three hours in advance, the manager of the Thiên Vương Tửu (Alcohol of the Gods) restaurant in Ho Chi Minh City will personally bring the live animal to your table and slit its throat to assure you of the unimpeachable freshness of the upcoming dish. At a neighbouring table you may catch fellow diners having their cobra's heart opened up, its medically touted blood or 'snake wine' pumping into a waiting decanter. Unlike sources of ivory, pangolins are pathetically easy to catch. Their scaly covering may pose a challenge for animal predators, but when they are shaken from a tree or a bushy hideaway, the perfectly curled-up ball into which they form themselves is a pangolin harvester's perfect convenience. Into the bag go the scaly balls and into the truck goes the bagful. Tens of thousands of these animals are caught this way every year, most of them merely for the scales which, when ground fine, are advertised as promoting lactation, helping to heal sores and rashes, banishing headaches, and curing anorexia, infertility and pretty much anything else that might ail you. The fact that since the scales are entirely keratin, and thus ingesting them is no more medical help than eating chewed fingernails, has no effect on the size and success of the pangolin market, which asks $3,000 for a kilo of scales plucked from the roasted animals. Ostensibly, and in response to their increasingly endangered status, there are heavy penalties for pangolin-smuggling. In January 2021, China jailed two smugglers for fourteen and thirteen years respectively. But the fact that, for a long period, companies in China and Vietnam have been permitted to manufacture more than sixty panaceas, all containing pangolin scales, has done little to constrain cross-frontier traffic. Overland routes from Cambodia, Thailand and Malaysia are still busy with truckloads of pangolins. In some cases they are even used for striking fashion statements, though it is not known if King George III, presented by the Marquess of Hastings, governor-general of Bengal in 1820, with an Indian coat and helmet made entirely from pangolin scales, ever got to wear that exotic outfit in the last year of his reign.[6]

An ironic consequence of the rise in demand for animal-sourced remedies is that they have ended up contributing to the ailments

Pangolin-scale coat made for George III (Indian, early nineteenth century).

they are thought to cure. In the spring of 2020 a group of Chinese scientists published analyses of coronavirus-infected pangolins confiscated from smugglers in 2017 and 2018 by customs officials at Guangdong.[7] The receptor binding domain of the virus was 97 per cent identical with that of SARS-CoV-2. Though this is not enough to clinch the case for pangolins being the intermediary host for the virus between a reservoir mammal like a bat and the end destination in humans, it adds to the growing evidence that the waves of terrifying diseases coming at the world faster and faster are almost always zoonotic. They are the direct result of what we have done to our planetary habitat.[8] Climate change has added to the witches' brew since the flooding that comes with extreme weather events has created more breeding pools for disease-carrying mosquitoes which, thanks to global warming, now also have an extended season in which to multiply. The massively extended disease ranges of West Nile fever and Zika virus are the result. In a disconcertingly Gothic footnote that Mary Shelley would have appreciated, the melting of glaciers on the Tibet–Qinghai border into a vast saline lake has revealed viruses dated to 15,000 years ago and said to be unlike any yet known to contemporary science.[9]

The years since 1980 have seen outbreaks of new infections at a rate of one every eight months in hot zones from Brazil to central Africa to south-east Asia, most of them viral. They include the catastrophes of HIV and Ebola, as well as SARS and H5N1 bird flu. The routinisation of long-distance trade in animals has speeded up the pace of these contagions. H5N1 originated in two mountain eagles illegally transported to Belgium from Thailand; chytridiomycosis, the fungal disease which made ninety species of amphibians extinct and another 124 species lose 90 per cent of their population, was spread by the international traffic in African clawed frogs. Sickness in animals has, inevitably, made its way into the human population transporting, marketing and consuming them.[10] M-pox (formerly known as monkeypox), first identified in 1958 in macaques, has reservoirs in striped mice, giant pouched rats, African rope squirrels and brush-tailed porcupines. A first American outbreak in 2003 has been traced to some of those exotic animals being housed with prairie dogs for the wild pet trade.[11] The jump of the disease from animal to human populations in Africa is itself a cascade of all the disruptions – demographic, social and environmental – that have stirred new contagions from dormancy.[12] For forty years no human cases of M-pox were recorded. But between 1970 and 2018 the population of Nigeria almost quadrupled from nearly 56 million to 195 million. The demographic explosion drove the conversion of rain forest to farmland and conurbations, along with the migration of reservoir species of animals into cities. A series of floods generated by climate change accelerated this migration, and, ironically, the termination of smallpox vaccination programmes due to the announced declaration of the extinction of the disease in 1979, weakened immunity to the closely related M-pox virus. From two African zones – west Africa and the chronically war-ravaged Democratic Republic of Congo – the international trade in wildlife exported the disease to the United States and beyond.

The SARS epidemic of 2003–4, only barely contained, has been traced back to the meat of masked palm civets, shredded and combined with chrysanthemum petals and minced snake to make the high-priced delicacy 'dragon-tiger-phoenix soup', served in up-market restaurants in south China. The virus jumped, not to civet-eaters, but to others in the supply chain leading to the dish: breeders of captive civets held in filthy cages in Guangdong, transporters, slaughterers and cooks. It gets worse (or better) for an opportunistic virus. In Thailand, captive populations of masked palm civets are fed exclusively on coffee bean 'cherries' which, as they travel through the gastro-intestinal tract, have the acidity extracted from them by the action of digestion-aiding enzymes. The neat piles of coffee cherries packed in civet excreta will then end up as your speciality java of the day, expensively priced on the market. Imagine how many opportunities there might be for a virus to make the jump from an infected animal to a civet-shit gatherer slaving on minimum wage. Venti latte, anyone?

Although a letter to *Nature Medicine* in March 2020 from Kristian Andersen and four microbiologist colleagues argued, on the basis of genomic analysis, that 'it is improbable that SARS-CoV-2 emerged from the laboratory manipulation of a related SARS-CoV-like coronavirus' and that it was more likely to have come from an animal reservoir – like *Rhinolophus affinis*, the intermediate horseshoe bat – there is, at the time of writing, no definitive verdict on the virus's aetiology.[13] Live mammals known to be susceptible to SARS, such as hog badgers, foxes and (especially) raccoon dogs sold for both fur and meat, were stored and sold in quantities at the Huanan Seafood Wholesale Market in Wuhan, and the first known infected person was a vendor there. In March 2023, raw genetic data taken from swabs around the stacked cages of raccoon dogs showed that an animal did indeed carry the SARS-CoV-2 virus, although whether it contracted the infection independently in the wild or was infected by a human remains as yet unproven.[14] In January 2023, 156 microbiologists joined a commentary by the editor of the *Journal of Virology*, Felicia Goodrum, asking, optimistically, for a less politicised 'rational discourse' on the subject, stating that 'at this time and based on the available data, there is no compelling evidence' supporting either 'an accident' or 'nefarious actors' at the Wuhan Institute of Virology.[15] That still remains the case, since data supporting the US Department of Energy's 'low confidence' opinion, made public in February 2023, that a lab leak was the likely origin of the virus, remains classified. Four 'intelligence community elements' and the National Intelligence Council have taken the opposite view (albeit also with gnomic 'low confidence') that exposure to an animal infected with SARS-CoV-2 or 'a close progenitor virus' was the more probable origin. Lab accidents, however, are not unknown and no one in the microbiological community takes issue with renewed, stringent attention being paid to safety issues in laboratories working with genetically manipulated viruses, especially those with potential transmissibility to humans.

Unfortunately, there may never be a definitive explanation of the origin of SARS-CoV-2, but there is no doubt that the closeness between human and wild animal populations has enabled 'reverse zoonosis': viral leaps from humans to non-humans, and then back again. It is thought by some epidemiologists that this is the route that the Omicron variant of COVID-19 took, mutation taking place in infected rats which then transmitted an adapted virus back to humans.[16] On 27 April 2022, it was reported that a virulent strain of avian influenza – H3N8 – had infected a four-year-old boy exposed to chickens and crows in the Chinese province of Henan. On 23 February 2023, Cambodian health authorities reported the death of an eleven-year-old girl in Prey Veng province from H5N1, the virus responsible for the pandemic of influenza in wild and domestic bird populations. The virus had already made the jump from avians to mammals including Peruvian seals and Spanish mink. At the time of writing, although the Cambodian girl's father also tested positive for H5N1 infection, there is as yet no clear evidence of human-to-human transmission. But the likely epidemiological implication of this news from Cambodia has already made the World Health Organization (WHO) state that the report is 'worrying'. Wildlife, intensively fed and bred livestock, and humans to all intents and purposes now constitute a common planetary reservoir of perpetually evolving and mutating micro-organisms, some of them baleful. The Global Virome Project, established, as its name suggests, to coordinate world-wide research, estimates that there are 1.6 million potential zoonotic viruses in the world with just 1 per cent of them currently identified and analysed.[17]

All this is happening at ever briefer intervals. Demography remakes geography, transforming – right now, and not for the better – the future of life on earth.

By the end of 2021, up to eighteen million people had died, world-wide, from COVID-19 infection.[18] You would suppose that in the face of a pandemic – an outbreak which by definition is global – together with a recognition of shared vulnerability, governments and politicians might have set aside the usual mutual suspicions and, under the aegis of the WHO, agreed on common approaches to containment, vaccination and control.[19] Needless to say, nothing remotely like that has happened. If anything, the reverse has been the case: responses to the pandemic sharply diverged, even within entities like the European Union, ostensibly committed to common policies. Decisions taken by individual American states on vaccination requirements and mask mandates thwarted federal guidelines, deepening the already bitter cultural divisions between 'red' and 'blue' America. Ron de Santis, the Republican governor of Florida, cast himself as the voice of Regular Folks' mistrust of expert opinion handed down from the Centers of Disease Control: the people's anti-Fauci.

To some extent the raising of walls, psychological and institutional, is understandable. The instinctive reaction to contagion breaking out somewhere distant is to erect barriers against its importation. For a while, geographically isolated countries like New Zealand benefited from the possibility of self-sealing. But two years' experience of the pandemic, in particular the unpredictable incidence of recurring outbreaks and viral mutations, has made the locking off of discrete zones of exclusion all but impossible. The need for an alternative, transnational approach to containment, mitigation and protection, coordinated by the WHO (since that is why it was established in the first place in 1948), has never been more urgent. The geographically uneven and glaringly unequal supply and delivery of vaccines and therapeutic drugs has only underlined this need. Because mutations arise most easily in thinly vaccinated populations, the comment of Tedros Adhenom Ghebreyesus, the WHO director-general, that 'until everyone is vaccinated no one will be safe' ought to have been an epidemiological truism.

This was not, however, the attitude of the then president of the United States. At the end of May 2020, during the most desperate early days of the pandemic, Donald Trump announced that the US would be withdrawing from the WHO.[20] His major justification was to complain that the organisation had become a pawn of the Chinese government and had, in effect, been an accomplice of Beijing's efforts to disguise the origin of the COVID outbreak. In Trump's view this meant that China and the WHO, working as collaborators, had knowingly unleashed the contagion on the world with the unpardonable consequence (if not actual intention) of damaging his re-election prospects. They had had the audacity to launch the embarrassment virus with millions of fatalities as collateral damage. Whether or not COVID-19 was the result of a lab leak at the Wuhan Institute of Virology, it is undeniable that China did initially play down the magnitude of the outbreak in Wuhan. The WIV was not at all transparent in making documentation of its experiments with genetically manipulated viruses available, yet the WHO was prepared to take on trust Chinese statements, such as they were, about the origin and spread of the disease. It was, however, far from alone in this incuriousness. In the early stages of the outbreak there was no more ardent cheerleader for Xi Jinping and his government's COVID measures than Donald Trump himself. 'China has been working very hard to contain the virus,' he said in January 2020, and a month later, 'I think China's handled it

[COVID] really well.' *Politico* found no less than fifteen such public comments, all in this generously appreciative vein.

Once, however, Trump concluded that China had weaponised its own epidemiological dishonesty and incompetence expressly to make him look bad, his mentions of the virus invariably came with a tag of culpability, as in 'the China virus' or more facetiously 'the Kung flu'. There is a history to attaching misleading nicknames to pandemics, the better to characterise them as an alien plague falling upon a vulnerable homeland. Although the first documented cases of the horrific influenza outbreak of 1918 occurred in a military establishment in Kansas, the pandemic became known as 'the Spanish flu', principally as a result of that country's willingness (unlike belligerents in Europe) to report candidly on the severity and extent of the contagion. The cholera which swept through most of Europe in the nineteenth century, and which in Britain's case arose from local sanitary contamination, became commonly known as 'the Asian cholera' or more offensively still as 'the yellow peril'.[21] In no time at all, discussion about the origin and transmission routes of COVID had likewise collapsed into the usual mire of military metaphors, so that its progress became an 'invasion' against which 'defences' had to be manned, battles fought, conquests pursued, to a decisive 'victory'.[22] Politically, it was all too easy for populist leaders, like Brazil's Jair Bolsonaro, infuriated by their impotence in the face of a microbial 'enemy', to emerge from an initial state of denial into a nationalist blame game; somehow, some other force, some other nation, was responsible for their country's predicament. Before long, any possibility of a clear and honest understanding of the common world-wide conditions that allowed such disasters to happen, not least the biological consequences of environmental degradation, became swallowed up by this default vocabulary of competitive nationalism. Astonishingly, the Johnson government of the United Kingdom was so intent on applying its new norms of Brexit isolation that it withdrew from the common European pandemic early warning information pool. Later, it made the claim that Brexit had allowed it to have the earliest and most successful vaccination programme, passing over the inconvenient fact that, as of July 2022, Britain nonetheless had the highest case and mortality rate of any state in western Europe.[23] In an impressive demonstration of cutting-nose-to-spite-face delusion of autarkic grandeur, heavily locked-down North Korea rejected an offer of 3 million doses of Sinovac vaccine from China as well as another major offer from AstraZeneca, and as of April 2022 was one of only two countries in the world (the other being Eritrea) to have no vaccines at all within its territory.[24]

Mercifully, it has not all been a zero-sum game. In late March 2021, twenty five world leaders, including Emmanuel Macron, Boris Johnson, Mario Draghi, Angela Merkel, Cyril Ramaphosa, Wolodymyr Zelensky and the head of the European Council, Charles Michel, as well as the prime ministers of South Korea, Fiji, Thailand, Chile, Senegal and Tunisia – but, depressingly, missing the leaders of the United States, Japan, Russia and China – issued a statement explicitly acknowledging the chain linking human and non-human lives and destinies. Invoking the multilateralist idealism of the years following the Second World War that sought a reconnected world through the United Nations and agencies like the WHO, they proposed a legally binding international treaty to deal with future pandemics. Such a treaty would embody 'an approach that connects the health of humans, animals and our planet'. This built on the impressive 2015 *Lancet*–Rockefeller Foundation joint report seeking to establish a 'One Health' globally indivisible approach to the environmental-epidemiological nexus that would necessarily 'transcend national boundaries'.[25] On 1 September 2021, Merkel and Todros Adhenom Ghebreyesus opened a Hub for Pandemic and Epidemic Intelligence in Berlin. In a gesture more appropriate for a country fair or the launch of an ocean liner, they cut a ribbon in two places. The ribbon was striped red and white as if simultaneously alerting visitors to peril and bidding them enter anyway. The Hub's mission brief says that it is meant to provide global data linkage and the sharing of advanced analytical tools and predictive models, the better to be armed against future outbreaks. 'No single institution or nation can do this alone,' Ghebreyesus declared. 'That's why we have coined the term "collaborative intelligence".' But there is already data-gathering at the WHO Academy in Lyon and preparations for the storage of infectious material at a secure bio-bank in – where else? – Switzerland. None of this, however, overcomes the immense disparity of resources, for both research and clinical trials, between richer countries and the regions of the world from which new infectious diseases often arise. The Global Health Network led by Professor Trudie Lang in Oxford is a promising effort to decentralise epidemiological and microbiological research and establish programmes of training in countries where it is desperately needed. This effort to supersede national self-interest with genuinely internationalist pooling of resources is uncontroversially commendable. But in some quarters there is uncharitable muttering about compulsive Hub-ification. It is easier, those critics say, to manufacture a hub than set it spinning.

This moment in world history is no less fraught for being so depressingly familiar: the immemorial conflict between 'is' and 'ought'; between short-term power plays and long-term security; between the habits of immediate gratification and the prospering of future generations; between the cult of individualism and the urgencies of common interest; between the drum beat of national tribalism and the bugle call of global peril; between native instinct and hard-earned knowledge. If it is a happy answer you want to the question as to which will prevail, it is probably best not to ask an historian. For history's findings are more often than not tragic, and its boneyard littered with the remains of high-minded internationalist projects. The appeals of idealists fill whole-page declarations in earnest broadsheets and win funds from far-sighted philanthropic foundations. But the plans and the planners are demonised by the tribunes of gut instinct as suspiciously alien, hatched by cosmopolitan elites: the work of foreign bodies.

Not invariably, though.

PART ONE

East to west: smallpox

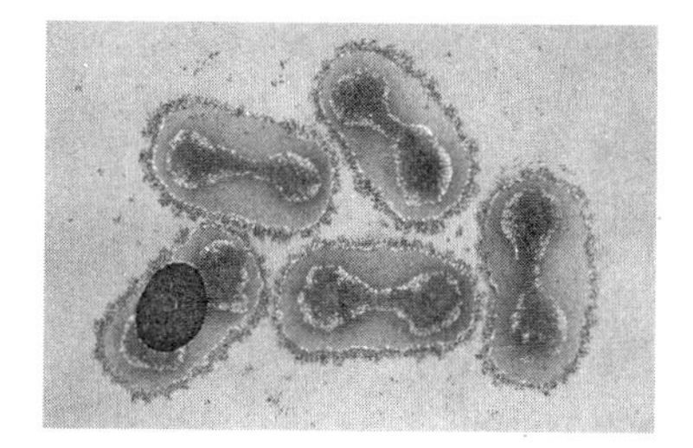

ii

'The fresh and kindly pock'

Nicolas de Largillière, portrait of Voltaire aged twenty three, 1728.

Two hundred pints of lemonade, was, Voltaire acknowledged, a surprising treatment for a victim of smallpox; nonetheless he was quite certain it had saved his life.[1] For some days, in early November 1723, it had been touch and go. For a week, his head and body burned with fever. Clarity deserted him. The sudden sickness wrecked a moment Voltaire had been eagerly anticipating. The marquis de Maisons, Jean-René de Longueil, as rich as Croesus, had invited Voltaire to his chateau on the edge of the Forêt de Saint-Germain-en-Laye, 9 miles north-west of Paris. A company from the cream of society and the republic of letters was gathered for a reading of Voltaire's latest play, a Herodian melodrama: *Mariamne*.[2] Its author, despite the *succès d'estime* of his earlier play, *Œdipe*, was not yet irreversibly established as a literary fixture. Everything seemed auspicious. Built by François Mansart in the previous century, the house was contrapuntal music in stone; what Versailles might have been had architectural monomania not set in. This reading, in the *salle des fêtes* with the star of the Comédie Française, Adrienne Lecouvreur, performing the title role, would surely make Voltaire. He would do his best to bat away the flattery with gestures of unconvincing modesty. It was what was expected.

But the virus gatecrashed the party and all was ruined. Fearing the worst, even before the first pustule had erupted on Voltaire's angular face, most of the guests fled. The marquis de Maisons, however, stayed put. At twenty four he was five years younger than Voltaire, not just the president of the Paris Parlement but a keen amateur scientist with an experimental chemistry lab. The marquis was also France's only grower of coffee, the beans said to compare favourably with the best mocha. Now, he walked through the enfiladed chambers of his beautiful house, distressed but practical. Money and power, of which he had plenty, meant he could call on the highest connections for help. So the personal physician of the Chevalier de Rohan, M. Gervasi, was summoned to attend the prostrate writer.[3] Understandably though unprofessionally, the doctor had needed some persuasion to hurry to the chateau, seeing that the patient was probably in the grip of a disease believed to have already killed 20,000 Parisians in twelve months. Voltaire might or might not be at death's door, but either way, he was bound to be infectious. Nonetheless, with whatever reluctance, Gervasi arrived and did his professional duty. An initial examination confirmed the doctor's pessimism, 'an opinion', Voltaire commented sardonically, 'which the servants made sure to communicate to me'. The usual prescriptions – a bleeding and a purge – were vigorously applied, but, as expected, pustules broke out on Voltaire's face and body, quickly filling with viscous gunk. At one stage, matters looked so dire that Maisons wondered out loud whether, notwithstanding his guest's famous scepticism in matters of faith, it might not be prudent for the *curé* to make a call. Expecting the worst, Voltaire agreed and duly made confession, which, he wrote to the baron de Breteuil, '[it] will not surprise you, did not take long'. Forever anxious about the state of what he called his 'delicate' constitution, and believing bodies to be merely the shabby casing for the soul, Voltaire resigned himself, 'tranquilly awaiting death', aggrieved only that it would prevent his putting finishing touches to *Mariamne* and the completion of his epic poem *La Henriade*, set in the sixteenth-century wars of religion. Most upsetting was the bitter thought that he would have to depart from his friends *de bonne heure*, much too early.

Would those friends grieve for him the way he had grieved for his bosom companion La Faluère de Genonville, who had been among those taken just two months before? Distraught when that had happened, Voltaire would write a poem lamenting the loss of the man with whom he had shared much, including a mistress: 'We who loved all three – reason, foolishness and love / The enchantment of tender errors.'[4] His condition rapidly deteriorated. As the blistering pustules turned purulent and broke, they gave off a rank stench. Servants crept about him, economical with timing, handkerchiefs pressed to their faces. The château was now a place of utmost danger. Still, the noble Maisons remained, watchful and touchingly concerned. That the customary, base instincts of self-preservation had yielded to hospitality and humanity moved Voltaire deeply. Two others also behaved well. The actress Adrienne Lecouvreur stayed with him until the arrival of his close friend Nicolas-Claude Thieriot, who had ridden 40 miles post-haste to be at the bedside. 'I was fortunate enough', Voltaire wrote to Breteuil, 'to have had beside me a man who can be counted among the very few virtuous men who truly understand friendship rather than merely know it as a word.'[5]

Despite (or because of) his ominous judgement, Gervasi persevered with his patient. Since Voltaire had long decided that he was his own best physician, this took some forbearance. Voltaire's view, common in his day, was that smallpox was nothing more than the vascular system's way of ridding itself of dangerous 'impurities' caused by an imbalance of humours. Thus, the whole sickness was generated internally and spontaneously rather than through any kind of invasive infection. The idea that smallpox was inherent in a 'natural' process of self-regulation in which the excessively 'moist' blood of childhood was evolving into the drier sanguine of adulthood had been rehearsed in 1676 in the *Observationes medicae* of Thomas Sydenham, ex-Parliamentary trooper and largely self-taught physician and biologist.[6] Despite his rough-and-ready medical education Sydenham had been befriended by the great eminences of the Royal Society, Robert Hooke and Robert Boyle – albeit not generously enough to get himself elected as Fellow. Most likely it was the influence of Boyle's 'corpuscular' theory of disease that persuaded Sydenham of the existence of innate 'morbific particles' residing within the human organism. Confessing ignorance of what exactly triggered their active corruption of the blood, Sydenham followed contemporary speculation that this might have something to do with atmospheric disruptions or untoward motions 'in the bowels of the earth'. Alongside these traditional speculations and his devotion to the doctrine of the humours, Sydenham believed that inflammation should be understood as a 'striving at the digestion and concoction of the inflamed particles with the intention of afterwards discharging them on the surface of the body and finally expelling them from her boundaries under the form of little abscesses'.[7] Logically, then, the visitation of smallpox (or measles, which also interested Sydenham) was to be seen not as a destructive, but as an ultimately salutary process: a violent refreshment of the metabolism. Of course, it might be so violent as to kill the subject of regeneration. But we now know that the immune system can, in some circumstances, deliver a cytokine storm that can do more harm than good to a SARS-CoV-2 infected body.[8]

Voltaire subscribed to many of Sydenham's assumptions, including the notion that smallpox, notwithstanding its lethal harvest of victims, was, as he wrote to Breteuil, 'merely a purification of the blood, favourable to nature, which, by cleansing the body from the agencies of impurity, prepares it for vigorous health'.[9] Accordingly, the worst thing one could do was to suppress or interrupt the 'natural' course of the disease. Despite the onset of feverishness, conventional treatment – described by the tenth-century Persian doctor Abu Bakr Muhammad Ibn Zakariyya al-Razi, called 'Rhazes' by European writers – counselled heat, as the best accelerator of the pustular exit of toxins. Patients were wrapped in blankets and confined to warm rooms to induce a sweating out of the infection. But Voltaire already knew that this was useless if not dangerous, as was covering erupted sites on the body with tight plasters or, as the physician Richard Morton recommended, rubbing the body with alcohol-rich 'Rabel Water'. Especially unhelpful, Voltaire thought, except in cases where 'sluggish blood' required encouragement to flow, was the ubiquitous 'Countess of Kent's Powder': a seventeenth-century concoction of 'crabs' eyes' (actually lime found in the stomach of crayfish), pulverised pearl and white coral, and the miracle substance 'contrajerva' (dorstenia) root, first brought from the New World by Francis Drake and imagined to be an antidote to, inter alia, syphilis, diarrhoea and tumours.[10] To Voltaire's disdain, many of the French aristocracy, including the duchesse d'Orléans, swore by 'Milady Kent's Powder'. He commented to Breteuil that survivors of a smallpox attack were in the habit of supposing their survival had something to do with these supposed remedies, when in fact they were lucky not to have been killed by them. The charlatans who made their reputation from dosing patients with such cordials and went around Paris boasting of their remedial talent should, he wrote, be summarily imprisoned as poisoners.

Sydenham had insisted that cooling, not warmth, was called for. It was thought to reduce blood pressure while somehow encouraging the coursing of foreign matter towards evacuation through the broken surface of the skin. Bring on the lemonade; again and again, together with another great bleed and eight emetic purges. The wonder is that Voltaire survived, but he did, and believed that the correct remedies had been applied to save his life as well as to preserve him from the blindness which struck many smallpox survivors along with the cratering of their skin. As his strength returned towards the end of November 1723, Voltaire longed to be off, not least to end his being a burden on the kindly Maisons and his household.

On 1 December, he was finally out of the house of his sickness and back in his Paris apartment, a stone's throw from the École de Médecine. But exhilaration was abruptly cut short. The following day brought the awful news that no sooner had Voltaire's carriage exited the grounds of the chateau than the floor of the room in which he had stayed had burst into flame, taking with it the apartments below and an entire wing of the majestically elegant house. Though Maisons was at pains to console his guest and insist that the disaster was in no way his fault, Voltaire was mortified and bewildered. When he had left the house, there was, he wrote, barely an ember glowing in the fireplace. But he learned that a wooden beam beneath the floor of the unevenly paved hearth must somehow have caught fire. 'I was not the cause', Voltaire wrote, 'but the unhappy occasion.' This, however, was enough to make him feel irrationally guilty; that somehow he had repaid the kindnesses of the man who had treated him 'like a brother' by incinerating his chateau. Along with the admiration he felt for Maisons, the pain of this consciousness would, he wrote, remain with him, his whole life long.

This may have been true, given the poignant postscript. In 1726, Voltaire's elder sister Catherine-Marguerite died of smallpox at the age of thirty nine. Five years later, it carried off the guardian angel of his own ordeal, René-Jean Longueil, the marquis de Maisons, at just thirty two. Remembering his solicitousness, Voltaire referred to him (a little oddly) as his 'father'. The shocking loss of Maisons swept Voltaire into a despair which, he wrote, 'has brought me close to senselessness'. What made it worse was that 'he died in my arms . . . by the negligence of doctors'. What he may have meant by this is the medical profession's ignorance of, or downright hostility to, inoculation. For while Voltaire was experiencing the terror of smallpox, the first efforts were being made to persuade the French nobility and their physicians of the wisdom of the procedure. In 1723, a Huguenot doctor, Jean Delacoste, who had seen the effectiveness of inoculation in England, wrote to a fellow physician, Claude-Jean-Baptiste Dodart, vouching for its success in saving lives. Dodart in turn asked the president of the Royal Society of Physicians in London, Sir Hans Sloane, for some corroboration. A year later, Delacoste published his version of the initiatives in England, but when the question of inoculation's safety and moral propriety was put to the Faculties of Medicine and Theology in Paris, it was judged a criminal act, both for introducing toxic matter into the bodies of otherwise perfectly healthy people and – more damningly – for usurping the judgement of the Almighty, in whose hands alone lay the arbitration of life and death, health and plague. In the face of such opposition, Delacoste concluded there would have to be a '*grande révolution*' before inoculation was accepted in France.[11]

It is possible, even likely, that Voltaire had read Delacoste and knew of the campaign to bring inoculation to France, cut short by the death in 1723 (in the arms of his mistress rather than on a bed of smallpox) of the Regent, Philippe of Orleans, who had declared himself a supporter. Delacoste's admiring report of English inoculation would explain why Voltaire's own advocacy – the very first in any language meant for a lay rather than a learned readership – appeared as the eleventh of twenty four *Letters Concerning the English Nation*, published in England in 1733, ten years after his illness.[12] The following year saw a French edition appear with the euphemistic title *Lettres philosophiques*.[13] If the French title was a half-hearted attempt to disguise the fact that the book was a comparison of English and French systems of government, attitudes to religion and freedom of conscience very much to the disadvantage of Voltaire's country of birth, it was a failure. On its appearance in France, the book was immediately banned and burned by the public hangman, thus reinforcing Voltaire's judgement about the correctness of his cross-Channel comparison. In August 1726, three months after his arrival in England (and though he had made a quick clandestine trip back to France a month earlier), he had written to Nicolas-Claude Thieriot that he was thinking of staying put in 'a land where the arts are honoured and rewarded'. He worried, though, whether his drastically shrinking fortune and chronically frail health would be robust enough for a life in the 'hurly burly' of Whitehall and Westminster.[14]

Besides philosophical principles, there were personal reasons for Voltaire's Anglophilia, not least that a stay in England would be preferable to a second spell in the Bastille. In April 1726, Voltaire had a brutal reminder that being lionised by well-read Parlementaires – 'nobility of the robe' – was no protection against malicious social contempt of the higher-born 'nobility of the sword'. Oddly, a peevish quarrel broke out with the aristocrat who had sent his personal physician to the stricken Voltaire's bedside. As grateful as Voltaire was for Gervasi's attention, he discovered that it had not been given gratis. Moreover, the sum he owed the doctor (along with other debts) was steep enough for the hard-pressed writer to have to sell off some of his Paris furniture. There were two rounds of mutual insults with the Chevalier de Rohan, both exchanged in full public hearing. The author of *Œdipe* had announced himself for the first time as 'Voltaire', a change from his actual name of François-Marie Arouet. In the foyer of the Opéra, Rohan had taunted him by greeting him as 'M. Arouet or Voltaire or whatever it is'. Voltaire of course rose to the bait, mocking the Chevalier's elaborate name, Guy Auguste de Rohan-Chabot, as 'Rohan or Chabot or whatever'. At the Comédie Française a week or so later, it got childishly worse. The Chevalier repeated the mock greeting while Voltaire shot back that 'at least I have chosen my name while you have dishonoured yours'. It was not over. When Voltaire was dining at the invitation (apparently) of the duc de Sully, a servant informed him there was someone at the door to see him, which indeed there was: a gang of Rohan's heavies, who proceeded to beat him up. Infuriated, Voltaire appealed to Sully and even the court at Versailles for an apologetic satisfaction. Getting nowhere, the rake-thin, perpetually hypochondriac writer bought pistols and swords and lessons on how to use them. Rohan vanished, which could be construed as a refusal to duel with a social inferior or (more likely) an onset of panic. Voltaire denied seeking a duel but he was nonetheless arrested as a threat to public order and taken to the Bastille. He had in any case been toying with the idea of moving to England and had become acquainted with the high Tory politician and philosopher Henry St John, Viscount Bolingbroke, exiled in Paris after making the mistake of supporting the Jacobite Pretender instead of the incoming Hanoverian king, George I. This, however, did not preclude Voltaire from grandly writing to the king for an invitation to come to England. Unsurprisingly, no reply was forthcoming from George, who was not celebrated for his interest in literature. But the French government had no objections at all to its gadfly crossing the Channel, whereupon Voltaire found himself at Calais on 9 May 1726 waiting for the packet boat.

He arrived in England barely able to speak a word of the language and knowing almost no one. Bolingbroke was now resettled in England but Voltaire was not confident enough to go and see him. He did, however, have letters of introduction from Bolingbroke as well as from Horace Walpole, the son of the Whig prime minister, so that both sides of politics were willing to help. Helped by lessons from a Quaker tutor, Voltaire mastered English impressively enough to be able to write to Alexander Pope in that tongue, which is just as well since Pope had no French. Before he returned to France three years later, he had been befriended by Swift, Congreve, Addison and Edward Young.

But the man who did most to open doors for the exile was a young silk merchant, Everard Fawkener.[15] Voltaire had met him in 1725 when Fawkener was returning from a nine-year stint in Aleppo as the commercial agent of the family import firm, Snelling and

Fawkener, trading in the Ottoman empire under the aegis of the chartered Levant Company. It is likely that the encounter first piqued Voltaire's interest in the Turkish Levant, and that this interest was further stimulated when Fawkener offered free lodging at his house in Wandsworth, on the south bank of the Thames a few miles upstream from London. Though Voltaire, hungry for an attentive audience, moved back to London at the end of 1726, he later returned to Wandsworth, lodging with a local scarlet-dyer who worked for Fawkener. It was there that Voltaire finally finished *La Henriade*, started to write his tragedy *Brutus* (dedicated to Bolingbroke), and began the research that would go into his history of the warrior king Charles XII of Sweden.

Fawkener's company and reports about the Levant also informed Voltaire's desperately needed literary success: the tragedy of *Zaïre*, written in three weeks in 1732 after he had returned to France and first performed a year later. Set in the medieval Holy Land of the Turkish–Crusader wars, the play features an enslaved Christian woman who falls in love with the sultan of Jerusalem. Due to marry the Muslim ruler, Zaïre is ordered by her father – a descendant of the Christian kings of Jerusalem – and her brother to be baptised. Mistaking family loyalty for romantic betrayal, the jealous sultan stabs Zaïre only to kill himself (Othello-like) when he discovers the truth. The play was an instant triumph, given another thirty performances that year and earning its author a six-week stay at Versailles so that it could be acted at court. Voltaire himself would sometimes play the role of Zaïre's father. In the fulsome dedication prefacing the play, he acknowledged his debt to Fawkener.

Voltaire had gone oriental. And the mix of curiosity and romance that coloured his orientalism continued in the *Letters Concerning the English Nation*, published that same year, 1733. Sitting exactly halfway through the volume (the first half given over to essays praising the superiority of British politics, government and religious freedom, the second to contemporary luminaries, above all Isaac Newton) is Voltaire's essay on smallpox inoculation. Nothing, he thought, exemplified the virtues of English modernity better than that procedure. But, paradoxically, that modernity lay in the commercially minded nation's openness to taking lessons not from the classical, but the oriental world. Was the received wisdom, that the orient was hopelessly mired in decadence and superstition, correct? Plainly this could not be the case if the Ottoman world had come up with something that would save lives more reliably than sweating sheets, gallons of lemonade or the Countess of Kent's panacea. And it had been those inveterate travellers the English who had recognised this, adopted it and established it as safe medical practice.

The turn to 'oriental' wisdom was all the more surprising because it involved what must have seemed a bewilderingly counter-intuitive procedure: the introduction of foreign matter, pus, drawn from an infected person – into the body of a perfectly healthy *child*! What kind of insanity was that? How could it possibly be that such an action would prevent, rather than guarantee, frightful, probably lethal, sickness? Ever the sly dramatist, Voltaire opens his essay pretending to withhold judgement:

> It is inadvertently affirmed in the christian [*sic*] countries of Europe, that the English are fools and madmen. Fools because they give their children the smallpox to prevent their catching it; and madmen because they wantonly communicate a certain and dreadful distemper to their children merely to prevent an uncertain evil. The English on the other side, call the rest of the Europeans cowardly and unnatural. Cowardly because they are afraid of putting their children to a little pain; unnatural because they expose them to die at one time or another of the smallpox. But that the reader may be able to judge whether the English, or those who differ from them in opinion are in the right, here follows the history of the famed inoculation.[16]

In addition to challenging one long-ingrained assumption – that Christian Europe had nothing to learn from the barbaric orient, Voltaire overthrows another: that male learning was superior to female intuition. In his narrative, the verifiable mitigation of smallpox mortality had been accomplished by folk practices of the east and the wisdom of women. What had been recognised and adopted by the scientists of the Royal Society – those British heralds of scientific modernity – turned out to have been practised 'immemorially' by . . . Circassians.

The source for the Circassian romance of inoculation which takes up most of Voltaire's essay must have been the *Travels* of the Huguenot Aubry de La Mottraye, published in London (where he had settled) in 1723–4 and including engraved plates of the Ottoman sultan's seraglio by the young William Hogarth.[17] It is conceivable that Voltaire first encountered La Mottraye's two volumes during his stay at Wandsworth since the Levant trader Fawkener is likely to have had them on the shelves of his compendious library. The fabled beauty of Circassian women, as well as their enslavement as Turkish odalisques, would become an erotic fixation of the Romantics, beginning with an entry in D'Alembert and Diderot's *Encyclopédie*. But the story of their preservation from smallpox disfigurement began in the second volume of La Mottraye's popular book. In 1711, travelling deep into the country north-east of the Black Sea marked 'Circassia' on the maps of the time, La Mottraye describes the people as 'the handsomest in the world' (while marvelling, mean-spiritedly, that their immediate neighbours turn out to be the ugliest).

> As I advanced amongst the Mountains and saw no body who was sear'd with the Small Pox, I bethought my self to ask them, if they had any Secret to preserve themselves from the Havock which that Distemper makes among so many other Nations; they inform'd me, that 'twas owing in a great measure to their inoculating them, whereupon I desir'd to be told their manner of doing it, which they explain'd to me exactly enough for me to comprehend it without seeing the Operation.[18]

Searching for an actual demonstration, La Mottraye finally found one in a village he calls Deglivad. There, he writes, he saw a girl of around four or five years old being 'carried to a young Boy of Three Years old who had the Distemper naturally and whose Pocks began to suppurate or were ripe, and an Old Woman performed the Operation for those who are of this Sex who are the most advanced in age are believed also in Wisdom and Knowledge . . . and exercise generally the Practice of Physick'. Using three needles fastened together, the woman made pricks in five places on the little girl's body: on the stomach, near the heart, and on the navel, wrist and ankle

> till the Blood came and at the same time [she] took some Matter from the pocks of the sick person and apply'd it to the bleeding Parts which she cover'd first with Angelica leaves, dry'd after with some of the youngest Lamb skins and having bound them all well . . . the Mother wrapped her daughter up in one of the Skin coverings which . . . compose the Circassian beds and carried her, thus pack'd up in her arms to her own Home where . . . she was continued to be kept warm, only a sort of pap made of Cummin flower [flour] with two thirds water and one third sheep's milk without either flesh or fish and drink a sort of tisane made of Angelica, Bugloss roots and Lickorish.

La Mottraye wrote that another 'easier' way was available to try to induce infection and that was to place a naked, healthy child in a bed with an infected one, though he added that inoculation was preferred as 'infallible' in bringing on an attack that left fewer marks. Such was the urgency with which parents sought that remedy for their children, especially their daughters, that they would think nothing of riding a whole day in search of a promising donor of pus.

La Mottraye's narrative may or may not be fanciful, but it is strikingly dominated by women: the little girls who were the primary (though not exclusive) receivers of inoculation; the 'old women' who were the surgeon-variolators; and not least Circassian mothers, who wanted to ensure that their daughters were marketable as concubines to Turkish pashas, viziers and even perhaps the sultan's own seraglio. Describing conduct which he knew would shock European opinion (even at its most marriage-market hypocritical), Voltaire, as was his wont, is simultaneously dry-eyed, breathily erotic and sentimental:

> The Circassians are poor and their daughters are beautiful and indeed it is in them that they chiefly trade. They furnish with beauties the seraglios of the Turkish sultan and of all those who are wealthy enough to purchase and maintain such precious merchandise. These maidens are very honourably and virtuously instructed to fondle and caress men, are taught dances of a very polite and effeminate kind; and how to heighten by the most voluptuous artifices the pleasures of their disdainful masters for whom they are designed. These unhappy creatures repeat their lesson to their mothers in the same manner as little girls among us repeat their catechism, without understanding one word they say.[19]

This, too, was from La Mottraye, who explained that although the Circassian sex traffic in daughters might seem, on the face of it, a mercenary outrage, their parents believed that

> at least the Girls will be happily provided for, by being advanced into the harem of the Grand Seignior where they may become Empresses or else into those of the Bashaws [pashas] and other rich persons where they will have fine clothes and rich Jewels and have everything that is most delicious in life. This prepossession which is generally received makes the Daughter when sold, part with her Mother without regret and the Mother on the other hand with her good fortune . . . a prosperous journey[20]

Rather than see the selling of daughters as an unnatural crime against family morality, Voltaire regards the instinct which led the Circassians to adopt childhood inoculation to protect their investment as entirely understandable and natural. His description of them as a 'commercial nation' is meant as a compliment: the preservation of family fortune. How different, after all, was their grooming from that of European aristocratic girls for the venal marriage market on which their fortunes depended? The brutal disfigurement wrought by smallpox on the faces and bodies of both sets of girls, Asian and European, was equally a catastrophe that would determine the whole course of their futures. It was only rational, then, Voltaire thought, for Circassian mothers to be the first to adopt inoculation for their children.

For Voltaire, smallpox inoculation was a sign of the common sense of 'intelligent women' like the poet Lady Mary Wortley Montagu, whose four-year-old daughter had, in 1718, been the first person to be inoculated in England under something like professional supervision and in a blaze of publicity. Lady Mary had many reasons to take that bold (some said foolhardy) step. Not long after coming into his own as Earl of Kingston-upon-Hull, her elder brother William had died from smallpox at the age of twenty, leaving behind an even younger widow and two small children. It was especially distressing for Mary, as William had been her defender against their martinet father's attempts to foist an unwanted husband on her. Instead, she had eloped with Edward Wortley Montagu, the brother of a friend, warning him on the eve of their elopement that she would bring nothing to the marriage other than 'a nightgown and petticoat'. Two years after the death of William, in December 1715, she herself suffered a life-altering attack of the infection. Though she survived the ordeal, smallpox took away her eyelashes, and the rest of the famous good looks that had had her toasted as the 'Beauty' of the Kit Cat Club, and bewitched the rather sober-sided Edward. But, as envious rivals and hard-hearted critics uncharitably pointed out, the facial cratering left by the distemper had scotched any prospects of a rising career at the court of the new king, George I.[21]

In 1717, during her husband's embassy to the Ottoman sultan Ahmed III, Mary noticed, not least while in the company of naked bathers at a Sofia hammam, that none of the women had been marked by smallpox. This miracle, she learned, was the result of inoculation. So struck was she by this information that in Constantinople, while Edward was away with the Ottoman court at Adrianople, Mary had her six-year-old son (also Edward) inoculated. When Voltaire met her in 1727 at Twickenham, she had become famous, or to the many belligerent critics notorious, as the champion of the life-preserving, face-saving procedure. This brave but eminently sensible advocacy, in Voltaire's view, made her – and her influential convert, Princess Caroline of Ansbach – peerless examples of practical reason and virtue. Mary was one of 'the most intelligent women' in England and Caroline, having in that same year of 1727 become queen on her husband's accession as George II, 'a delightful philosopher on the throne'.[22] That women should have been the principal agents of amelioration was, Voltaire thought, inconceivable in his own country, for all its vaunted obsession with beauty and noisy lust for life. That those women should have successfully introduced and popularised an alien, oriental practice only made the whole business doubly confounding.

But then the abatement of smallpox mortality had become a matter of political urgency. The disease kept on disrupting the dynastic continuity that was the condition of royal power and stable government. In 1694, it had carried off Queen Mary II of England. Louis XIV's heir, the Grand Dauphin, and the Habsburg Holy Roman Emperor, Joseph I, both died of smallpox in 1711. And it had ended the seven-month reign of Louis I of Spain in 1724. Louis' father, Philip V, who had abdicated in favour of his son, was forced to return to the Spanish throne. In 1730, another child-ruler, the fourteen-year-old Tsar Peter II perished from smallpox. In 1700, the sole surviving heir to Queen Anne (after seventeen pregnancies), Prince William of Gloucester, died aged eleven. There was talk of smallpox, though an autopsy also discovered fluid on the brain that was a symptom of encephalitis. Whatever had killed the boy, a constitutional crisis was triggered by the unavoidable end of a Protestant Stuart ruling house. In 1701, an act of Parliament bestowed the succession on the elector of Hanover.

Around this time, Queen Anne's personal physician, the prodigious polymath Dr Martin Lister – among other things the world's first arachnologist (specialist in spiders) and conchologist (shells) – received a letter from a distant relative, Joseph Lister, a merchant trading with the East India Company at Amoy in Fukien province, describing the measures that the Chinese took to bring on 'mild' smallpox the better to avert the more lethal kind. Given the immediate crisis brought on by infection, it seems likely that Martin Lister had, perhaps on hearsay, sought out that information. In February 1700, the anatomist and osteologist Dr Clopton Havers, like Lister a fellow of the Royal Society, read a communication on Chinese inoculation, which was then the subject of discussion.

As early as 1683, Lister, then the Society's vice-president, had already classified the new, lethally virulent strain of smallpox as 'an exotic disease of Oriental people, not known to Europe or even Asia Minor'. He was both wrong and right. In some form or other, an orthopoxvirus had been around since late antiquity and possibly earlier. A fourth-century Chinese text had described symptoms of an infection which brought on suppurating pustules. In 582, Bishop Gregory of Tours, in an extensive survey of pandemics including a terrifying wave of bubonic plague, wrote of a disease which began with high fever and proceeded to an outbreak of spots that then turned into pustules 'the size of grains of millet' (around 2.0–2.5 mm),[23] an image which the twelfth-century poet Theodore Prodromus changed with equally vivid force into a rain of hailstones peppering face and body. 'I almost had my soul spat out on account of this disease,' he wrote in painful recollection.[24] But the family of orthopoxviruses is broad and constantly evolving and those early infections may have been severe chickenpox.[25] What is not in doubt is that the 'pocks' featuring in literature before the seventeenth century did not kill at anything like the rate of modern epidemics. As a result, the appearance of the pustules in children was treated as an unexceptional episode of their growth.

In the middle of the seventeenth century, the virus mutated and it did so most dramatically in China. Mortality rates climbed steeply, becoming a matter of grave concern for the Qing imperial government. Their invading armies, which had penetrated the Great Wall and put an end to the Ming dynasty, were for the most part ethnically Manchu without acquired immunity to the much more lethal version of smallpox already making its way through central China. As a result, the Kiangxi Emperor made inoculation, already traditionally practised, a state policy. The imperial family, including the emperor himself, were inoculated, but the common way of engendering a mild form of the infection as defence against the deadlier form was, as Havers had noted and the Jesuit missionary Père d'Entrecolles reported in 1712, insufflation: the blowing of dried, powdered pus up the noses of children and adults. D'Entrecolles reported that swaddling children in the clothing of a smallpox victim was also used in some villages, but that insufflation was thought – and proven – to be more effective as a prophylactic.[26] Even to those Europeans championing inoculation this seemed eccentric, and was documented as having been attempted just once in eighteenth-century England: by Dr Richard Mead on an eighteen-year-old convict, Ruth Jones. Whether nasal inhalation would be as effective as subcutaneous inoculation remains an open question. Currently, billions of dollars are being invested in developing an over-the-counter nasal spray against COVID-19 which could dispense with the need for nurse-administered vaccinations, though early clinical trials by AstraZeneca have shown, at best, mixed results.[27] But for Voltaire, who read d'Entrecolles's printed letters (not least because they revealed in print for the first time the exciting secrets of Chinese mastery of hard-paste porcelain), their early adoption of inoculation was yet more proof that the Chinese were the 'wisest and best governed [people] in the world'.[28]

The sudden interest in what the Chinese did about the new lethal variety of smallpox was not disinterestedly scientific. In his remarks to the Royal Society in 1683, Martin Lister identified the paradox of Europe's colonial future. Smallpox in its mortal form had been unknown in the west and Asia Minor until a 'Spice Trade was opened by the late Princes of Egypt to the remotest part of the

East Indies where it rages to this day'. It is unclear who those 'Princes of Egypt' were – possibly the Mamluks – but the East Indies had indeed been the theatre of colonial ambition and conflict in the seventeenth century. Lister's implication – doubtless supported by what information he had from his kinsman Joseph Lister – was that commercial appetite came with a price tag of heavy mortality. Those who paid were not just the freebooters of the orient. Since infections travelled along with merchandise and those who traded it, at some point, it was feared, Europeans at home would be on the receiving end. So the dawning of ruthless colonial enterprise generated, at the same time, profit and infection, which thereafter would walk hand in hand down the imperial centuries. With them would go the search for cures and mitigations from wherever they could be found; which, at the beginning of the history of inoculation, turned out to lie within the very realms on which the colonial powers were setting their hungry eyes.

Occasionally, there were Europeans who recognised and reported this paradox. Almost always they were the rare birds who lived in two worlds at once: as the agents, servants and narrators of imperial power and plunder, but also as the sympathetic and often learned multilingual chroniclers of the cultures and traditions in which they found themselves. 'Orientalism' does not begin to do justice to their complex identity; certainly not to John Zephaniah Holwell, the most famous survivor of the 'Black Hole of Calcutta', whose published narrative of the ordeal established in the British mind an image of Bengali Indians and their ruler, Siraj ud-Daula, as brutally inhuman and barbaric.[29] But, a decade later, in retirement at his country house at Pinner, Middlesex, Holwell, an ardent vegetarian, also published the first account of inoculation, practised by high-caste itinerant Brahmins.[30] His fascinated description is detailed and largely free from colonial condescension. The inoculators, he tells us, went from house to house, ensuring first that patients had abided by a preparatory regime of abstention from ghee, flesh and fish. Punctures were preceded by an eight- to ten-minute friction on the arm with a dry cloth, then a slight wound 'the size of a silver groat' was made, to which a pledgit treated with last year's pus, diluted in some drops of Ganges water, was then applied. The procedure was accompanied throughout by chanting from Sanskrit sacred texts and the wound was covered for six hours. During the time of fever, patients were to be doused over their heads morning and evening with 2 gallons of cold water. So far from dismissing it as dangerous, Holwell thought this treatment to be an Indian version of Thomas Sydenham's cooling prescriptions and concludes that since the procedure seemed to be universally successful 'it must have been originally founded on rational principles'. He even considers that the Brahmin explanation that smallpox and other infectious diseases are caused by 'multitudes of imperceptible *animulculae* floating in the air', transmitting sickness through respiration or surface contact, ought to be taken seriously.

Among British colonials in Bengal, the original and open-minded Holwell was atypical. Most of the earliest information about successful smallpox inoculation came from nearer at hand: 'factory' communities of Europeans embedded within Levantine port cities like Constantinople and Smyrna or the great entrepots like Aleppo in Ottoman Syria. Those who answered enquiries from Paris and London were Levantine hands combining the roles of commercial agent, diplomatic secretary and interpreter – and sometimes were physicians into the bargain. In the age of smallpox anxiety, institutions like the Royal Society acted as clearing houses for incoming bulletins from abroad: arbiters of what information was scientifically reliable and what was not – and, often enough, battlegrounds of opposing judgements. In 1706, Dr Edward Tarry, physician to the English factory in Aleppo, but who had also resided for years in the European trading colony of Pera in Constantinople, reported that an 'old Greek woman' had assured him that she had personally inoculated upwards of 4,000 patients all without any ill effects.[31]

This figure of the 'old Greek woman' became the protagonist of nearly all the early smallpox inoculation narratives. That itself was a remarkable turn on the part of communities of the learned, like the fellows of the Royal Society, whose mission was to demote, if not altogether eradicate, folk wisdom and replace it with modern, empirically demonstrated science. That the two kinds of knowledge might not, after all, be irreconcilable, but complementary, was an unexpected illumination. And this was not just a matter of European learning becoming receptive to what non-European cultures might teach. For during the campaign for inoculation in Britain in the 1720s, it was 'discovered' that the adoption of the 'kindly pock' had long been practised in Wales and the remoter parts of highland Scotland.[32] Early ethnography in other countries, too, 'discovered' customary inoculation in regions commonly dismissed as backward: villages in the Auvergne and Périgord in France, Danish Jutland, and further afield in the Kabyl mountains of the Maghreb.[33]

As the tide of death rose, belief in 'philosophical' circles grew that foreign wisdom was, after all, worth considering, given that traditional European remedies had seemed of little avail. But mediators between the science of the learned and the practitioners of custom were indispensable. They were travellers along the shifting boundaries of empires and religions, peripatetic frontier-crossers: versed in many tongues, educated in many places and in many disciplines, from botany and geology to herpetology; equally at home in the khans where European merchants, along with their inventory, were housed and in the courts of emperors to whom they ministered as doctors, surgeons and dispensaries. Two of those frontier-crossing virtuosi were Giacomo Pylarini and Emanuele Timoni, and their cosmopolitanism was reflected in their multiple names: Jacobus Pylarinos and Emmanuel Timony, Timones or Timonis. Both had been educated at the medical school of Padua University, which drew its students from across Europe and the Levant. They both wrote in Latin, Italian, French and German, and lived in the Adriatic-Aegean-Ottoman world where Latin, Greek and Turkish cultures flowed together, acting in effect as medical interpreters, translating for western Europeans what seemed on the face of it a recklessly unnatural act into something valued as a life-saver.

Timoni's family, long established in what had once been the Genoese island colony of Chios, were translators: holders of *berat* patents, granted to non-Muslim subjects of the Ottoman empire, entitling them to the protection of European law and jurisdictions.[34] As such, they served many masters but since the early seventeenth century had had a particular attachment to the British embassy where they were employed as secretaries as well as interpreters. Emanuele's father Demetrio and his older brother Giorgio were both dragomans for the British embassy in Constantinople. But body physicians like Timoni were also figures of trust with personal access not just to the sultans but to the seraglio, where almost as much politics was carried on as at court. At one point, Timoni was offered the post of chief physician to the seraglio, which, for whatever reason, he prudently declined. He did, however, work as informal adviser to the British ambassador in Constantinople, William, Lord Paget, while the latter was helping negotiate the Peace of Carlowitz, which ended (at a high territorial price for the Ottomans) the war between Turkey and the Holy Roman Empire. (Paget had previously served as ambassador in Vienna, which positioned him as arbitrator in the painful negotiations.) Timoni must have been in Paget's good graces, since three years later, on the ambassador's return to England, he took the physician-surgeon with him. In 1703, Timoni added a degree from Oxford to his Padua credentials. In that same year, Paget proposed his protégé for election to the Royal Society, where he was duly installed as a fellow in his early thirties, the only non-British member from the Levant and north Africa to make the grade in the entire eighteenth century.

Not everyone in the British embassy and commercial companies in Turkey shared Paget's high opinion of the physician. His successor in the embassy, Robert Sutton, thought Timoni gave himself altogether too many airs and graces, and suspected him, as did the commercial consul in Smyrna, William Sherard, of playing both ends against the middle, as a triangular conflict between Russia, Sweden and Turkey dragged on into the new century. Timoni was plausible enough as a doctor-diplomat to serve, at the same time, both Sultan Ahmed III and the Swedish king, Charles XII. The

'Northern War' between Sweden and Russia came to a southern climax at Poltava in Ukraine where Tsar Peter I destroyed the Swedish army before handing responsibility for the fate of the captive king to the sultan. Timoni had come to know Dr Samuel Skraggenstierna, Charles XII's personal physician, and thus found himself amid the military remnant of Swedes holed up in the citadel of Bender, keeping the Turks at arm's length, courtesy of the local Tatar prince, Antioh Cantemir. Some of that Balkan experience would return to Scandinavia with the eventually liberated fragment of Charles's army, which is why, as you have probably asked yourself, 'Swedish meatballs' taste exactly like Turkish *kofte*.

Ten years older than Timoni, Giacomo Pylarini was from the Ionian island of Cephalonia, one of the last possessions of the Venetian *stato da mar*.[35] Pylarini was even more of an inexhaustible cosmopolitan than Timoni: the path of his career went through the Greek and Slavic realms of the Ottoman empire – to Crete where he was physician to the Ottoman governor, Ismail Pasha, then to Prince Cantacuzanus in Wallachia and across the Ottoman–Habsburg frontier to the vice-voivode of Serbia, Jovan Monasterlija. It may have been this familiarity with the shifting frontiers of the Balkan Slavic world – at a time when Peter the Great's ambitions were facing south – that gave him entry, around 1690, to the tsar's court as personal physician. Staying in Moscow until his Mediterranean constitution could no longer tolerate the bitter cold of Russian winters, Pylarini returned to Constantinople in 1701, but then, as a native subject of Venetian Cephalonia, he spent five years in Smyrna as the consul of the republic.

On 3 June 1714, with the gouty Queen Anne literally on her last legs, the editor of the Royal Society's *Philosophical Transactions*, Richard Waller, read a communication by Timoni on smallpox inoculation: the first publication in a European language on that counter-intuitive subject. Those observations, however, had not been addressed directly to Waller or to the Society but to Dr John Woodward, geologist and professor of physick, or medicine, at Gresham College in London. Woodward extracted what he judged to be the essence of Timoni's report but, having been expelled from the Society three years earlier for 'conduct unbecoming to a gentleman', was disbarred from reading the report himself. That unacceptable conduct had been the vocal barracking of the target of his obsessive contempt, the naturalist and physician Hans Sloane, then in the middle of addressing the fellows on bezoars, the creamily variegated stones found lodged in the digestive tract of animals and humans. This was just the kind of dilettante curiosity that irked the combative Woodward. The proceedings of the Society, he thought, ought to be confined to some coherent programme of analysis and synthesis rather than allowing a random display of chunks of disparate information. But Woodward's grievances against Sloane were also personal and political. The Caribbean plantation-owner was in the process, he thought, of colonising the Society, preferring his friends to fellowships, the better to position himself as successor in the presidency to Sir Isaac Newton. It was to the elderly Newton that Sloane complained about Woodward's behaviour, the most recent outburst not being the first such indignity inflicted on him. A council was called and the unrepentant Woodward ejected from the Society. Sloane would indeed succeed Newton on the latter's death in 1727 and later go on to found the British Museum.

Woodward had a particular stake in passing on the findings of Emanuele Timoni. The speciality that had made his own reputation was his *Essay Towards a Natural History of the Earth*, based on his rich fossil collection. But he was also a fellow of the Royal College of Physicians with patients including Richard Steele, the co-founder and editor of *The Spectator*, and he had strong opinions about the best way to treat smallpox when the 'secondary fever' came on. This consisted essentially in heavy vomiting, triggered if necessary by a tonsil-tickling feather Woodward had designed to bring on the needful retch. This prescription only reinforced Woodward's reputation as nothing more than an eccentric amateur. Two of the fellowship in particular, John Freind and Richard Mead, ridiculed Woodward so mercilessly that the dispute over smallpox treatment turned from a duel of words into the actual thing. Swords were drawn and might have shed blood, had not Woodward, true to his farcical reputation, lost his footing and stumbled to the ground. 'Take your life,' shouted Mead, to which Woodward gamely shot back that he would take anything at all, just so long as it was not Mead's physick.[36]

Woodward must have hoped that his part in publicising Timoni's account of inoculation (together with a Latin aetiology of smallpox) would restore his reputation and overturn the expulsion from the Royal Society against which he had unsuccessfully appealed. The document was certainly dramatic news for a Europe terrified by smallpox. What he would describe, Timoni wrote, was common among 'Circassians, Georgians and other Asiaticks' and had been introduced among the Turks for forty years.[37] It had been so successful that the inoculated were 'scarce sensible they're ill or sick' and it was especially 'valued by the Fair since it never leaves Scars or Pits'. The operational method was very close to Aubry de La Mottraye's description of Circassian practice. A 'boy or healthy young lad' with smallpox was sought; at around twelve days following the first appearance of symptoms, 'ripe' pustules on the donor's leg were punctured. The infectious matter was then drawn into a glass vessel which needed to be kept warm, 'close to the bosom', while it was being taken to its intended recipient. The healthy patient would then have their skin 'ripped up a little' by a surgeon's three-edged needle or lancet and the pus muddled with the blood using 'an ear pick' or some such implement. Just where on the body the subcutaneous scratches or incisions were to be made varied a good deal according to the operator's custom, though Timoni expressed the view that the muscles of the arm were most promising for the infection to take. Wherever it was administered, the inoculation site was then covered by half an empty walnut shell. In a week or less, the pustules would dry up and fall off by themselves leaving no discernible scars behind: a truly miraculous thing. In the eight years that he had been observing inoculations, Timoni claimed he had never witnessed any 'mischievous accident', dispelling the reports of fateful disasters and deaths which 'have been sometimes spread among the Vulgar'. Should the inoculated die, it was invariably because they had been suffering independently from other maladies. He was not, however, he emphasised, claiming that inoculation was a panacea.

Timoni's was the first account of smallpox inoculation to be published in English, but although he writes that he had been observing it for eight years, the truth may be that he was actually introduced to the procedure by his fellow Greek physician Pylarini when they met in Smyrna in 1712. Their accounts both claim success for inoculation, but the close detail in Pylarini's version suggests the experience of a veteran practitioner familiar with the procedure since the severe epidemic of 1701. Pylarini's report came to the Royal Society, also edited by Richard Waller, albeit in Latin, which, even in that classically learned age, limited its readership. But it was the product of an affinity of naturalists linked by a chain of Anglo-Greek-Ottoman knowledge. In his own right, Waller was a close observer and illustrator of animals and plants; he had published research on fireflies, frog spawn and the bills and tongues of woodpeckers. He had also produced fine drawings intended to illustrate the work of the great parson-naturalist John Ray. And it was Waller who took the initiative in making enquiries about inoculation in the Ottoman world. But when he wrote to William Sherard, the British consul in Smyrna, it was as one 'natural philosopher' addressing another. Like a striking number of Europeans in the ports and entrepots of Asia Minor and the Levant, Sherard was not just a commercial agent but an enthusiastic adept at many kinds of knowledge, old and new. From a modest Leicestershire background, he had studied botany with Joseph Pitton de Tournefort at the Jardin des Plantes in Paris. There he had met the Dutch botanist Paul Hermann, who as director of the Botanical Garden in Leiden had Sherard come there for further collegial study. But Sherard was never just a gentleman-scholar of means. Surviving inevitably meant the thankless grind of tutoring the sons of aristocrats in England and Ireland. When that became insupportable, he escaped, adventurously, following the example of his teacher, Pitton de Tournefort, who had sailed to thirty eight islands in the Aegean in search of species to discover and inventory before making his way through Anatolia, Armenia and Georgia all the way to Tiflis. In his more modest way, Sherard also wanted to build his own personal empire of knowledge, sailing in small boats through the Greek archipelago, riding with mule trains and dromedary caravans deep into Asia Minor and the Caucasus. Ancient history excited him as much as natural history so when he became consul for the Levant Company, housed within its khan in Smyrna, Sherard took advantage of its location to set off on archaeological expeditions to Halicarnassus and Ephesus, and to make pilgrimages to the Seven Churches of Asia Minor. His country house at Sedi-keui, 7 miles from Smyrna, was filled with learned loot: plant and seashell fossils from the lacustrine basin of the region, the product of primordial volcanic eruptions. There were cabinets full of such objects together with inscriptions from antiquity and the centuries of Christian Byzantium.

The two ostensibly commercial consuls at Smyrna – Pylarini for Venice, Sherard for the English – were like minds whose itineraries of knowledge criss-crossed each other. The much-travelled Pylarini had returned to Constantinople from Russia during the height of the epidemic of 1701 and had himself and others inoculated, beginning in 1704. Most of his early patients belonged to the Rum Greek urban nobility, but it had been in Thessaly that he first encountered the matriarchy of inoculators. His description of their practice was originally given to Sherard, who sent it on to his brother James, a fellow of the Royal Society as well as an apothecary. Almost hidden in the text was the startling information that a Mr Hefferman, secretary to the British ambassador in Constantinople, Sir Robert Sutton, had already had his two sons inoculated. The letter was first published in its original Latin form in the *Philosophical Transactions* of 1716.[38] It was unusual (and very different from the Woodward–Timoni communication) in its richness of social information. In Pylarini's narrative, the fear of inoculation is softened by the occasion being turned into a semi-ceremonious occasion in which 'buying the pocks' was formalised by a donation of money to the child donors. The matrons, who would have been well known to villagers, waited until the fierce summer heat had abated in September before enquiring of local families who might wish to have their children inoculated. Sometimes the procedure was no more than abrasive rubbing with cloth soaked in infected matter, but more usually shallow subcutaneous incisions were made in four or five places on the body of the recipients. Pylarini described the visit of one of these matronly inoculators – elderly, 'simple and honourable' – to the house of a well-to-do Greek who wanted to have his children inoculated but wondered whether the procedure was to be trusted. After reassuring the father, the inoculator made light punctures on the middle of the child's forehead, the chin and both cheeks. That was just the start of the operation, which also pricked the backs of hands, the interstices of toes, everywhere, in fact, but those fleshier areas of the body that Pylarini believed more likely to achieve a good result.[39] Unlike Timoni, who gave the impression (followed by Voltaire) that Turks in abundant numbers had adopted the practice, Pylarini knew that within the core of the Ottoman empire, inoculation was mostly confined to non-Islamic peoples: Jews, as well as Christians. The women inoculators were exclusively Greek Christian and Lady Mary Wortley Montagu described some of them making a pattern of incisions in the sign of the cross.[40] Islamic clergy were adamant that inoculation was a sacrilegious interference in Allah's judgement on the sinful and their vehemence on the subject was enough to deter the Turkish faithful from adopting the practice.[41] Seraglio concubines who had been inoculated to preserve their looks were most often from populations which had not yet been Islamicised.[42]

But the Muslim world was no more a monolithic bloc of belief and practice in the eighteenth century than it is now, and there is ample evidence that in some Islamic societies – Persia (where something like Chinese nasal inhalation seems to have been the norm) and the Maghrebi states of north Africa, and some parts of Syria – notwithstanding clerical condemnation, inoculation had long been widely accepted. In 1700, in a response to enquiries from the Royal Society, Cassem al Gaida Aga, the ambassador of the bey of Tripoli to the Court of St James's, remembered his father

> carrying us five brothers and three sisters to the house of a girl that lay ill of the Small pox and had us all inoculated the same day. Now he that had most of all, had not above twenty pustules . . . Otherwise this practice is so innocent and so sure that out of a 100 persons inoculated not two die whereas on the contrary out of a hundred persons that are infected with the small pox the natural way there die commonly about thirty. It is withal so ancient in the kingdoms of Tripoli, Tunis and Algier that no body remembers its first rise and it is practiced not only by the Inhabitants of the towns but also by the wild Arabs.[43]

Those 'wild Arabs' are likely to have been Bedouin or Berbers. Later in the eighteenth century, Patrick Russell, the Levant Company agent in Aleppo (whose side passion was the useful science of herpetology), wrote to his brother Alexander, a fellow of the Royal Society, that despite those solemn denunciations at Friday prayers, inoculation was widespread not just in the country around Aleppo but in Baghdad, Basra and Damascus. Kurds, Jews (after some rabbinical resistance) and Druze, the minorities of the region, as well as desert Bedouin, all practised it pre-emptively. In Russell's description, the little formalities of 'buying the pox' observed by Pylarini in Ottoman Greece were designed to reassure anxious children. Families of the recipients would come to the house of the donor bearing dates, raisins and sugar plums as a gift, so that an exchange of health would then take place between the two children; if the recipient child was too young, the role would be taken by his or her mother.[44]

In 1715, the Scottish ophthalmic surgeon Peter Kennedy arrived in Constantinople to research eye diseases in the Levant and Asia Minor. But, after he was introduced to Timoni, he became interested in smallpox inoculation and included a chapter devoted to the subject in his compendious work *External Remedies*.[45] His short account largely followed Timoni's and Pylarini's descriptions of the procedure, though, unusually, he addressed himself to whether inoculation would prevent a second infection. His answer was that in most cases this could indeed be assumed, but should there be a reinfection it would be relatively harmless and was known by dismissive nicknames like 'the bastard pox' or 'hogpox'. And then there was something more momentous to which Kennedy turned his mind. Could it be that a habit that was practised by Turks (as he thought) and certainly by Persians would find acceptance in his own country? His was the first mention that taking 'the fresh and kindly pock' was known in the Scottish highlands, where infectious matter was rubbed on to prospective takers, and so should be thought of not as an alien, but as a native, practice. But when he considered metropolitan Britain, Kennedy was more pessimistic. Despite the fact that in Constantinople there were 'several merchants' who knew that the account of inoculation's effectiveness was true, 'we in Britain probably being more Timorous and fearful of our lives in this case because of the great Mortalities which accompany this Distemper with us' were unlikely to adopt the habit even though it was maintained by its champions to be 'so innocent, it need be no more minded than giving or taking the Itch'.[46]

As it turned out, not everyone was quite that fatalistic. On 1 April 1717, Mary Wortley Montagu, smallpox survivor, wrote from Adrianople to her friend Sarah Chiswell, whom she had invited to accompany her to Turkey, that she would relate 'a thing that will make you wish yourself here. The small pox so fatal and so general among us is here entirely harmless by the invention of engrafting which is the term they give it.' She then describes the 'set of old women' and their September inoculating visits, which were organised as 'parties' of fifteen or sixteen families. The inoculator asked 'which vein you please' and then 'immediately rips open that you offer her with a large needle (which gives you no more pain than a common scratch) and puts into the vein as much venom as can lie upon the head of her needle and after binds up the little wound with a hollow shell and in this manner opens four or five veins'. Afterwards the children played together and remained 'in perfect health' until the eighth day when symptoms began. 'They have very rarely above twenty or thirty which never mark and in eight days time are as well as they were before their illness.'[47]

It seems likely that Wortley Montagu had already decided to have her six-year-old son inoculated but told her friend that her interest would not stop with the safety of her own family. It was not the English in general who would stand in the way of this 'Turkish' innovation, but, as she anticipated, the medical profession, which stood to lose financially once all their potions and powders, cordials and compresses, purges, plasters and bleedings could all be done away with.

> I am patriot enough to take pains to bring this useful invention into fashion in England and I should not fail to write to some of our doctors very particularly about it, if I knew any one of them that I thought had virtue enough to destroy such a considerable branch of their revenue for the good of mankind . . . Perhaps if I live to return I may, however, have the courage to war with them. Upon this occasion admire the heroism in the heart of your friend.[48]

But then Wortley Montagu was not just receptive to what she saw of Ottoman Turkey. Her 'Embassy Letters', published only after her death, make it plain that, almost as soon as she crossed the frontier from the Habsburg Holy Roman Empire, she fell headlong in love with almost everything about it.[49] An exception were the brutal raids janissary troops inflicted on country villages. In Belgrade she discussed poetry and the Qur'an, and contrary to strict Islamic law, sipped wine with the local governor, Ahmet-Beg, deciding that anyone so refined had, somehow, to be that surprising thing (for Europeans), a Muslim free thinker. In Sofia she rode in a Turkish *araba* carriage, happily hidden by a scarlet drape which could be raised to peer through the grille. A visit to a hammam was a thrilling revelation: two hundred women, their skin 'shiningly white', unmarked by either pocks or the welts of tight lacing, sipping coffee or sherbet as they lounged on divans, all distinctions of rank banished since they were all 'in a state of nature, or in plain English stark naked'. Their nudity, however, seemed the opposite of lascivious. 'I know no European court where the ladies would have behaved themselves in so polite a manner to a stranger.' While not joining the women in their bare liberty, Wortley Montagu satisfied their curiosity by opening the skirts of her riding dress to display her stays. The reaction of the women was to assume that the confining underwear must have been imposed by her husband, which in turn gave rise to an unsettling thought that perhaps they were right, and that it was she and her kind, and not the occupants of the seraglio, who were the real prisoners of male expectations. The enforced confinement described in books like Aaron Hill's *Present State of the Ottoman Empire* was, she thought, wholly misleading. The Turkish women she encountered were the freest people in that realm and indeed beyond. In Adrianople and Constantinople, she exchanged western dress for the coverings of the *asmak* veil and the *ferige* robe which allowed her to 'ramble all day' in the souk or even a mosque without detection. But she also ordered fancy costume: a muslin chemise, brocaded jacket, soft Turkish slippers and turbans.

Wortley Montagu was also taking lessons in Turkish and Arabic and in a larger sense was open to what the Ottoman world could teach Europeans rather than learn from them, hence the enthusiasm of her letter to Sarah Chiswell on discovering smallpox inoculation. Although she does not seem to have read John Woodward's account in the Royal Society's *Transactions* of 1714, it is likely she heard about the procedure from Emanuele Timoni in person since during the time the Wortley Montagus were in Adrianople, he was also in residence, acting as the sultan's physician and dragoman-interpreter for the British. It must also have been Timoni who found and hired the Greek woman who inoculated the six-year-old Edward in Constantinople the following year. If so, it would have been one of the last things Timoni did. Not long afterwards, according to the clergyman-traveller John Covel, a friend of his father's, driven to 'shame and despair' by the machinations of Mary's husband Edward (about what, exactly, is unknown, but it might have been something to do with Timoni's ambiguous dealings between diplomatic camps), the physician-dragoman-diplomat took his own life.[50]

This sad end did not hold up the momentous inoculation. Along with twenty liveried servants and a chaplain, Edward Wortley Montagu had hired a middle-aged Scottish bachelor surgeon, Charles Maitland, as embassy physician. In 1722, amid the heat of inoculation controversy in Britain, Maitland published his account of what had taken place on 19 March 1718. But unlike the descriptions of Pylarini and Mary herself of the 'honourable' and 'simple' women inoculators, Maitland paints a spectacle of rough elderly clumsiness from which the Scot had to rescue the boy. 'The good woman went to work; but so awkwardly by the shaking of her hand and put the Child to so much torture with her blunt and rusty needle that I pitied his Cries who had ever been of such Spirit and Courage that hardly any Thing of Pain could make him cry before; and therefore I Inoculated the other Arm with my own instrument and with so little Pain that he did not in the least complain of it.'[51]

The operation completed, little Edward was kept in a warm room (which in the Turkish spring would have been warm indeed). Five days later, Mary informed her husband, then at the camp of the grand vizier at Sofia, that 'the boy was engrafted last Tuesday and is at this time singing and playing and very impatient for his supper. I pray God my next [letter] may give as good an account of him.'[52] It did. The fever came and went; then a hundred or so pustules which swiftly oozed, dried and dropped, leaving no mark behind, other than scars from the needle jabs. Mary's next letter reported that 'your son is as well as can be expected and I hope past all manner of danger'. Since she heard nothing back, Mary sent a third letter, dated 9 April, repeating that 'your son is very well' but then adding, tartly, 'I cannot forbear telling you so, tho' you do not so much as ask after him.'

Three months earlier, at the end of January 1718, Mary had given birth to a daughter, also called Mary. In 1721, back in England after Edward had been recalled from the embassy having failed to achieve the peace between the Holy Roman Empire and Turkey that had been the whole point of his mission, his wife decided it was now time to inoculate their small daughter. A smallpox epidemic had swept over Britain, bringing terrifying mortality. Maitland would call it 'a destroying angel'. A number of Mary's friends, Twickenham neighbours and relatives, including her sixteen-year-old cousin Hester, had perished in the early months of the year. But there was no reason to assume what she was about to do for her three-year-old daughter would be uncontroversial; quite the opposite. For as the tide of terror and sickness rose, and the cemeteries and death pits filled, disputes over treatment became angrier. But they were, for the most part, confined to the old alternatives: hot or cold – blankets and fires or chilled wet sheets and lemonade; emetic purges or bleeding or both. What Mary Wortley Montagu was about to do was a first, at least in England, the folk inoculation in parts of Scotland and Wales being as yet unheard of in the metropolis. Moreover, the Royal College of Physicians was waging a forceful campaign against unlicensed surgeons. In this agitated atmosphere, for Wortley Montagu to import an untried foreign innovation – 'Turkish' inoculation, the deliberate infection of a healthy child who might never otherwise catch smallpox – seemed to her many critics to be tempting fate; worse, an 'unnatural act' for a mother, bent on some sort of experimental self-indulgence.

Only a woman as exceptional as Mary Wortley Montagu, someone who combined motherly tenderness with coolly reasoning intelligence, someone not confined by insular and traditional prejudice, could have risked this and faced down all the hostility that came her way. She did, of course, have the experience in Constantinople to bolster her confidence, but that was no guarantee that a second inoculation would be as harmlessly effective. On the other hand, it was obvious that Wortley Montagu did not wish her own terrible experience, recalled in every pock mark, to be visited on her daughter and namesake. So, she proceeded.

Charles Maitland was summoned from Hertford to perform the operation. He did not relish the prospect. Maitland was a fellow of neither the Royal Society nor the Royal College of Physicians. He had nothing of the esteem of Hans Sloane, John Freind or even John Woodward. He was a provincial nobody; not even a physician, just a lower-ranked common surgeon, and this was Twickenham, not Constantinople. Moreover, the stakes were incomparably higher. Lady Mary was a Figure: a close friend of the dramatists William Congreve and John Gay; an even closer friend of the passionately enamoured, but unrequited, Alexander Pope. Her own poems were published and much reviewed; Godfrey Kneller was painting her portrait *à la Turque*. Should things not go well it would be the end of Maitland's reputation and, with it, his livelihood. He was hard put to it, then. He did not say no, but he laid down conditions: first, that there should be a waiting time until the spring; second, that there must be knowledgeable witnesses attending the event – whether in a spirit of counsel or potential exoneration was left unsaid. He also wished an old friend of his from his home town, Dr James Keith, to be there, since that poor father had already lost two of his three sons to smallpox, and was anxious to see if there might be a way to preserve the last surviving boy.

The 'engrafting' of little Mary went ahead in April and was as much a success as it had been for her brother. The 'preparations' of purges and special diet much favoured by doctors were ignored given the little girl's robust constitution. She too was seen 'playing' not long after the operation and was also left unmarked when the dried pustules disappeared. Keith was so impressed (and emotionally grateful) that he had his surviving son swiftly inoculated. Wortley Montagu herself became not just a personal but a public champion of the procedure, inviting visitors to come and see her daughter, no worse for the inoculation. It was celebrity medicine: a combination of family sentiment and scientific reason, enacted

Godfrey Kneller, portrait of
Lady Mary Wortley Montagu, *c.* 1715–20.

and publicised to disarm public criticism about what, to many, still seemed a shockingly counter-intuitive procedure of dubious alien provenance. The campaign – for it immediately became that – could not have been launched without three decisive forces that were present together only in eighteenth-century Britain, two of them in the heart of the establishment, one very much outside it. First was the willingness, in the midst of a terrifyingly lethal epidemic, of senior luminaries of the Royal Society and the Royal College of Physicians to take a radical innovation seriously. Some of them, such as Dr John Arbuthnot, physician to Queen Anne, were not esoteric academic scientists but polymathic virtuosi. In addition to practising medicine, Arbuthnot (who had rejected the dangerously Jacobite side of his family) was an eminent mathematician and statistician, a satirist, inventor of the mythical 'John Bull', and a friend of Jonathan Swift; and he published, inter alia, on ancient weights and measures, and the first work on the rhetoric and uses of political lying. As an old Tory in what the Hanoverian succession had turned into a Whig world, Arbuthnot knew the ropes of noisy public contention and could act as a feeder for the second decisive agency in campaigning for a medical revolution: a freewheeling, sensation-hungry, prolific press. Many of the fly-by-night newspapers and journals were local as well as metropolitan, but for all of them, novelties such as inoculation were the catnip of profit. Third, and perhaps most decisive, was the active interest of the Hanoverian court. The affable Arbuthnot (Swift commented that the only poor thing about him was his waddle) was welcome at Leicester House, the residence of the Prince and Princess of Wales, and along with Wortley Montagu may well have played a part in prompting the interest of the princess, Caroline of Ansbach, who had nearly lost a child to the 'distemper'.

Relations between the king and his son ranged from frostiness to hearty contempt and hatred; at one point, George I was so displeased by his son and daughter-in-law's choice of god-parents for their new-born child that he had them put under house arrest in St James's Palace. But the royal heart might have been slightly mollified when Anne, the Princess Royal, contracted a bad case of smallpox in 1720, aged ten. The same year saw a tentative reconciliation of father and son, enough at any rate for Caroline to persuade the king to use his royal pardon in an experiment designed to demonstrate the effectiveness of inoculation. Even by the standards of the eighteenth century, the project was cold-blooded. Condemned convicts in Newgate prison were to be offered their freedom in exchange for undergoing inoculation. Initially it was thought that as few as two would suffice, but since that would scarcely have been a trial at all, the number was raised to six; three of each sex. All but one of the volunteers were awaiting the gallows. Mary North, a perennial shoplifter who had previously been transported but returned to steal again, was thirty six; the remainder were all young petty criminals. Anne Tompion, twenty five, and a possible relative of the famous watchmaker Thomas Tompion, picked pockets and had been caught taking eleven guineas from a couple she and her husband, a keeper of a bawdy house, had tricked into taking a pleasure trip on the Thames. Elizabeth Harrison, nineteen, had stolen the impressive sum of sixty two guineas from her mistress; John Allcock, twenty, made off with horses as well as silk handkerchiefs and cravats; Richard Evans, nineteen, had stolen 19 yards of Persian silk; and John Cauthery, twenty five, had taken all of three wigs from a barber, the sum of which barely but unambiguously qualified him for the death sentence, though at the last minute his sentence was reduced to transportation.[53]

Prior to their date with the lancet, the six were removed from condemned cells reeking of ordure and crawling with lice into quarters where they could be somewhat cleaned up. At nine o'clock on the morning of 9 August 1721, they were ushered into the inoculation chamber. There, they were unexpectedly confronted by a staring audience: twenty five physicians and men of science, most of them fellows of the Royal Society and Royal College of Physicians, including Sir Hans Sloane, Dr Johann Steigerthal, George I's personal physician, and, answering to Maitland's request that there be someone with actual experience of the procedure in Turkey, Dr Edward Tarry. But there was also a German onlooker who reported that when Maitland unsheathed his operational blade, the convicts could be seen to shake with fear. The six received nicks on both arms and their right leg, and then had the infected matter delivered into the shallow wounds. Mary North the shoplifter, much given to vapours, came over faint on this occasion; but the rest got through it with a modicum of calm, cheered, no doubt, by the prospect of their imminent liberty. This was just as well, since a few days later, impatient for the infection to take and disappointed that symptoms were not yet showing, Maitland inoculated them all over again with a fresh and, as he hoped, more effective batch of pus.

The experimental subjects were inspected daily, no part of their bodies, male or female, being left unexamined.[54] Maitland recorded the appearance of pustules on the thighs and breasts of the women convicts, as well as on faces and limbs, and noted their number, appearance and suppuration. Everything seemed to be going as well as could be hoped, though 'unaccountably' every so often, the prisoners would take matters into their own hands. On 18 August, John Allcock 'prick'd open all the pustules he could, (and there were sixty of them) with a pin'. Somehow the result was not disastrous. At the end of August, North abruptly doused herself in cold water (probably to cool the oppressive summer heat in their confined quarters), 'thence catching a violent colic' which lasted two days, though that too seemed not to jeopardise the experiment. (It might have helped.) On 6 September, nearly a month after the original inoculation, with all the prisoners judged to have fully recovered, they were 'dismissed to their several Counties and Habitations'. Some of them lost no time to revert. Six weeks later John Cauthery, the young wig-robber, was caught at it again, tried and sentenced once more to transportation. But as the temporary custodian of their physical, not social, condition, Charles Maitland was able to conclude, 'The Thing has been successful . . . far beyond my expectation.'

Princess Caroline must have agreed since, according to Sloane, she made an astonishing proposal to inoculate all the orphans of the parish of St James in Westminster. In the end, however, just six were treated in March 1722, but so successfully that they were turned into a living display on behalf of inoculation. Sceptics and admirers alike were invited to inspect the orphans at 'Mr Foster's House' between ten and noon of a morning and two and four in the afternoon. The two trials, together with inoculations taking place around England and reported to the secretary of the Royal Society, James Jurin, were enough for Sloane and the serjeant-surgeon of the king, Claudius Amyand, to recommend the procedure to Princess Caroline. On 17 April 1722, Amyand made incisions on the arms and legs of her two younger daughters, Amelia and Caroline, while Maitland applied the infected matter. The two girls caught the desired mild case of smallpox and, to Maitland's great relief, recovered perfectly.

Widespread press coverage of the trials on the convicts, the orphans and not least two royal children did not mean, however, that anything like a consensus had been established in favour of inoculation. That year, 1722, saw an outpouring of hostile opinion: fearful, indignant and deeply suspicious. In July, Edmund Massey preached a sermon 'Against the Dangerous and Sinful Practice of Inoculation' at St Andrew's Church, Holborn. The choice of pulpit was not fortuitous. St Andrew's was the bastion of High Church Toryism and had been the parish of Henry Sacheverell, impeached by Parliament for preaching a sermon against the infiltration of the true Church by 'False Brethren', a broad tent of heretics including Dissenters, Unitarians, Muslims and, needless to say, Jews (especially those masquerading as 'New Christians'). Since the Hanoverian accession, and the abysmal failure of the Jacobite rebellion in 1715, High Toryism had been reduced to political impotence. But it was still possible, through astute indirection, for Tories to preach against new-fangled, especially foreign, innovation; the Hanoverians; the Whigs; and, since absolutely everything was fair game for politicisation, the 'dangerous experiment' of inoculation, so obviously a monstrous interference with God's plan for the world. That one of the very latest promotions of inoculation, published just the previous year in London, was the work of a Portuguese Marrano Jew, Jacob (Henrique) Castro Sarmento, who had hawked his quinine-based 'Agua da Inglaterra' and was now settled in London, would have done nothing to dissuade High Tory churchmen like Massey that the procedure was dubious, alien and un-Christian: a mountebank miracle.[55]

Massey's chosen text for his sermon was a passage from the book of Job: 'So went Satan forth from the Presence of the Lord and Smote Job with sore Boils from the sole of his foot to his crown.' Lest any in the congregation suppose that any relief from the current affliction ought to be welcomed, Massey was adamant that any such act of mercy was the Almighty's alone to bestow. Anything else was the grossest presumption, by implication an act of faithlessness and a (literally) diabolical challenge to God's scourging of iniquity. Satan, the patron of inoculation, knew what he was doing. For when feckless men told themselves they were protected against infection, they became eager to commit the very sins which smallpox was punishing. Those who were the Devil's accomplices in this 'strange' (meaning alien) practice were 'foolish and unskilful men'. Together they compounded transgression.[56]

In June 1722, a month before Massey's sermon, another recalcitrant Tory, William Wagstaffe, physician at St Bartholomew's Hospital and a fellow of both the Royal Society and the College of Physicians, published a long 'Letter showing the Danger and Uncertainty of Inoculating Against the Small Pox'. Besides his medical credentials, Wagstaffe was something of a Character: 'much in social company' for his wit, though famously 'indolent', whose 'irregular habits' sent him to Bath for a cure. Whatever ailed the doctor got the better of him, for Wagstaffe died at the spa aged forty. But he had been one of the witnesses at the Newgate inoculations and the lessons he drew from the experience were the opposite of Maitland's and those of his friend and fellow Tory John Arbuthnot.[57] Whenever patients survived inoculation, Wagstaffe wrote, it was because they had not in fact had been given smallpox at all, their 'pimples' being much more likely to have been chickenpox or some relatively harmless infection. It was commonly known, he argued, that different people had quite different qualities of blood. Any prophylactic not taking into account those distinctions was dangerously indiscriminate.

But Wagstaffe's most serious objection, of a piece with High Tory insularity, was racial and truculently national.

> The countrey [*sic*] from whence we deriv'd this experiment will have but very little influence on our Faith [in its safety and effectiveness] if we consider either the nature of the Climate or the Capacity of the inhabitants . . . Posterity will scarcely be brought to believe that an experiment practic'd by a few ignorant Women among an illiterate and unthinking People should of a sudden and upon slender evidence so far obtain in one of the Politest Nations in the world as to be receiv'd to the Royal Palace.[58]

In other words, what might succeed in far-off warm climes was not universally transferable and very unlikely to work in Britain where what Wagstaffe characterised as the 'National' blood, not to mention the British air, was incommensurably distinct. Treatments of oriental origin would never take, not least because of differences in diet. British national blood was 'the product of the richest diet' consumed by even 'the meanest of our people . . . not famous for their abstinence'. And in any case the whole notion of exchanges of blood and body fluids was itself 'repugnant' and wholly untested, except in the unfortunate experiments done in the previous century with 'mangy dogs' and 'sound' hounds with predictably unfortunate results. Wagstaffe also shared the belief that apart from being 'useless' at what it purported to achieve, inoculation was an invitation to other maladies to invade an otherwise healthy body. 'New-fangled notions of infusing such a malignity into the blood is the foundation of many terrible Diseases', especially disastrous in tender, vulnerable children. 'Operators' claiming success could then, unchallenged, take their pernicious procedure into the provinces unless there was a way to prevent 'such an artificial way of depopulating a Country'.

Other critics in the same High Tory vein, like the surgeon Legard Sparham, posed as the defenders of Old England against such a destructive novelty, a medical version of the South Sea Bubble; at best a diabolical gamble, at worst a fraud perpetrated on monied aristocrats, the fashionable, the gullible and the easily 'infatuated' populace. Until now, Sparham wrote, one could 'never dream that Mankind would industriously plot to their ruin and Barter Health for Disease, nay what is worse give up little innocents a Prey to the most calamitous Ill from a wrong judgement of averting worse'.[59]

There was one notable, and eloquent exception, to the Tory chorus of disparagement: the amiable John Arbuthnot who, in a pamphlet defending – and recycling – Maitland and attacking Wagstaffe for misrepresenting him, showed the reading public that old men had polemical teeth. Arbuthnot turned upside down the warnings about the exotic origins of inoculation. The 'method . . .

utterly unknown here till of late [but] practised all over Turkey and other parts of the East an hundred or for ought we know some hundreds of years . . . obtained amongst an illiterate sort of People I here offer to recommend.'[60] What kind of Tory *was* this? One untroubled by the example of foreign bodies; also a shameless lyricist of the family romance which, in the publicising of the Wortley Montagu and Princess Caroline mother-child stories, had already struck a sentimental chord with the public. Lest people be affected by the prejudiced and the fearful, Arbuthnot wrote, let them consider 'what would not tender parents give to secure to them the lives and Features of their beloved offspring when they behold them disfigured by the loathsome disease, pittings, seams and scars! What films and Fistulae and sometimes Blindness in their eyes. What ulcers and imposthumes on their bodies, contractions of the nerves and even lameness of Life.' There was nothing to be scared of, he wrote, just consider the facts.

But when the disease suddenly appeared in a place where it had hitherto been unknown, especially in a culture founded on attention to the punishing dispensations of the Almighty, neither romantic endorsements of inoculation nor empirically recorded data were a match for fear and suspicion.[61] In the spring of 1721, smallpox arrived in Boston aboard a British merchant vessel. The outbreak was immediate and severe. By the time it receded the following summer, 5,759 of the city's population of 11,000 had been infected and 844 had died, a rate of one in six, nearly identical with the uninoculated death rate in London. One man at least, the Puritan minister at North Church, Cotton Mather, anticipated what had to be done. In 1716, he had sent a letter to John Woodward at the Royal Society reporting a conversation with his enslaved servant Onesimus.

> Inquiring of my Negro Man, Onesimus who is a pretty intelligent Fellow, whether he ever had the Small-Pox he answered both yes and No and then told me that he had undergone an Operation which had given him something of the Small Pox, and would

> forever preserve him from it, adding that it was often used among the Guranatese and whoever had the courage to use it, was forever free from the Fear of the Contagion. He described the Operation to me and showed me in his Arm the Scar.[62]

'Onesimus', meaning 'Useful', was the name Mather had given to the African servant ten years earlier in 1706; it was taken from Paul's Epistle to Philemon, relating the story of a runaway slave who had been converted by the apostle when they encountered each other in prison. Likewise Reverend Mather believed it his duty to Christianise the African, but apparently failed to overcome the latter's resistance. That same year, 1716, Onesimus bought his freedom by purchasing a replacement for Mather's household. Five years later, when smallpox was raging, the instructional relationship between enslaved man and master was in effect reversed by Onesimus's report. What was more, in *The Angel of Bethesda*, the book Mather wrote after the epidemic had abated, he reported speaking to a number of enslaved men in Boston from the west African Guaramantee region 'who all agree on one story': the practice of taking 'juice' from the pustules of the pox and infusing it into blood shed from cut skin. Mather was struck by the description exactly echoing the reports from Timoni and Pylarini that he had recently read in the *Transactions* of the Royal Society.

When the epidemic struck Boston, Mather made this known to the surgeons and physicians of the city and local towns such as Cambridge and Roxbury. His reward was to be the immediate target of vituperation, not least from two physicians, William Douglass and Lawrence Dolhond, who claimed to have personally seen soldiers, Muscovite and French, die from inoculation thirty years earlier. Mather's standing as a church minister did not prevent threats turning violent. One evening, 'my wife and children sitting in the parlor', a grenade was thrown into Mather's house. Fortunately for the family, it struck some solid Boston furniture, dislodging the fuse and thus preventing detonation. But the experience was terrifying.

Among the medical men, only the surgeon Zabdiel Boylston (also a surgeon's son), who had, he said, 'barely escaped' death from the disease, was persuaded by Mather's personal appeal, inoculating his six-year-old son Thomas, his enslaved servant Jack and Jack's 2½-year-old boy Jackey.[63] Though hostile physicians sneered at him as a mere cutter for (gall)stones, Boylston proselytised by distributing copies of Timoni's and Pylarini's reports to the Royal Society, and inviting citizens to visit inoculated patients, including those in his own household. But with stories circulating that the inoculated could pass on the disease, there were no takers. The town's governing selectmen adamantly refused to read the Timoni–Pylarini reports, much less make them available to the public. Heroically, none of this deterred Boylston from inoculating 242 patients in Boston, Charlestown, Cambridge and Roxbury. Many of them were children and adolescents, and a good number were unfree servants in white households, both adults and children. Without detailed guidance other than what he had read from the Royal Society, his techniques were 'the Turky way' as he understood it: using a sharpened toothpick to open a pustule for extraction, a lancet to make a quarter-inch incision; cabbage or colewort leaves, though sometimes a half-nutshell, to cover the site; and depending on the condition of the patient, bleedings, emetics and vomit encouragements, plus a variety of cordial concoctions including oil of almonds, syrup of marshmallow, black cherry water, 'plague water', syrup of violets, and tisanes from Virginia snakeroot and sheep dung.

Boylston attended all 242 inoculations personally and his descriptions of calming fearful families, middling and poor alike, are often touchingly vivid. Nine, mostly small children, together with Black servants and an Indian maid, were inoculated in the household of Harbottle Dorr. After the procedure, all were kept in a single room where 'the poor children in their sickness and the winter's cold' were gathered, 'one crying, another complaining, one wanted Drink, another to do his Needs, one to get up, another to go to bed so that together with opening and shutting of the Doors, the gingling [*sic*] of the warming pan, Fire shovel and tongs there was scarce a Minute in twenty four hours that all was still and quiet.'[64] Of Boylston's 242, none died or suffered the blindness that was almost as terrifying. The pox brought on by inoculation was what he called 'kind and distinct'; the number of pustules was limited, fevers relatively light and short, and scabs fell off leaving no pits or scars to speak of. But his evident success was no match for the combination of clerical and medical denunciation. Physicians like Dolhond and Douglass were going to have no truck with remedies recommended by enslaved Africans and passed on by their masters as 'virtuoso amusements', nor for that matter anything at all coming from heathen Turkey and the Levant. It was, they insisted, self-evident that 'the natural tendency of infusing with such malignant filth into the mass of blood is to corrupt and putrify it'.[65] Heaven forbid that such alien practices should be countenanced by good Christians. But Boylston, like his English counterparts, believed that his work was indispensable, as lives were saved, 'to the praise and glory of God'.[66]

Matters went better in England. During the spring and early summer of 1722, polemical sceptics could hardly hope to prevail against the likes of Sir Hans Sloane, Princess Caroline and John Arbuthnot. But in one sense, the critics had a point: after the treatment of the young princesses, inoculation did become *à la mode*, especially in London. But the publicity that came with fashion also risked disrepute should something go wrong, as inevitably it did. It mattered not that Maitland had defensively warned that some of the inoculated might die – though not, he emphasised, from inoculation. The servant of Lord Bathurst, whom his patron had had inoculated as a test, died, followed by the small son of one of the grandees of English politics, Henry Spencer, Earl of Sunderland, who himself had died just a few weeks before. Sudden doubts, of course, were a godsend to the papers and journals, and they made the most of it, stoking anxieties. Enthusiasm for inoculation, at least in London and among the well-to-do, cooled.

Not all the critics, for that matter, were choleric old-school Tories. In September 1722, a 'Plain Account of the Inoculating of the Small Pox by a Turkey Merchant' appeared in *The Flying-Post*, not questioning the value of inoculation itself, but fiercely attacking the ways in which it had been administered in England. In contrast to the simple and expeditious way in which the operation was executed in Turkey, the 'Merchant' believed that unnecessary preparations, potions and regimens had been concocted by physicians with greater regard to lining their pockets than to the welfare of their patients. Indeed those 'preparations', along with purges and bleedings, almost certainly ended up 'weakening bodies that were to go through a distemper', sometimes with fatal results.[67] The author – who was, of course, Mary Wortley Montagu – roundly denounced 'the Knavery and Ignorance of Physicians', declaring that she on the other hand 'would get nothing by it but the private Satisfaction of having done good for Mankind'. Her contrasts were dramatically invidious. In Turkey, just so much 'venom' was applied as would rest on the tip of a needle; in England, so much infected matter was brought to the incision as to be dangerous to the subject. In Turkey, a slight opening of the skin sufficed; in England, the 'miserable gashes' were so excessive that they actually endangered the loss of an arm. In Turkey, patients were not obliged to down quantities of useless cordials; in England, they were more likely to actually increase rather than diminish fever by doing so, and 'to such a degree as may end their lives'.

What Wortley Montagu was really complaining of was that a benign innovation from Turkey had been appropriated and exploited by English physicians and surgeons, clumsily and greedily, to the detriment of their patients. The more it had been assimilated into customary English practices, the more dangerous inoculation had become. Casualties of the deadly combination of ignorance and greed were, then, due to the mistaken belief that the British way was superior to anything ignorant matrons from the east could possibly provide. The pointed irony, however, was that, in order to launch this scathing attack, slightly softened though it was by the *Flying-Post* editor deleting some of her bitterest sarcasm, Wortley Montagu had to disguise her gender. Not long after her letter was

Jean-Étienne Liotard, *Lady Montagu in Turkish Dress*, c. 1756.

published, the terrier-like William Wagstaffe attacked the 'sham Turky-Merchant' but failed to identify either the name or the sex of the author.

Wortley Montagu's animus was against metropolitan doctors who she thought were exploiting the 'distemper' for their own profit and bringing inoculation into disrepute by compromising its effects with needless and excessive bleedings, purges, vomits and cordials. But her anger short-changed the work of physicians in the counties and country who adopted and championed the 'method' in the teeth of overwhelming local hostility. No less than her, they too, were heroes.

iii

SAFELY! QUICKLY! PLEASANTLY!

In the early spring of 1722, a Yorkshire physician, William Whitaker, passed on to the fellows of the Royal Society a letter he had received from a county colleague, Thomas Nettleton, about the latter's experience of attempting to establish inoculation in his home town of Halifax. The letter was so vividly revealing it was read to the Society in May of that year and subsequently published in its *Transactions* the following year.[1] As eighteenth-century Yorkshire physicians went, Thomas Nettleton was out of the ordinary. Though he came from a modest background in Dewsbury and was schooled at Bradford Grammar School, he got his medical education in Edinburgh and then at Leiden University, where Herman Boerhaave was transforming the study of physiology. Armed with his Leiden medical degree Nettleton returned to west Yorkshire. He would have been familiar with the severity of smallpox, but the toll it took in the winter of 1721 and the spring of 1722 shook him. In Halifax, 43 people out of 276 cases died; 38 out of 177 in Rochdale, Lancashire; and 189 deaths out of 972 in Leeds. Overall, the West Riding had a 20 per cent mortality rate.[2] Reading the accounts of Emanuele Timoni and Giacomo Pylarini, Nettleton had become intrigued by inoculation, which he believed had long been practised 'in those [Turkish] parts of the world with constant success'. When 'called to many in the Small Pox whose Cases were so deplorable as to admit of no relief', he decided to take the possibility of the 'method' seriously. To foes who called inoculation a crime, Nettleton replied that 'I was so far from knowing it was a Crime that I always thought it the duty of our Profession to do whatever we could to preserve the lives of those who Commit themselves to our Care. And I knew no reason why we ought not, with all humble thankfulness to Almighty God to make use of any means which his good providence shall bring to light conducing to that end.' Inoculation, in other words, was a gift from God, not a defiance of His will.[3]

Though he knew he was engaged in a prophylactic experiment, Nettleton wrote that he 'had succeeded so much beyond . . . expectations', his patients getting through the sickness 'with so much ease' that he was able to offer it both in the town of Halifax and in the villages in the surrounding countryside. Before long, he was being besieged by local families clamouring to have themselves and their children inoculated. As of the date of writing in December 1721, he had inoculated forty and with the exception of one death, 'all got well through the Distemper and are at this time thro God's blessing in very good health'. In contrast to the elaborate procedures Mary Wortley Montagu criticised in the practice of surgeons and physicians in London, Nettleton followed 'Turkish' simplicity, applying just two or three drops of infected matter into incisions made with a lancet. Sometimes he found it easier and just as effective to soak 'small pledgets of cotton' from the donor's pustule and apply them to the shallow wounds. Affectingly and unusually, Nettleton paid almost as much attention to his patients' state of mind as to the state of their bodies, endeavouring to 'to use all proper means to drive away all Fear and Concern'. As to their being made to abstain from all meat and liquor, the common-sense Nettleton thought such precautions overdone, irrelevant to the outcome, and potentially harmful if the patients were 'oblig'd to live too low'. Better to 'let them eat and drink as usual'. He also refrained from plying his patients with all kinds of medicines post-op, especially alcohol-infused 'cordials', preferring instead to let the fever and the pustules take their course.

In almost all cases of child deaths, Nettleton scrupulously

recorded the details of their sickness to make the distinction between those who died having been inoculated and those who could be said to have died *of* inoculation, of which there were, year to year, hardly any. But he was honest enough to describe touch-and-go cases, often of very young children, and recording the anxiety of parents (whom he names), many of them already having lost children to the disease and frantic to find a way of preserving those that survived. A Mr Turner had 'buried three' of his four children; the last was inoculated and survived. What to do about the children of John Symons's family put Nettleton in deep perplexity. The first to be inoculated ran a violent fever, was convulsive and died, though not necessarily from the inoculation itself. If Nettleton went ahead with others and they perished too, not least from having contracted the disease naturally from their brother, he 'would be charged . . . But if I did nothing I did very much fear they might all dye.' Being a decent man, Nettleton took the only decision he could 'wherefore I was willing to run the risk of my reputation rather than that the children should perish'. He did, however, warn the mother and father that 'I could not answer for the Success in case they had already catch'd the Infection'. In the event, one of the children, a girl, did die but the remaining two survived.

Nettleton's fears for his reputation were understandable. Though forty inoculations in 1721 was not a small number, he wrote that it 'might perhaps have been greater if I would have press'd it . . . I only took such as desir'd it of themselves, being cautious of persuading any Body to it because I had but little Authority hereabouts to support me tho' I sought to acknowledge the kindness of many of my friends who being concern'd that this Method would be of some use were very zealous to promote it.' Yet all of those supporters were up against 'the vigorous opposition' they met 'from many honest, well-meaning persons who could not but fancy that it is an unlawful and unsurmountable Practice. They have gain'd a great Majority on their side here as well as in other Places where it has been Practic'd; I only wish that they would have been less busy in raising and spreading false groundless Reports whereby the

Matter has been very much misrepresented.' As a result many had been deterred from inoculation, and they and their children 'unhappily . . . taken off by Small pox' were the victims of their own folly. But Nettleton was not discouraged and remained optimistically convinced 'that when this Affair is set in a true light and found to be always safe and effectual I believe all the objections raised against it will fall'.

What could be done immediately? It would help if 'those Gentlemen who have justly gain'd the greatest Honour and Reputation in our Profession' might be persuaded to declare publicly in its favour. This meant getting into the publicity fray; moving beyond a discussion of the like-minded and learned to a counter-campaign in the press against hostile publications like *Applebee's Original Weekly Journal*. This would start with comparative data about the chances of surviving smallpox uninoculated or with the 'method'. Publicising those numbers, Nettleton believed, would be 'the greatest means to be found in the Country to remove the unreasonable prejudices which do generally prevail against a Method which I believe has nowhere been put into practice with any other Aim than to do some service to Mankind'. There was a clear road to that promotional campaign. William Whitaker, the recipient of Nettleton's first report on Yorkshire cases, was a friend of James Jurin: mathematician, physician and, crucially, for a time both secretary of the Royal Society and editor of its published *Transactions*. Though he came to enjoy great esteem, and the favour of the Society's president, Isaac Newton, whom he revered and made the subject of his own publications and lectures, Jurin was not an establishment natural philosopher. The son of a Huguenot cloth-dyer, he had been a scholarship boy at Trinity College, Cambridge, and tutor to another Trinity scholar, Mordecai Cary, with whom he went to Leiden to study with Herman Boerhaave. Jurin then was another traveller on the frontiers of different kinds of knowledge and was open to exactly the kind of evidence-grounded campaign Nettleton was proposing. In May 1722, he read Nettleton's letter to Whitaker at a Royal Society meeting but then moved beyond it to ensure that their correspondence was regularly published in journals and newspapers. What followed over the next seven years was a unique collaboration in early epidemiology: however rough and ready, the first systematic accumulation of comparative data about the effectiveness of medical intervention.[4] Before it was over when the epidemic receded in the late 1720s, all kinds of information-gathering techniques had been pioneered and widely and unapologetically published. Because Jurin's own calculations of deaths from smallpox were tabulated from annual Bills of Mortality – and unreliable in disaggregating specifically smallpox-caused deaths in the sick, especially among infants likely to die of any number of maladies not connected with the virus – the only sure way was to initiate local house-to-house enquiries. In the 1660s John Graunt, who has a claim to being the first epidemiologist, scrupulously compiled statistics on plague deaths, but Nettleton was the first to pioneer methodical micro-surveys, sending 'a careful person, once a year' to every street in Halifax enquiring and recording cases of smallpox together with deaths and survivals.[5] By 1725, a printed requirement asked 'all persons concern'd in the Practice of inoculating the Small Pox to keep register of the Names, Age and Habitation of every person inoculated, manner of operation, days of sickening, eruptions, [and] the sort of Small Pox produced' and to submit information for the previous year not later than February so that Jurin and the Society could publish their updated record in the spring. Before long there were even preprinted forms to make all this simpler for reporting inoculators. Driven by the shared need to demonstrate their faith in inoculation's effectiveness and safety, especially in the face of accusations that the inoculated had no protection against catching the disease 'naturally', local inoculators were eager to report. Dr George Lynch of Canterbury expressly added to his account that it could be taken as 'faithfull' since it had been compiled by a 'proper person going house to house'. But as Andrea Rusnock has pointed out, those domestic visits could themselves trigger hostility and suspicion for their intrusiveness. Reporting a high number of cases was not what local gentry, merchants and shopkeepers wished to have aired in public.[6]

The assumption of the inoculators was that, however fierce the resistance, there was a potentially persuadable public to be reached: first, among those families who had already lost members to the disease and were frightened for the survival of their remaining children; second, within the ranks not just of enlightened physicians but the much larger category of surgeons and apothecaries in the provincial towns and country.

The division of labour and rank into which the medical profession had always split was assumed to hold for this new 'method' of inoculation. And to a great extent this was indeed the case. Inoculator surgeons, in London especially, worked under the supervising authority of physicians. Surgeons, who were still looked on by many of their superiors as the artisans of medicine, did the cutting and dressing; physicians were responsible for the regimes of preparation and recovery on which, so it was still believed, good outcomes crucially depended. It was the physicians (in London almost always fellows of the Royal College of Physicians) who were credited with the learning, enabling them to judge dominant humours – whether a patient, even (or especially) a child, was 'sanguine' or 'bilious' – and who prescribed diet and any restrictions on activity they thought necessary. So even Claudius Amyand, serjeant-surgeon to the court, who alone counted for almost a third of the total inoculations in Britain in 1722 and 1723, necessarily worked under the authority of the likes of Sir Hans Sloane, or Drs John Arbuthnot and John Freind. His patients, overwhelmingly drawn from the elite, would have expected nothing less.

But as the practice moved out into the provincial towns and country, this specialisation broke down. Physicians, concerned not to lose status and livelihood over a still highly controversial operation, and acutely aware of the few deaths of the inoculated publicised in the anti-inoculation press, must have been shy of lending their authority to what was still thought of as a risky procedure. Thomas Nettleton in Yorkshire, who both presided over and personally practised inoculation regardless of the possible cost to his standing, was exceptional in this respect. On the other hand, even the very limited demand in the earliest years meant that often surgeons worked on their own, and still more controversially, as Jurin's reports document, apothecaries took over the whole business. In Salisbury, ninety nine people were inoculated by two surgeons, Geldwyer and Foulks, together with a Mr Elderton, who, since he was not identified as 'Dr', was equally unlikely to have been a physician.[7] In Portsmouth, the naval garrison physician, Dr Brady, supervised the inoculation of all of six officers, but Mr Waller, an apothecary, is listed as having inoculated fourteen. More surprisingly, a Mrs Roberts 'near Leicester' and a Mrs Dorothy Ringe at Shaftesbury are listed as operators, an outlandish thing, but then an anonymous London woman is reported as having 'inoculated by herself', the operation vouched for – as if they were standing back as she did it – by her father, 'a person of credit', and a (presumably cautious) attending apothecary.

And there were places in Britain where none of the physicians' learning was needed, for inoculation had been practised since 'time out of mind'.[8] One region of folk inoculation was south Wales, in particular the counties of Pembrokeshire and Carmarthenshire. Dr Perrot Williams, well aware of accusations that a practice so foreign to British ways could not possibly be trusted, reported to the Royal Society from Haverfordwest that two villages to the west of Milford Haven, Marloes and St Ishmael, were perfectly familiar with the operation and so sanctioned by custom that no one knew or cared when it had begun. One nonagenarian fellow remembered not only his own inoculation but his aged mother telling him about receiving hers, thus welding together a chain of memory stretching back at least a century and half and possibly longer, to the time of Queen Elizabeth. It was a natural, native procedure then and called for little skill and no fancy lancets. The Welsh way involved abrasive, concentrated rubbing, at several points on the arms until the skin showed 'an excoriation or soreness', at which point the infected matter would be applied, sometimes on a wad of cloth. But there were also villages where it was more customary to puncture the skin with just the slightest pricks of a pin. Improvisation abounded. George Owen, the lawyer son of the Bishop of St David's, told Williams that he and five or six (he couldn't quite recall the exact number) of his school fellows used the back of a pen-knife on themselves or each other, until satisfyingly they bled a little. Williams, to whom nothing much came as a surprise, reported that this all went off perfectly well, producing mild smallpox and no lasting marks, without any of the elaborate preparations and post-op dietary regime that physicians insisted were indispensable. Very often too, the process of obtaining poxy matter involved buying it, so the most anciently British corner of Britain turned out to share customs practised on the 'Barbary Coast' of north Africa and in Syria. The more Celtic the region, the more likely folk inoculations were to be discovered: in the Hebridean islands for example or the remoter Scottish highlands.

Elsewhere, the reports to Jurin (more than a hundred of them) came from inoculators standing ready to perform wherever fearful people asked for their services. Most often they were gentry but not always; tradesmen, shopkeepers and farmers also got inoculated if they could afford it. The map of treatment spread throughout England – to Liverpool and Bristol, Bedford and Nottingham, Stratford-upon-Avon and Winchester; villages in Dorset, Hampshire, Sussex, Kent, Lancashire and Northumberland. Some of those who desperately wanted the treatment were prepared to travel long distances – as much as 40 miles in one case – to get it for themselves or their children. Likewise, dedicated inoculators like Nettleton rode the rounds of their county, and in his case beyond Yorkshire to the peaks and dales of neighbouring Derbyshire. It was an extraordinary breakthrough but not, as yet, anything like a mass phenomenon. In 1729, the last of the Royal Society reports, edited by Johann Gaspar Scheuchzer, Jurin's successor as secretary, counted just 1,087 persons receiving inoculation from around seventy operators over six years.

James Worsdale, *Portrait of James Jurin*, 1740s. He holds an edition of Isaac Newton's *Principia Mathematica*.

Given what Jurin delicately characterised as the 'warmth and zeal' of opponents, the hostility of local gentry in the provinces and London, and newspapers sensationalising the occasional death, especially when it occurred in aristocratic households, the limited uptake is not surprising. It was still an extraordinary leap of faith for a healthy person or a parent of a healthy child to expose themselves or their offspring to what was commonly called a 'poison', trusting that this would avert death or disfigurement. As James Kirkpatrick, the author of the encyclopaedic *Analysis of Inoculation* (1754), put it, 'Seeking security from a Distemper by rushing into the embraces of it could naturally have very little tendency to procure it a good Reception on its first Appearance.'[9] Nonetheless, Jurin, Nettleton and Williams, along with statistically minded colleagues like John Arbuthnot, were from the outset determined to prove numerically, beyond all possible refutation, that the risks involved in inoculation were minimal compared to those of catching the disease 'naturally'. In a letter to Dr Caleb Cotesworth, published first by the Royal Society in December 1722 and then more generally for the public a year later, Jurin posed the essential question which, once settled by numerical evidence, he believed must surely disarm both fear and scepticism: How much more likely was it that the uninoculated smallpox patient would die than the inoculated? And to those who believed that the chances of catching the disease in the first place did not warrant the risk of inoculation, he added the question: How likely was it that one could get reinfected?[10] Analysing the numbers he had received from Nettleton in Yorkshire, Williams in Haverfordwest and a surgeon in Chichester, Jurin calculated that of those who caught smallpox while uninoculated, one in six would die of it. On the other hand, any deaths among the inoculated were exceptionally rare (though not unknown). Nettleton reported none from the eighty he had inoculated the first year. Over six years, the average mortality rate of the inoculated was 2 per cent, that of the uninoculated at least 16 per cent (and for the most severe years 18 per cent). In contrast, the last Royal Society report in 1729 had just 26 of the 1,087 inoculated dying, by no means all of them clearly attributable to the treatment itself.[11]

None of these figures dissuaded hardened sceptics. In a reply to Charles Maitland's defence of inoculation, Edmund Massey's uncle, Isaac, a London apothecary, stated that the comparison was tendentious since it failed to take into account that the vast majority of those perishing from smallpox were also the poorest, living in conditions that made them likelier to catch the disease and suffer more severely from its ravages, thus skewing the figures of the uninoculated deceased. Had those of modest means been able to afford the kind of attentive surveillance available to the rich, the numbers of those dying would be reduced.[12] It was a fair point, but while he could have, as yet, no notion of the behaviour of viruses (though the word itself was in common use in the medical profession), Jurin had an educated instinct, and plenty of evidence from fatalities in the propertied classes that smallpox mortality, like plague, took no account of rank and circumstance. He remained resolutely optimistic that should the 'test of experience plainly declare for the affirmative side' of his two questions – comparative mortality and whether or not inoculation immunised against further infection – 'every gentleman who has the Honour to serve his Country in the capacity of a physician will have Integrity and Humanity enough to declare himself honestly and openly in Favour of the Practice . . . for if the Practice of Inoculation be really found to be a Means of Preserving Life it will not be easy to make the World believe that it is criminal to use it'.[13]

Wishful thinking. The power of numbers, especially since they were suspected and contested, did not sweep away opposition. For the time being, sermons beat statistics. Smallpox receded in the 1730s and with it the urgency to try what, to many, still seemed the voluntary courting of 'distemper'. Mighty tomes appeared, the mightiest of all Thomas Fuller's *Exanthematologia* (1730) and Kirkpatrick's *Essay on Inoculation* (1743) and *Analysis of Inoculation*, offering for the first time an aetiology of smallpox, the comparative history of past pandemics, and a strong endorsement of inoculation.[14] Those volumes were meant for a learned readership especially within the medical profession. But, cumulatively, they created a textual canon of inoculation lore: forerunners appearing in scriptural lands – Syria, north Africa, the Caucasus; wandering prophets and apostles (Timoni, Pylarini and Wortley Montagu) spreading the new gospel, performing miracles of pre-emptive mercy for the mighty and the humble, lords and prisoners, children, even suckling babes, all in the teeth of derision and enmity, while all around them, unbelievers perished.

Something of this scripture stuck in the public mind, its potential for salvation reinforced whenever smallpox pandemics returned across Europe, as they did in the early 1740s, '50s and '60s. With each fresh wave of death and disfigurement, the line between resistance and acceptance shifted. Little by little, inoculation, initially condemned as dangerously alien and an offence in the sight of God, became naturalised, nationalised, even an act of Christian charity. What was coming to be seen as an unavoidable cycle of epidemics encouraged the dawning of a civic consciousness in which protective self-interest was married to genuine social altruism and Christian duty. One of the principal benefactors of the London

Small Pox Hospital, which opened its doors in 1746 on Windmill Street, off Tottenham Court Road, was the much-travelled Dr Robert Poole, who, as well as working as the hospital's first physician, was an ardent Methodist and, under the name of Theophilus Philanthropos, the author of a series of devotional works. The hospital offered preparation and inoculation gratis to the poor, though £1 6*d*. was required up front to cover the costs of a possible funeral, a condition unlikely to have persuaded the nervous to occupy one of its thirteen beds. Another house in Lower Street, Islington, not far from Poole's own home, was opened purely for inoculation. But in 1752, the institution moved to Cold Bath Fields in Clerkenwell, where it could accommodate 130 patients in six separate wards. What had become a major institution now received the blessing of at least some members of the clergy. On its opening, Isaac Maddox, the Bishop of Worcester, preached a sermon insisting that true Christians should think of inoculation not as a presumptuous interference with the will of the Almighty, but on the contrary as another of His merciful blessings. Bishop Maddox's choice of pulpit – St Andrew's, Holborn, where Edmund Massey had condemned inoculation – was again not fortuitous. The 'method' had been anointed. That same year, 1752, the Foundling Hospital in Bloomsbury announced that it, too, would henceforth inoculate all children over three taken into its care.

This social expansiveness was not purely altruistic. The propertied classes, who had been most open to the new operation, realised that it was in their own interest to persuade ordinary folk, with whom, even in the highest quarters, they perforce had to rub shoulders, of the wisdom of inoculation. Household servants were among the first, outside the circle of their own children, to be inoculated.

Ultimately, though, in entrepreneurial England, it was profit rather than piety or civic duty which transformed inoculation into a mass phenomenon. Once inoculation made money for its providers, what had been seen as a suspect foreign import was transformed into a national habit. In April 1757, a Suffolk surgeon,

Robert Sutton, whose oldest child, also called Robert, had suffered from a brutal attack of the disease, advertised in the *Ipswich Journal* that he had rented a 'large house' in his village of Kenton to receive people 'disposed to be inoculated by him for the small-pox'.[15] Board, nursing, fish, fowl and tea would cost a patient seven guineas, five for farmers, the fee to be paid up front. Out-patients visited by Sutton or one of his sons on their circuit rides would be charged a guinea in the event of a trip to treat twenty or fewer, but only half a guinea if thirty could be bundled together. Sutton was betting that the demand for inoculation had made a leap from the elites of town and country to the much broader middle, trade and farming classes. The immediate briskness of his business proved him right. A second 'inoculation house' was rented in the autumn of 1757 and then a third custom-built as an in-patient clinic. Sutton took his treatment on the road to Framlingham, Harleston and Halesworth, and across the border to Diss, in Norfolk, bustling little market towns where herdsmen and shepherds switch-drove their flocks over the cobbles. In 1759, he created specialised houses respectively for 'reception and preparation', a third for treatment and nursing and a fourth for 'airing'. Word got around in the inns and churches, passed by families living in the rose-washed cottages of Suffolk, and Sutton made sure with repeated advertisements in East Anglian papers that it spread beyond the immediate county. In 1760, he announced in print that he had inoculated 200 patients in that year, none of whom had suffered any untoward effects.

Demand was so strong that it rapidly outpaced what Sutton and his sons could personally service. But the shortage was turned to commercial advantage. For what was in effect a franchising fee, surgeons and apothecaries were licensed in other towns to deliver the treatment using the shallow incision, fresh 'airing' and patent medicine promised to minimise side effects. The charge tariff was lowered to expand the custom. Five-guinea boarding was now the general rule, but Sutton also offered 'easier terms' for folk of more modest means, accommodated in a separate, economy-class house.

In 1762, patient-friendly improvements were announced. Sutton now claimed to inoculate with no incision at all. The number of pustules erupting on face and body would be less than a hundred, and bed rest would be minimised so that patients might speedily return to their farm or counter. The Secret Sutton Remedy would ensure that the distemper would make only the lightest visit. By the mid-1760s, there were sixty three Suttonian operators working across England.

Suttonian inoculation became a family business. Six of Robert's eight sons fanned out across England, dividing the rapidly expanding market. Robert Junior served broader East Anglia, moving up to bigger towns like Bury St Edmunds, and into Norfolk, and by 1763 (so his father advertised) had inoculated 600 with no ill effects.

Thomas opened a house in the Isle of Wight, James in Yorkshire. But it was the second son, Daniel, who took the Sutton brand to an altogether new commercial level. Location, he knew, was everything. His first house was at Ingatestone in Essex, on the main road (now the A12) between the port of Harwich and London. Not everyone in the small town was happy with this, but in 1764, when smallpox swept through Essex, plenty of local people lined up for treatment and through traffic, as Sutton had guessed, was continuous. Bulk inoculation was offered. That same year, Daniel contracted with the village of Maldon to inoculate its entire population: seventy 'gentlemen and tradesmen' and, in a recognition of the dangers of contagiousness, 417 of the village poor. The brand now had a catchy promotional slogan – SAFELY! QUICKLY! PLEASANTLY! – and, in an astute pre-emptive move, Daniel hired an Oxford clergyman, Robert Houlter, to preach the blessings of Suttonian inoculation in a specially built chapel. His sermon, 'The Practice of Inoculation Justified', was duly published and circulated while Houlter's son, Robert Junior, handled the promotional end of the business, later taking the practice to Ireland where he in turn licensed sixty 'partners' to act throughout the country. In 1762, Robert Sutton Senior inoculated 400 – by the standards of the time a good number. But where his father and brothers were inoculating hundreds in a year, Daniel treated thousands: 7,618 in 1766 alone, 22,000 since 1763. Speedy turnover was of the essence: 487 treated in one single day, an entire village of 700 pricked before and after lunch; the first documented clinical production line in medical history. In the late 1760s, Suttonian treatment went metropolitan. Daniel, together with his younger brother William, moved to London, opening a house for society's elite on Kensington Gore. Robert Junior operated in Paris for ten years, and the Sutton brand was sought after by the courts and elites of Europe. Emissaries came from Poland and the German states seeking to bring one of the Suttons to deliver treatments on the Continent. A rival inoculator, the Quaker Thomas Dimsdale, had since the 1740s been practising in Hertford (where he might have learned something from Charles Maitland) using a more intrusive technique involving an incision kept open with finger and thumb while a thread soaked in matter was drawn through it. Dimsdale's practice was successful enough that, in 1767, he was invited by Tsarina Catherine the Great to inoculate her and her son Tsarevich Paul and was rewarded with not just a great fortune of a cool £10,000 plus an annual pension of £500, but the title of baron of the Russian Empire.[16]

Not all of the mass inoculators made fortunes, but those who did ascended swiftly and impressively up the ranks. With profits from his treatments, Robert Sutton Senior bought a grand house at Framingham Earl in Norfolk and lorded it in local society. Baron Dimsdale multiplied his Russian fortune by turning to banking as senior partner in Dimsdale, Archer and Byde, and then more grandly in the bank of Baron Dimsdale and Son. That fortune in turn staked him a place in politics, and he sat as MP for Hertford during the 1780s.

Inoculation had become serious big business. But this had only come about by creating elaborate and lengthy preparatory and post-operative treatments, customised for individuals according, so the inoculators said, to their particular physical constitutions and dominant humours. Patients governed by sanguine humour, for example, would require regular and copious bleeding lest their 'high blood' aggravate their tendency to inflammatory fevers and make inoculation perilous. Bilious subjects on the other hand needed purges and emetics to ease their costive and clotting tendencies. Optimising receptiveness was a finely judged business, but the whole success or failure of inoculation was said to turn on these nice judgements. Only physicians who could call on their accumulated treasury of wisdom and who had access to a substantial pharmacopoeia could be trusted with the all-important preparation (usually lasting ten days to two weeks). They alone would know what adjustments and dosages should be prescribed when the fever broke, pustules erupted and filled; what cordials and tonics patients might need to aid recovery. This expertise, diagnostic and pharmaceutical, was what provided inoculators with their authority and their money. Even the more forward-looking – including those who followed Thomas Sydenham in prescribing a cold, rather than warm, care during the course of the illness – assumed preparation was indispensable to success. James Kirkpatrick's *The Analysis of Inoculation*, the most commonly consulted bible of procedure, called for emetics, consisting in some cases of ipecacuanha, oxymel and extract of squills; in others the 'Aethiope' combination of powdered tin, calomel and grains of rhubarb in tansy water was supposed to act as a 'vermifuge' (wormer).[17] Dimsdale, in so many respects a shrewd and careful operator, prescribed for many cases a ten-day preparatory diet of 'pudding', gruel, sago, something known as 'Gruber's Salt', powdered crab claws, calomel and 'emetic tartar'.[18]

Inoculators might differ on the contents and selection of these prescriptions, but the humoral assumptions on which they were based went unchallenged until a work appeared in 1764 which denounced them as spuriously unscientific and the customised preparations as completely pointless, based more on the self-interest of physicians than the welfare of patients, and in many cases, actually harmful. *Réflexions sur les préjugés qui s'opposent aux progrès et à la perfection de l'inoculation* was a revelatory and revolutionary thunderbolt which overthrew in 200-odd pages every major assumption governing inoculation, except the sovereign principle that introducing the material of infection into a healthy body would pre-empt, rather than invite, lethal illness.

The author was Angelo Gatti, to this day an unsung visionary of the Enlightenment.[19] A professor of medicine at Pisa University, Gatti had been born in 1724 in Ronta in the rolling country of Mugello, 40 kilometres north-east of Florence. Church schooling had the not uncommon effect of persuading him to abandon the seminary for studies in mathematics and medicine. His apprenticeship in the latter was at the ancient Ospedale della Santa Maria Nuova in Florence, but Gatti's horizons were always expansive. In the early 1750s, he travelled to Algiers, Constantinople and, intriguingly, England, thus following an itinerary that was becoming something of a medical grand tour for any young physician interested in inoculation. Along with his boyhood friend Giovanni Targioni Tozzetti, Gatti had become a convinced champion of inoculation, as yet hardly practised in the Italian states. But the British mercantile community trading in the booming port of Livorno introduced the procedure, which was just as well since they had probably brought the disease with them when it struck the city in 1756. The sudden spread of smallpox made Tozzetti and Gatti determined to bring inoculation to Tuscany, where Grand Duke Francis and his regent, the reform-minded comte de Richecourt, were sympathetic, as were, surprisingly and helpfully, leading members of the Faculty of Theology.

But what turned Gatti into the figure he himself drolly described as a '*petite célébrité*', the object of both admiration and execration, was a French connection. In January 1755, the most famous champion of inoculation outside of England, Charles-Marie de La Condamine – ex-soldier, scientific virtuoso, mathematician, intercontinental voyager and traveller and in every respect one of the more swaggeringly outsize figures of eighteenth-century culture – sailed into Genoa aboard the *felucca* made available to him at Antibes by Prince Corsini.

Charles-Nicolas Cochin, portrait of Charles-Marie de La Condamine, late eighteenth century.

Ostensibly La Condamine was in Italy for his health, but the real purpose of his trip was to secure a dispensation from Pope Benedict XIV for a marriage to his much younger niece and god-daughter.[20] Badly scarred by a childhood attack of the disease, La Condamine had been a schoolmate and good friend of that other famous smallpox survivor turned polemicist, Voltaire. In the late 1720s, they had been partners in a lucrative and not wholly legal scheme. A keen student of probability, La Condamine had figured out (though this did not require much in the way of advanced mathematical theory) that the recently established French state lottery paid out a greater sum than the entirety of ticket purchases. Pooling their investment, Voltaire and La Condamine made a handsome killing by buying up almost an entire issue. Official attempts to prosecute the two of them failed, but La Condamine sensibly decided it would be a good idea to use his winnings to finance a sudden trip through the Middle East to study the engineering of Egyptian obelisks and pyramids. The journey also took him to Constantinople where he remained for five months. In the Ottoman capital he picked up the lore of the pioneer inoculators – Timoni, Pylarini and Wortley Montagu – but also observed Marseillais traders inoculating their own families. While he was on his next major expedition, to Peru – commissioned by the Royal Academy of Science to measure the length of a degree of meridian at the equator – La Condamine witnessed the success of a Jesuit settlement at Grão Pará inoculating local Indians: a tiny remnant of the millions of indigenous people he knew had been annihilated by smallpox, imported by the Spanish conquistadors.[21] La Condamine returned from his Andean-Amazonian odyssey (one of the most extraordinary voyages of the century) with specimens of cinchona bark as an antidote to malaria, and *caoutchouc* – rubber – (both ignored or rejected) as well as a passion to convert France to inoculation. What Jean Delacoste and Voltaire had begun decades earlier, he aimed to finish. In April 1754, La Condamine delivered a lecture to the French Royal Academy of Sciences refuting objections to inoculation, the 'vampire' that drank the blood of vulnerable multitudes. In a dramatic analogy he was certainly qualified to invoke, La Condamine compared contracting smallpox to a lottery in which the chances of drawing a fatal ticket might be one in seven. In any given year in Paris, 1,400 people would draw a black ticket, the lot of which meant death. Exaggerating outrageously, he asked, rhetorically: What happens if inoculation is accepted? The number of those fatal tickets is reduced to 'one in three hundred, or five hundred or a thousand'! Varying his metaphorical mathematics, La Condamine rousingly declared, 'All future ages will envy us this discovery. Nature decimated but Art millesimated [*sic*] us.'[22]

Published as *Mémoire sur l'inoculation de la petite vérole*, La Condamine's lecture, heady with romantic hyperbole, was an instant sensation, going through five quick editions. In 1756, the first Frenchman was inoculated: the marquis de Chastellux, who like La Condamine combined a military career with a scientific bent. A minor vogue followed among the liberal nobility and intellectual elite. In an act of exemplary public significance, the duc d'Orléans, cousin to Louis XV, imported the Genevan physician Théodore Tronchin to inoculate his two children, the duc de Chartres and Mlle Montpensier. Inoculation was suddenly, fashionably, *de rigueur.* '*Bonnets à l'inoculation*' – decorated, creepily, with red-spotted ribbons – were worn to high-minded salons. The procedure now acted as a cultural sorting mechanism. To be personally inoculated or, more dramatically, to have one's small children treated, became a sign that one belonged to the enlightened classes; partisans of scientific knowledge, undaunted by the irrational superstitions of clergy. Inoculation was the badge of modernity; a sign of tender concern for children (the new cult); a determination to spare them the terrors of horrifying disfigurement or death.

The manifesto of inoculation, La Condamine's *Mémoire* was swiftly translated into English, Dutch, Spanish, Portuguese and, in 1756, Italian. One of its translators, the Florentine Filippo Venuti, met La Condamine in Livorno during the worst visitation of the disease, and became his publicist and promoter on what turned into an eighteenth-century approximation of a triumphant book tour. The circle of admirers and campaigners to whom La Condamine was introduced during his progress through the peninsula included Gatti's close friend Tozzetti who, with the blessing of the Tuscan government, had begun to inoculate poor children in Florence at the Ospedale degli Innocenti. So although no direct documentary evidence exists of a meeting between La Condamine and Gatti, it seems improbable that this did not happen, either in Florence or in Pisa where La Condamine went in 1755 to measure and ponder the marvellous inclination of its campanile.

La Condamine, though, was not the only French *philosophe* in Italy in the 1750s. The marquis de Choiseul, arbiter of French foreign policy and friend of the Enlightenment, got Gatti to inoculate his grandchildren in Rome and at some point Gatti met the materialist philosopher Claude-Adrien Helvétius. In 1758, Helvétius's *De l'esprit* argued, shockingly, that 'we are all the product of what surrounds us' and (taking John Locke further than he would ever have gone) that mentality was nothing more than the input of physical sensation. A year later, the royal licence to publish Helvétius's book was revoked and copies were burned by the public executioner. Three years later, in 1761, the now scandalous materialist invited his friend Gatti to come to France to inoculate his children.[23] That visit turned into a decade-long stay in France where Gatti was patronised and protected by Choiseul, who successfully recommended him to the king as a consultant physician. At the same time, Gatti quickly became the inoculator of choice for the radical fringe of the Enlightenment, operating on the three children of the atheist baron d'Holbach and being helped, edited and translated into French by the Abbé Morellet.[24] A good part of his attractiveness to the boldest of the *lumières* was Gatti's faith in the power of 'nature' (rather than a discredited pharmacopoeia and complicated preparations and treatments) to take care of the inoculated. Nature as healer had been the watchword of Tronchin, who had returned to Geneva and stayed put to rub shoulders, sometimes abrasively, with both Voltaire and Rousseau. But Angelo Gatti, persuasive, passionate, unguarded in the manner of the first generation of Romantics, seemingly a force of nature himself, was there, in Paris, at the centre of the inoculation storm.

In June 1763, with smallpox racing through the capital, Joseph Omer Joly de Fleury, the president of the Paris Parlement (a judicial not legislative body), banned inoculation pending a report from a committee of inquiry. It was to have twelve members, six known to be in favour, six against, and would deliver its results to the Faculty of Medicine. Threatened by the independent, royally appointed Academy of Science, the Faculty was an ancient professional medical guild, jealous of its monopoly and authority and for the most part bitterly hostile to what it considered the dangerous, not to say English, novelty of inoculation. Joly de Fleury was himself an adversary of the *philosophes*; he had ordered the burning of Helvétius's *De l'esprit*, and banned the *Encyclopédie* itself in 1759. As such, he was all too ready to believe the fallacy, derided by Gatti, that inoculation was the cause of, not the answer to, smallpox epidemics. Gatti quickly discovered that the patronage of Choiseul, the court and the leading lights of the salons did not protect him against ferocious public abuse. Expecting gratitude, he had stirred a hornet's nest of vituperation.

In response, he began to write *Réflexions*, which was published the following year under a 'Brussels' pseudo-imprimatur that barely disguised the fact that it had actually been printed in Paris.[25] The little book of some 200 pages is an explosive act of medical iconoclasm and, as such, almost calculated to make as many enemies as possible. Gatti's basic French was not up to the task in hand, so the translation was undertaken by his *philosophe* friend Morellet, a radical political economist in his own right. In a second book published three years later under the same pseudo-imprint, Gatti self-consciously revelled in his contrarianism. 'My ideas are entirely different from currently prevailing notions; the rules that I set out are diametrically opposed to those that have been followed up to this moment.' And indeed his targets were not, as they had been for Jurin, Nettleton and La Condamine, hardened anti-inoculators, but the majority of the *practitioners* of the 'art' who Gatti thought had brought it into disrepute and were fleecing patients, leaving the operation to 'empirics' as well as high-class but woefully misinformed physicians, most of whom based their treatments on half-digested gobbets of pseudo-biology serving their mercenary self-interest. There was only one other publication about inoculation that had launched a comparable attack: the fiery letter against expensively incompetent doctors, published forty years earlier in the *Flying-Post* by the 'Turkey merchant' who turned out to be Mary Wortley Montagu.

Unlike every other book written about inoculation, Gatti's belongs to a distinctively mid-eighteenth-century literary genre: personal, impassioned, even confessional. Its pages have touches of Voltaire's polemical mockery and Diderot's family-friendly *sensibilité*. It begins with a mock apologia, an insincere expression of regret for having to write it in the first place. Gatti then reproaches himself for naivety. He had 'always thought that it was enough to let inoculation be justified by its evident success during [a time when] the ravages inflicted by the pox assure everyone of its necessity'.[26] It never occurred to him, he goes on, that such a simple and effective practice would be met with such mistrust and so many obstacles to its acceptance. He had foolishly assumed that a treatment so widely practised 'for centuries' and endorsed in so many countries could not possibly be opposed in the enlightened society that was France, nor that it would meet with such 'bitterness and hatred' when something that 'concerned the good of humanity' was being discussed. Then he admits, flatly, 'I was mistaken in all that.'

Reluctant to believe that foes of inoculation would not yield to the evidence of case histories, Gatti decided to publish a list of all those – some 200 – he had inoculated since coming to France, none of them with adverse effects. Sceptics were free to consult the individuals. None did.

Gatti's dismay only deepened when he began to realise that in 'the inoculation war' he was opposed not just by dug-in enemies of the procedure, but by many of those touting it but who rejected a 'natural' treatment that dispensed with all the dietary regimes, potions and powders, emetics and blood-lettings, the lengthy confinements to bed, that had become standard 'preparation'. Those prescriptions, Gatti wrote, were the result of a complete misunderstanding of the true origin and nature of the disease itself. Worse, all the errors were perpetuated by a descriptive terminology which had no relationship whatsoever to the biological reality of smallpox. A term commonly used to characterise the nature of smallpox, understood as seething in the blood so that it could drive costive toxins to exits through skin pustules, was 'fermentation', sometimes also known as 'effervescence'. But, wrote Gatti, 'fermentation' (as distinct from fever) was a fantasy, since smallpox was not generated spontaneously within the metabolism as the result of some obstructive imbalance of the humours. The disease did not lie dormant within an otherwise healthy body until some atmospheric abnormality, bad diet or vapours arising from polluted environment triggered its onset. Had that been the case, even the most cursory reflection on the history of smallpox would have raised obvious doubts about humoral aetiology, since you would have expected smallpox to occur universally and regularly instead of locally and with intermissions of centuries. The disease had been unknown in the Americas until the Spanish took it there and altogether missing from Greenland until the arrival of Danish settlements. The reality was that smallpox was a virus (Gatti used the term) and was to be understood as 'foreign matter', entirely external to the body, transmitted contagiously from one infected human, or clothing contaminated by infection, to another. The viral carriers he characterised as 'atoms of poison' of 'prodigious subtlety', the 'invisible enemy that attacks us' and once within the body, capable of swift and copious reproduction. Anticipating by 130 years the pioneering research of Elie Metchnikoff, Gatti wrote that inflammation (often used synonymously with 'fermentation') was not, as was commonly held, the agent of the disease, but a salutary response; a 'necessary symptom' of the natural progress of the sickness.

All the elaborate matching of preparations to patients; the selection of optimal seasonal timing depending on age or thickness or thinness of blood; disputes over the quantity of pocky matter to be used; whether 'ichthorious' watery pus drawn from early pustules or thickly viscous material from 'ripe' ones was likely to promote a better outcome; the depth or shallowness, the angle of an incision; the number to be administered and their location on the body – *all* were utterly futile and beside the point. 'All the doctors have said, prepare the subject . . . And me? I say, do not prepare.' If anything, preparation was an obstacle to the essential goal of making inoculation expeditious, uncomplicated, inexpensive, safe and widely available. Examining prospective patients to see if their humoral or blood characteristics indicated a necessity for purges or bleedings or both, supposed to evacuate from the body the non-existent malignant substances that were thought to be lurking in the gut or veins, was a waste of time. Likewise, the elaborate diets customised according to the fatty or skinny appearance of the patient. All this was the product of '*l'art*': the body of accumulated wisdom confined to the medical profession and which Gatti in the wildness of his candour was saying was a spurious, ignorant fraud. All this elaborate performance did was to bring needless anxiety and expense to the already nervous patients, and often enough would have the reverse effect from what it advertised: the undermining of healthy bodies.[27] How did he know? Gatti writes that he himself had conducted modest but controlled experiments employing, for one group, emetics, dietary prescriptions and blood-lettings and, for the other, socially identical, group, no preparations whatsoever. The results showed conclusively that there was no difference in outcome.

At least in his two publications, Gatti did not supply any details about those comparative tests. But that such comparative trials, complete with controls, happened and almost certainly with his direct encouragement and cognisance is documented. The place, though, of this methodological breakthrough was London, not Paris. In 1768, four years after Gatti's *Réflexions* appeared, an English doctor, William Watson, published *An Account of a Series of Experiments Instituted with a View of Ascertaining the Most Successful Method of Inoculating the Small-Pox,* corroborating, from the evidence of a systematic comparative trial, all of Gatti's conclusions.[28] Watson was another unorthodox outsider in the medical community: the son of a Smithfield trader, a scholarship boy at Merchant Taylors' School, and for many years an apothecary with a sensational sideline in electrical experiments. His house at Aldgate became a walk-in demonstration theatre for those excited and curious about electricity, including the 'Butcher' Duke of Cumberland, who enjoyed a shock by touching Watson's apparatus with the point of the sword he claimed to have wielded at the recent battle of Culloden. In the middle of this lightning fame, Watson suddenly changed career, moving to Halle in Saxony-Anhalt to take a medical degree. When he returned to London, he moved to Lincoln's Inn Fields, a short walk away from the Foundling Hospital where admitted infants had been inoculated when they reached the age of three. Watson was appointed the hospital's senior physician in 1762 and five years later, in October 1767, carried out the first properly comparative trials testing whether preparation made any difference to the outcome of smallpox inoculation and whether mercurial and antimonial doses assisted post-op recovery.

Those trials might well have been triggered by discussions with Gatti, who was in London in 1767; he had already published his trenchant dismissal of the usefulness of preparation and may well have been among the 'eminent men' Watson mentions attending his 'experiments'. Given that this is the very first documented occasion, anywhere, of a recognisable clinical trial, albeit on a modest scale, Watson's protocols were impressively conscientious. Thirty one children were selected for inoculation, divided almost evenly between the sexes (one more boy), all with the same history of diet and dress and all in good health. All were inoculated in the same manner, with an obliquely angled lancet in two places on the left arm. But their preparation differed: one group of ten were given emetics and laxatives of calomel and jalap twice before the operation and twice after; another group of ten received senna laxative and syrup of roses before and after; and the third group of eleven received no preparatory doses at all. Results from the prepared groups – measured by the number of pustules, carefully counted by Watson's assistants (except on scalps covered by hair) – were, on average, precisely the same as in the unprepared group, and in all cases were in small clusters on face and body, none on eyelids or situated so close to the eyes as to cause the blindness which was a common misery for the uninoculated. The seventy four children who were inoculated during the course of the trials, including a second, eventually showed just 2,362 pustules in total, 'an inconsiderable number, indeed which physicians daily see in one limb only of an adult [uninoculated] person'.[29] The conclusions confirmed the heretical conviction that Watson and Gatti shared: that 'preparation' was entirely gratuitous to the success of inoculation, and that another fixation of inoculators as to whether pus should be taken from early watery or more developed pustules was equally beside the point: in all the studied cases 'though the treatment was so different the small pox was so slight as scarce to deserve the name of a disease'. Dosing patients with mercury- or antimony-based potions which were alleged to help recovery made no difference whatsoever. 'The boasted effect of the medical nostrums of . . . inoculators . . . are, in my opinion very little to be regarded . . . the most valuable nostrum of all, I apprehend, is not to do too much.'

None of the inoculators, including the Suttonians, changed their operational practices as a result of Watson's published trial results. And Gatti, who had certainly encouraged his English colleague, could not forbear from attacking fellow inoculators as well as anti-inoculators, comparing the attachment to 'preparation' to the imaginary fantasies of other cults to which the ignorant and prejudiced were uncritically devoted. All that was necessary for a good result, in both adults and children, was to ensure that they were not suffering from other, independently serious, maladies, which could indeed compromise the safety of 'artificial' infection. Even then, Gatti warned, given that there were no such humans who could claim to be in perfect health, it was unwise to wait until they could be declared altogether sound while an epidemic was raging. Once the subjects were judged in good *enough* health, all that was needed (as the Greek women knew) was the tiny amount of pus that could be held at the point of a pin or needle. And a minimally subcutaneous prick, between epidermis and dermis, not a stab into muscle, would invariably trigger the controlled infection. Puncture not incision should be the watchword. Moreover, attempting to maximise the number of pustules for an abundant ooze, so that patients would know that they did, for sure, have the real thing (and not, say, chickenpox) and have multiple exits for the toxins, was ridiculous and if anything dangerous. Just *one* pustule, Gatti writes, could be a sure sign of infection as much as hundreds.

Gatti's pages anticipate many of the techniques and norms of modern vaccination, though our own time stops short of the family home vaccination he thought perfectly practicable. Why? Gatti also offers, from sheer intuition, another piece of surprising advice. It would be preferable, he writes, if matter could be taken from the pustules of an *inoculated* patient, rather than someone who had come by the disease naturally, as the former was much more likely to convey a milder degree of infection, along with fewer, gentler symptoms. Flying in the face of inoculation practice based on the idea that the more pustules the better (short of fatal confluence) since the pocks were supposed to draw toxins from blood, Gatti insists (not least from first-hand observation, especially of children) that, on the contrary, the fewer pustules the better. And then without any benefit of bacteriology, he delivers a speculation, startling in its prescience, and which would become significant for modern defences against infectious diseases. 'I think', he writes, 'that the variolic matter which has passed through several bodies, being used for several inoculations . . . has less malignancy than matter from natural smallpox . . . perhaps one day . . . the poison going through several bodies may acquire an improved nature . . . [and] weaken, and denature insensibly through successive transplantations and finally cease to play a role so considerable among the several contagious diseases that have afflicted mankind.'[30]

The eye-opening radicalism of Angelo Gatti's two books is easy to miss. Their dull, formal titles seem to situate them as conventional additions to the already overloaded shelves of inoculation polemics. He wrote them under indignant stress and with a sense of wounded dismay that self-evident truth had such a hard time of defeating error. But running through them is also a strain of romantic victimisation we usually associate with Jean-Jacques Rousseau. Gatti's first, 1764, book was a shocked response to the bitter controversies surrounding a procedure which he had supposed, once the data had been digested, must achieve consensus, at least among the 'enlightened classes'. When the duchesse de Boufflers caught smallpox, three years after he had inoculated her, Gatti found himself on the receiving end of malicious personal abuse and recrimination, at the same time as the committee of inquiry was debating whether or not the 1763 ban could be safely lifted. Once the cynosure of Enlightenment fashion, Gatti, to his consternation, became the symbol of what were denounced as the false promises of inoculation, above all its guarantee against reinfection. In fact, none of its champions, including Gatti, claimed reinfection was impossible, merely that it was very rare, and when it did occur, as in the case of the duchesse de Boufflers, was never fatal and seldom severe. Gatti also pointed out that, usually, 'reinfections' were caught naturally after operators had botched the original treatment, failing to deliver the prophylactic mild dose. Nonetheless the abuse took its toll. Gatti found himself singled out as an upstart foreigner presuming to instruct native French physicians on their failings while the infection he had promised to eliminate continued its devastation. This would not be the last time unscientific opinion, including that of medical professionals suspicious of, or even hostile to, mass inoculation, turned on an innovator challenging their monopoly of authority. Falsely accused scapegoats recur with depressingly predictable regularity in the long history of inoculation. Like Gatti, they are often demonised as the bringers of false hope, the reckless spreaders of contagion, sometimes even secret spies or enemies of a nation's health. The inoculated, moving freely through the population, in this mind-set were a biological fifth column, the instigators (even when they were unaware of it) of new waves of deadly infection. In response, Gatti pointed out that while there were perhaps 300 inoculated in Paris, 'thousands of [uninoculated] people live, eat, play among us . . . walking in the streets, [congregating] in churches with the crust of smallpox still on their faces; scabs casually thrown away with no precautions . . . people dying of smallpox are exposed and buried in churches', their kith and kin attending the funerals. He himself had made proposals for the strict isolation of the infected but had been ignored; instead he was further attacked as an overbearing disrupter of piety, pleasure and the daily round of business.

Former friends and allies in the fashionable *monde* now shrank from him, though not Louis XV, who, despite failing to have himself inoculated (a personally fatal error, as it would turn out), appointed Gatti his consultant physician. The marquis de Choiseul, too, stood by Gatti, and brought him to live at his country estate. In 1769, in an act of scientifically grounded faith, the king and his minister ordered Gatti to inoculate the entirety of the cadet corps at the École Militaire. This was the first time that inoculation was seen to be a military necessity until, in response to an British act of biological warfare – the knowing distribution of smallpox-infected blankets to Shawnee and Lenape Indian allies of the French during the War of Independence – George Washington had his troops inoculated.[31] In 1770, Choiseul fell from favour and a year later Gatti – accused now of disloyalty to his protector – returned to Italy, practising and teaching in Naples, which, against the odds, and for the moment, had become the hub of forward-thinking ideas.

Gatti's two books would be forgotten by all except the occasional work of epidemiological history, along with his own presence and reputation. But not immediately. Matthieu Maty, the Huguenot secretary of the Royal Society, and sometime editor of the *Journal britannique*, which had brought English scientific and critical literature to a Francophone readership, translated Gatti's *New Observations*, albeit, as he explains in a preface, in abridged form and with its polemical exclamations heavily watered down. Maty's unsolicited judgement was that it had been Gatti's tactlessness, his impassioned determination to take on whoever stood in the way of the 'progress and perfection' of inoculation, that had cost him and his cause dear.

The prudent Maty had a point. Gatti did have a gift for alienating the medical profession whose support he needed were inoculation to be widely accepted in France. But he came by his contempt honestly and scientifically, convinced that in good conscience he must do everything in his power to sweep away the whole farrago of 'preparation' and mercurial potions as a mass of wretched 'absurdity' (a favourite word). The closed guild of '*l'art*', its spurious knowledge, was perpetuated only because it served the interests of the profession, not the people it purported to help. The three weeks' prep and three weeks' recovery (with mercurial doses), the emphasis on informed vigilance, the esoteric concoctions, were either a folly or worse a collective conspiracy to maintain the mystery of '*l'art*': the authority of its vaunted knowledge, and the gross income of its practitioners. But the truth, where inoculation was concerned, Gatti insists, was that there was no mystery and no artful practitioners needed to administer it. In this case, at least, 'nature, a language truer and safer than that of medicine' would see patients through. But, physicians feared, were nature alone to be trusted, both before and after the operation, what need would there be for doctors? Gatti goes further still. Doctors, he writes in the stirring, fiery polemical climax of the 1767 *Réflexions*, had an interest in making inoculation against smallpox almost as fearful, painful, lengthy and profusely poxy as naturally acquired infection, so that patients, once rescued from death, could be duly, and deeply, grateful to medical wisdom for their preservation. If the whole business was reduced to a minimal, home-care, family-administered procedure, gratitude would instead be due to mothers or wet-nurses, and where would doctors be then? Not needed, that's where. Perish the thought.

Angelo Gatti was that rare and dangerous thing: a physician with something of an egalitarian temper, who set his face against medical hierarchy. It is no accident that his publications appeared in the same decade as Rousseau's literary raptures – *La Nouvelle Héloïse* and *Emile* – proclaiming learning from nature over learning from books, intuitive love over institutional authority, innocence over fashion, and transparent simplicity over social convention. For Gatti, the success of public health, especially in so grave a matter as smallpox, depended on universal accessibility and domestic self-sufficiency. In such matters there was much to learn from folk who lived furthest from urban culture. It was, after all, those commonly condescended to as 'barbarous people who . . . for aught we know . . . were the inventors of inoculation', and who had passed on their practical wisdom from generation to generation. Some details, to be sure, the elderly Greek women reported by Timoni and Pylarini got wrong. There was no need to make multiple incisions in many parts of the body, nor to cut so deep as to yield blood that could be mixed with pus. Covering the site of incision with any sort of plaster, bandage or half-nutshell was likewise a bad idea since that was more likely to encourage septic ulceration than prevent it. But there was one quality which made those women good operators: their sex.

After all, it was likely, Gatti thought, that the whole procedure had originated with 'tender, fearful mothers desirous of preserving their children from a cruel distemper by hurting them as little as possible'. He remembered the sudden, brutal outbreak of a smallpox epidemic in Urbino in 1746 when, without benefit of physicians or even surgeons and apothecaries, mothers took it upon themselves to inoculate their children. 'Such was the voice of nature and reason; so it is that women have always gone about it.' In his first book, Gatti anticipated being accused of holding a 'bizarre' opinion in this matter but replied that, after all, people were perfectly prepared to allow women to act as midwives in what was a much more complicated and perilous undertaking than the mere prick of a needle. Mothers or nurses with no learned training, no instruments that would terrify children (or indeed adults), no lancets, no forbidding rules and regimens and orders for bed rest, could accomplish the operation 'as well as the best physicians', if not better. For 'who is a better judge than the mother of her child's health, who more dextrous to perform the operation, who less likely to frighten the child and more fit to divert it'? Such a parental act 'requires no assistance, no operators and no expense'. The 'business may be done in an instant on a sleeping child with little or no pain if you prick with a needle without acquainting him you are going to give him a distemper'.

Centre stage, then, in Angelo Gatti's drama of inoculation were those first darlings of the Romantics: mothers and children. Mothers – in Greece and Turkey, north Africa, Syria, Circassia, Georgia and Armenia, most likely in China and south Asia too – had understood what was needed to prevent the death of their children or the horrible disfigurement that would blight their future. Mothers – Lady Mary Wortley Montagu and Princess Caroline – had been the first disciples of inoculation. Then came the royal women: Empress Maria Theresa (who lost a daughter to smallpox and had another scarred by it); her daughter Marie Antoinette, who had seen Louis XV die horribly of the disease, and insisted on becoming queen that her husband Louis XVI and his two brothers become inoculated; the Russian tsarina-mother Catherine, who made sure Tsarevich Paul had the same treatment. Gatti, the father and husband, wanted the same informed solicitousness to move all families. 'The only recompense I wish for', he wrote, 'is the joy of a father and mother to see their children sheltered from danger.' But he also knew that 'inoculation would never become universal unless it has . . . simplicity, ease and safety'. One day that would happen, 'truth will triumph over error' and the whole matter would at last be settled.

But not yet.

PART TWO

West to east: cholera

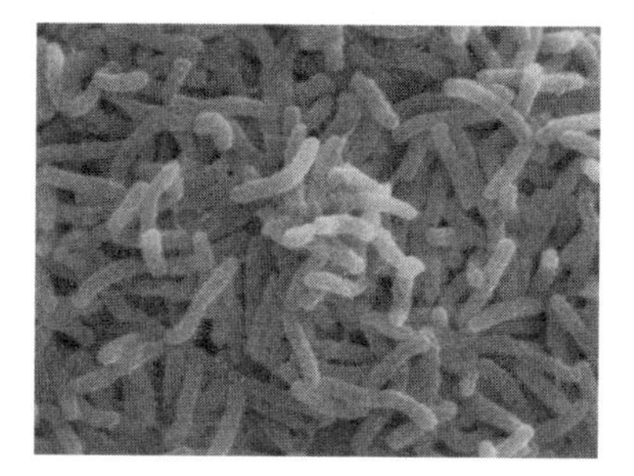

iv

Proust's travels

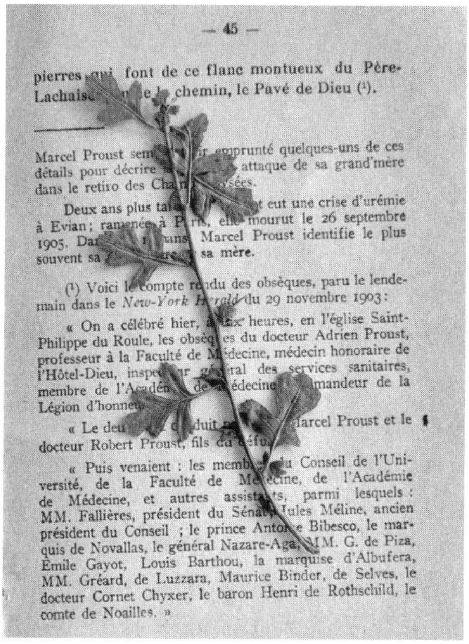

Some years ago in Paris, in a moment of idle curiosity, and for very little money, I bought a short book about Marcel Proust's father, Adrien.[1] Its author was the physician Dr Robert le Masle (also a translator, writer and illustrator) who was a friend of Marcel's younger brother, the urologist, Robert. The slight volume was meant as a commemorative tribute to the father of the brothers Proust thirty years after his death in 1903, but was published two years later in 1935. The delay, however, carried with it a stab of funereal poetry. That same year, le Masle and Robert Proust had made a pilgrimage to Illiers, the country town Marcel had turned into the idyllic 'Combray' of his childhood in a conscious revisitation of the expedition the family had made together in 1880. But in May 1935, Robert died, so the book became, between the lines, a double elegy.[2]

I knew nothing of this at the time that I bought the book. It seemed to me then merely a literary gesture of high-pitched reverence. The vaguely antiseptic associations of Adrien's career put me off doing anything more than taking a cursory glance at its pages. The most arresting thing about it, so I then supposed, was the commandment inscribed in English on the fly leaf, unsigned but dated July, vii, 1936: 'He who would not live in hell / Should read the books of Proust, Marcel.' Who would disagree? Away it went, sentenced to captivity in the chaotic obscurity of my shelves. But in the 'month of Mary', as Proust called it, May 2020, a whole new interest in public health led me to search for and recover the book, which, as sometimes happens as if self-animated, opened to a page that revealed something I had missed on first acquaintance. Between its leaves, as if inserted as a Proustian memory-prompt, was a small sprig of hawthorn, the *aubépine* whose heady spring blossom switched on the sensual wiring of the boy Marcel. Appearing in church, hawthorn seemed to the tingling youth a synecdoche of all nature. But hawthorn also features prominently in any pharmacopeia as a stimulant of sturdy cardiac health. Which is why it seemed ironic that this particular dried but exquisitely preserved sprig, retaining on its stalk minute thorns, lay over the passage where le Masle describes Adrien's death.

That demise on 20 November 1903 was, it was said, '*inopportune – untimely*' – a Proustian word, but fitting nonetheless. Adrien had been on the verge of a triumph, *finalement,* the realisation of his plan for a permanent international agency for public health. He had been attending the International Conference on Sanitation, the eleventh since its inauguration in 1851.[3] The 1903 meeting was Proust's seventh; he had sat through them all in Vienna, Dresden, Venice and Rome. He had bided his time though not always minded his temper with the frock coats, the bespectacled and opinionated, the epidemiologists and – alas – the diplomats: the intense Germans, the aggravatingly oblique, dependably disingenuous, English. Washington, in 1881, had been a steamship too far, even though Proust knew yellow fever would be the focus of discussion and he had been developing a serious interest in that particular disease along with all the other epidemics and pandemics. At international conferences on pandemics, he had become an indispensable fixture. *Le Figaro* had called him 'the founder of international public health'. That perhaps was a little much, but Dr Proust would not have brushed the compliment away.[4]

Nadar, photograph of Adrien Proust, 1886.

Some, the British especially, thought him pompous, arrogant, tiresome; others flattered his tirelessness. At sixty nine, on occasion he would succumb to moments of exhaustion, lowering himself on to the sofas of friends. He would rally, of course; it was expected of him. At the opening of the Paris conference in October 1903, he had given yet another speech on the *absolute necessity* of a permanent international public health agency. But while at all the other conferences he had registered the raised eyebrows, the rolled eyes, the theatrically stifled yawns, on this occasion it seemed possible that his vision of how the world would come together to manage pandemics might actually become fact. If such an institutional miracle was accomplished, it would be the capstone to his long career in public health. It had taken a mere forty one years of incessant campaigning, he joked to the gathering. But the long haul had worn him out. A beckoning finishing line can slow one down just as easily as trigger a sprint. It was undeniable, in his seventieth year, that he was noticeably more laborious: stouter; the broad, handsome face a little jowly; his brow creased; the carefully trimmed whiskers, silver. All the same, on 23 November, without taking his eye off the Eleventh International Conference on Sanitation elsewhere in the city, Proust got to his feet to address the annual meeting on tuberculosis, another of his special interests, along with infections of the brain, especially aphasia, about which he had written one of his many books. Aphasia – when a damaged brain makes vocalised words evaporate between thought and utterance – was a truly Proustian terror. He had also written and published on spinal damage, the involuntary writhing of athetosis, trichinosis brought on by parasitic intestinal worms and, lately, occupational diseases of the nervous system among metal workers, especially those handling copper.

But it was as a sentinel against the pandemics of the age – cholera, yellow fever and obstinately persistent, terrifying outbreaks of bubonic plague – that Adrien Proust had become famous in the medical establishment of the Third Republic, and beyond into the *monde* of science and letters. Or rather, he was a campaigner against the many inside the medical profession still denying that those infections were contagious. Instead, they argued that epidemics were discrete phenomena, occurrence and virulence entirely conditional on local qualities in the soil and the air. Sir Joseph Fayrer, the conference delegate representing British India (they demanded *two* representatives, of course, one from Britain, another from the Raj), insisted that cholera was just a variant of malaria, as indeed were yellow fever and plague, and, moreover, that the cause was to be found in a disturbance of atmospheric electricity. Early warning of a pandemic might be folded into a weather forecast.

On the morning following his speech to the lung people, Dr Proust paid a call to his younger son. Dr Robert Proust was already a successful urologist, a silver medallist in his professional examination: much for his father to be proud of, not that Proust *père* had exerted the least persuasion. Robert had studied with the charismatic founder of French gynaecology, Samuel Jean Pozzi, a recruit to Sarah Bernhardt's sizeable stable of lovers. In the last year of the Great War, Pozzi would be fatally shot by a patient who blamed the amputation the surgeon had performed on him for the onset of impotence. Not all wearers of prosthetic legs were as insouciant as Sarah, apparently. Robert Proust himself added gynaecology and urology to his standing – admirers joked that his pioneering procedure of prostatectomy ought to be known as proustatectomy – and was successful enough to be able to afford an elegant apartment at 136 boulevard Saint-Germain. When his father arrived for a morning coffee on 20 November, Robert was immediately struck by his unusual pallor. Acting on family instinct, he insisted on accompanying Adrien to the Faculty of Medicine where the professor was to chair a committee examining a medical student's dissertation. Just prior to that meeting, Dr Proust Senior went to his lab and then, anticipating a long academic examination, paid a prudent visit to the lavatory. He failed to emerge. When Robert broke through the door, he found his father lying on the floor, paralysed by a stroke. Adrien was taken to his apartment on the rue de Courcelles near the Parc Monceau where, two days later and without recovering consciousness, he died.[5]

The funeral was an Event; *le tout Paris*, or, at any rate, all those who mattered, showed up at the Eglise Saint-Philippe-du-Roule on the rue du Faubourg Saint-Honoré. Clinicians, epidemiologists, public health officials paid their respects, along with a former prime minister, Jules Méline, foreign ambassadors Scandinavian and Persian, French diplomats from the Quai d'Orsay, epauletted generals, and, notwithstanding the grieving Jewish widow who would survive her husband by just two years, a scattering of the *amitiés amoureuses* with whom (it was no secret) the deceased had been gallantly close (Laure Hayman, Marie van Zandt); also some of his elder son Marcel's friends whom the doctor-professor had beamingly indulged in his expansive middle age: a Prince Antoine Bibesco, Mathieu, comte de Noailles, the poet Robert de Montesquiou and, of course, a Rothschild. Marcel and Robert were pall-bearers. As it had always been, their burden was weighty.[6]

Adrien Proust had habitually called his elder son '*mon pauvre Marcel*' and, aggravatingly, continued to do so as the bronchial boy turned into the asthmatic man. There had been a time when the father, believing the asthma to be at least partly 'neurasthenic', prescribed the kind of tonics from which Marcel recoiled: gales of fresh air, blinding sunlight, shutters and windows thrown open to admit both. Marcel waited until his father was safely entombed at Père Lachaise cemetery before lining his study with cork and fitting impenetrably heavy curtains, so that his immense act of sensory and cerebral recollection, the visions, inhalations and dialogues, would become illuminated in the gloom, like radiantly focussed images thrown on a darkened wall by a camera obscura.

Not all of Adrien's medical doctrines had been sound or helpful. Convinced that compulsive masturbation led to homosexuality, he gave the sixteen-year-old Marcel ten francs for a therapeutic visit to a local brothel.[7] Unsurprisingly this did not go well. Reporting to his maternal grandfather, Nathé Weil (whom he implored to help with a loan so he could try a second visit), the anguished teenager confessed that 'in my agitation I broke a chamber pot, 3 francs, then, still flustered, I was unable to screw. I dare not ask Papa for more money so soon, so I hoped you could come to my aid . . . [Surely] it cannot happen twice in one lifetime that a person is too flustered to screw. I kiss you a thousand times and dare to thank you in advance.'[8] Then there was politics. Adrien was a middle-of-the-road republican, but this did not preclude falling out with his sons on the defining matter of the day: the innocence or guilt of Captain Alfred Dreyfus. The sons of their Jewish mother Jeanne were adamant Dreyfusards; Adrien, inexplicably, thought him guilty.

Nonetheless, as father and son grew older, they also grew closer, more mutually tolerant. After Jeanne and Adrien moved into their fine apartment in the 17th arrondissement, Marcel became a frequent visitor. This is not to say there were not bitter rows. After his father died, wanting to take back something harsh he had said in the heat of a political argument, Marcel made a comparison, unfavourable to himself, on how each of them behaved when feeling unwell. 'He was a lot better than me . . . all I do is complain . . . his only concern was that others not know he was sick.' Still more improbably, in the summer before Adrien's death, they became collaborators in public rhetoric.[9] Chartres, where Adrien had gone to school as a scholarship boy, decided to commemorate the twentieth anniversary of Louis Pasteur's vaccination against anthrax. As one of the school's most celebrated graduates, Adrien was asked to organise the ceremonies, for the occasion on 3 June, and to make a speech. Marcel was summoned to help. He had been spending the years since John Ruskin's death in 1900 attempting, with the help of his mother and two friends, to translate the Englishman's feverish prose. As if challenging himself, Marcel had selected one of Ruskin's definitively untranslatable outpourings, *The Bible of Amiens*, a Gothically serpentine chunk of a larger book about Frankish Christianity, intended (amazingly) for the young. But from Ruskin's communings with Gothic architecture Marcel lifted the detail of the 'stone of the Magus' which he claimed to see amid the sculpture of one of the door jambs of Chartres Cathedral. This allowed father and son, within the speech delivered by Adrien alone, to engage in a dialectical family two-step. Adrien deplored the barbaric superstitions encumbering medieval people's blinkered efforts to grapple with plague; Marcel invoked the sacred alchemy of the 'Magus'. Adrien celebrated Pasteur and modern medicine; Marcel reconciled faith with science.

The second occasion was still more unlikely, and more touching for being so. Adrien had agreed to distribute school prizes in his birthplace, Illiers, 25 kilometres south-west of Chartres. Standing on a platform, celebrated as the adventurous, much-travelled paragon of international public health, the doctor-professor was caught off-guard by a rush of unexpected sentiment. Prompted by Marcel's exercises in immersive recall, he found time arcing backwards. There, on the quai du Marché, opposite the church of St Jacques, was the house where, sixty nine years before, he had been born to Louis Proust and Virginie-Catherine Torcheux. The ground floor beneath their flat had been a general dry goods store, selling clogs, paper, sugar and home-made candles for both domestic and clerical use. Though the shop was long gone, sold by his mother after his father's death in 1855, Adrien thought he could still smell the wax, the honey, the spices on the well-stocked shelves. He was expected to praise medicine and the mission of public health in France and in the world beyond as the noblest of vocations, and so indeed he did, not stinting on the civic rhetoric. But for every allusion to the need for Modern Improvements – the shocking state of the river Loire, the unhealthy interiors of ancient dwellings – there was a dip into poetic vision. Weeds clogging the Loire turning stretches of it stagnant were, all the same, 'a beautiful tapestry'. If unwholesomely airless houses had to be knocked down, one should feel sadness alongside relief. Then followed a moment so surprising that it could have served as an overture to the enormous experiment that would consume Marcel. Looking around his little town, Adrien let a tear or two escape his spectacles as he went on in deeply Proustian vein:

> This emotion I feel coming back after sixty years [the count of time away from Illiers was not altogether accurate] you cannot be expected to understand. Not that I believe that fifteen-year-olds are in any way less intelligent than sixty-year-olds or less able to understand all manner of things . . . no, probably you are *more* apt. But there is one thing that is closed to young people, and which can only make itself known by an odd sensation of *pressentiment* . . . and that is . . . poetry; . . . the melancholy of memory.[10]

At that moment, the father was ventriloquising the son.

Marcel's realm of memory, if we are to believe the opening of *Swann's Way*, was, above all, a heightened sense of being. The intensity of Adrien's life, on the other hand, was driven by an impatient need for action. It had, after all, been shaped by growing up not maternally soothed with madeleines dissolved in tisane but in a storm of mass death. Half a century earlier, when he had been a young intern at the Charité Hospital in Paris and with little thought for his own safety, Proust had been responsible for triage, determining who among the incoming cholera patients could stay in the general ward, the Salle de Charles, and who had to be swiftly moved to the wards of the terminally sick. It had not been hard to make those brutal decisions. The infected arrived at the Charité already displaying the shadowed blue-grey complexion that augured a bad end. Most were racked by spasms that made them incapable of retaining any fluid. Futile top-ups of water, tendered by well-meaning nurses, would trigger further convulsive evacuations. All too soon, their metabolism would dehydrate to the point where vital organs stalled out, collapsing in on themselves. Half of them would die within three days. If the fatality rate at the Charité could be held at just 40 per cent, that was judged a great victory.

Eighteen thousand souls had perished in Paris alone in the pandemic of 1832; 100,000 in France. In 1865, a further 11,000 died in the city, which felt it had got off comparatively lightly. Adrien Proust had qualified as a doctor two years before and had developed special interests in brain disorders, but there was no escaping cholera, even though there was no medical consensus on how it arose or how, if at all, it was passed on. Pondering the pandemic of 1832 that had swept across Europe, Asia Minor and America, but as yet without any understanding of microbial pathogenesis, the majority of epidemiologists from India to America and throughout Europe classed typhus, typhoid, cholera and plague as variants of the same disease, arising in particular places for entirely local reasons. In doing so, they were remaining faithful to a tradition stretching back to inquiries into the cause of the Black Death.

Five centuries had passed since bubonic plague first arrived from Asia in Europe, and still the culprit was thought to be 'miasma': toxic emanations rising from the decomposing detritus of organic matter – vegetable, animal, human. In 1855, Henry Hartshorne published *On Animal Decomposition as the Chief Promotive Cause of Cholera*, in which he insisted that wind-blown particles of remains were the decisive agency of the disease. Those who may have found this eccentric still subscribed to the belief that local conditions – insanitary crowding, proximity to poorly maintained cemeteries, boggy soil, a concentration of foul air – could, in any unhappy combination, or just by themselves, create conditions hospitable to infection. The solution was not to inflict damaging interruptions of trade and business that were likely, it was said, especially in busy entrepots like Hamburg, Bristol and London, to cause more harm than the disease itself, much less to impose quarantines. Rather, it was to bring the full force of central and local governments to bear on eradicating local pollutants, whether they lurked in industrial tenements or the morass at the edge of a Levantine village. Local boards of health, accountable to a central administration, had been doing just that in Britain since the coming of the first dramatic visitation of cholera in 1832.[11]

But Adrien Proust the novice already suspected this catch-all explanation was an error. Cholera – and, for that matter, bubonic plague, which still periodically raged in the Middle East, Asia and Russia – had legs, he thought.[12] Infectious epidemics arose in places where they had been immemorially endemic, but to which the local population had developed what we now call natural immunity. But at some point the diseases penetrated populations of the healthy. Cholera had only entered Europe in 1817, and then with a vengeance in 1832 and 1854. The fact that the disease was not transmitted by the mere presence of more than one person in the same space did not mean it should not be classified as contagious, especially if there was shaking of hands. Even before there was any understanding of germ theory, it was plain to Proust and many of his contemporaries that there were countless social routes in which the infection could spread: through contaminated open sewers, ditches and canals; public and street fountains; soiled fabric and laundry. Proust thought he had already seen this in microcosm in Paris. As he followed Professor Guillot on his rounds through the wards of the Charité, the evidence of his eyes told him that the disease did not simply strike discretely, confining itself to districts where sanitation was poor to non-existent. Rich and poor alike occupied beds, just as moralising frescos of the Black Death had featured, the mighty taken along with the humble, all dispatched in their complacent prime. Though the life habits of the well-off, not to mention the access to a better (or at any rate a better-paid) class of physicians, may have stood them in good stead in fighting off the disease, it did not guarantee an escape. Prosperous citizens in the surrounding neighbourhood of Saint-Germain-des-Prés were laid low along with those who dwelled in the impoverished tenements of rue Saint-Denis or the swarming warrens of the Île de la Cité. Carriages drove from worse-off to better-off quarters of the city, the soiled fabric of a seat going with them. A little pitiless in his forensic curiosity, Proust would make enquiries of the addresses of the patients, observing especially details of their work habits and the routes they took to go about their daily grind. Daily life in Paris until the 1870s was unimaginable without *porteurs d'eaux* bringing water from the Seine or city fountains to street addresses. Wet-nurses coming into the city (still a profession in the second Empire) or market gardeners trundling churns of milk through city gates would take the disease back home with them, so that village churchyards gaped with graves. Cesspool collectors of human urine and excreta, sold for fertiliser, employed thousands of Parisians since the agricultural viability of an entire region around the metropolis depended on abundant crops. Huge deposits of steaming human waste were carted to depots outside the city where they stayed, ageing into the slowly solidifying black sludge that was one of Paris's major exports, sent as far afield as Normandy and over the Channel to England. The thriving trade assumed that excreta was not something to be terminally disposed of but profitably recycled. Even after the deadly epidemics of 1832 and 1849, epidemiologists had a hard time persuading governments that the business was not worth the danger.[13]

Convinced that data could change minds, the young Adrien Proust was already learning the social observation methods he would use far beyond Paris and France and which would win him recognition as 'the geographer of epidemics'. His provisional mapping, he liked to believe, was grounded in evidence. It was undeniable, he argued, that infections migrated contagiously, from arrondissement to arrondissement, from port cities to industrial towns, from the towns to the metropolis; for that matter, from country to country. By 1873, when the 39-year-old Proust published his exhaustive historical and geographical survey of the world's pandemics, the *Essay on International Hygiene*, he was already unambiguous. 'Cholera follows the routes of human travel . . . it is imported through human agency.'[14] As the railways and steamships shrank time and distance, the easier it became for infection to hitch a ride. Sanitary arrangements for the majority of passengers on a packet boat or a river paddle steamer, if they existed at all, were rudimentary and public, still more so for the crew. Disease, Proust thought, lay hidden in the waste, lurked in tainted drinking water.

Microbial pathogenesis was first revealed by the Italian anatomist Filippo Pacini some thirty years before the German Robert Koch, who, probably without knowing Pacini's work, took credit for the discovery. In 1854, the same year that Dr John Snow published his own determination that the cholera epidemic of 1848–9 in London could be traced to the single source of faecally contaminated water issuing from the Broad Street pump, Pacini published his 'Microscopic Observations' in the *Gazzetta Medica Italiana*.[15] Snow's forensic epidemiology was scrupulous, but once he had detached the infected pump handle, cholera bacilli disappeared, showing only generalised impurities as causal agents of the disease. Pacini, on the other hand, did autopsies on the bodies of victims in Florence and under the microscope actually saw in the intestinal fluids dense swarms of the microbes. The findings of both were initially dismissed and the two men might have bonded in the pain of their rejection. Evidence has come to light that they may well have corresponded with each other, perhaps sharing their baffled indignation.[16] Snow was furiously contradicted by much of the medical and epidemiological establishment in Britain. More scandalously, given repeated publications in 1865, 1866, 1871, 1876 and 1880, all of them copiously and sharply illustrated from his microscope slides, Pacini's breakthrough work was written off by those clinging to the belief that 'miasma' was the matrix of infection, his revelatory plates discounted as no more than entertaining illustration. The one exception was the British registrar-general, William Farr, who made a point of visiting Pacini while attending an international conference of statisticians in Florence in 1867. In the Hospital of Santa Maria Nuova, he found Pacini examining the intestines of a Danish artist who had just died of cholera.[17] So impressed was Farr that he devoted a special supplement to Pacini's work in his annual report on data in 1869. Two years later, Pacini declared himself 'amply compensated' by Farr's recognition, but when he died in 1883, no one else among the aetiologists of cholera had acknowledged his breakthrough.

A year later, Robert Koch, to immense acclaim in Germany, published his own revelation of the 'komma bacillus'. But a decade before, when Proust was writing his *Essay* on the history and aetiology of pandemics, he had an inkling of the truth, remembering what had been called 'microphytes' present in the body tissue of smallpox victims. The implication was that comparably independent organisms could be live 'germs of infection'. Self-evidently, when such germs travelled they were no respecters of frontiers. If they could turn epidemics into pandemics, then they had to be dealt with as an international, not national, crisis. In such circumstances, Proust wrote, what sense did it make for nations to deal with the contagion within their own territory while paying no attention to what was happening to their neighbours? If governments failed to understand the logic of this, then medical men would be obliged to turn diplomat to persuade them.

Which is what Proust trained himself to be. In 1851, in the aftermath of the 1848–9 outbreak of cholera, the government of the French Second Republic (not yet liquidated by its president, Louis-Napoleon) had taken the first steps to internationalise discussion of the character of the pandemic with the aim of agreeing measures of quarantine. Russia, which in 1849 had also suffered a major visitation of cholera, crossing its frontier with Persia, had an urgent interest in seeing something enforceable done. But nothing practical materialised. A second International Sanitation Conference was convened in 1859 with similarly inconclusive results. The third wave of cholera in 1865 was so brutal, and its progress from east to west so apparent, that a year later, with the disease not yet arrested, the conference moved, with surprising directness, to one of the worst-affected cities: Constantinople.[18]

It was February, a chilly month on the Bosphorus. Delegates from Sweden and the Netherlands, Britain and Portugal, the Papal States (still not incorporated into the new Kingdom of Italy), Prussia and Persia could comfortably wear their frock coats and overcoats as they sat discussing matters of life and death while their Turkish hosts smoked black tobacco, and waiters wearing tarbooshes on their heads and slippers on their bare feet brought trays laden with tulip-shaped glasses of tea. The city of perfect mosques might have been an awkward place to discuss the physical dangers posed by the hajj, the massed Muslim pilgrimage to the holy cities of Mecca and Medina. But the Ottoman Sultan Abdulaziz had travelled to Europe, and believed ardently in 'western modernisation', so he and his government were ready to discuss the health risks of the hajj, especially since the prevailing view was that it was Muslims from south Asia – Indonesia, India, Malaya and Singapore – who might carry cholera with them, transmit it while on board the crowded boats or in the squalid lodgings which were all most of them could afford. The infected might then take the disease with them, back to Egypt, for example, from where non-Muslims might in turn ship it west to the Mediterranean ports: Piraeus and Messina, Naples, Genoa and Marseille.

Whatever could be done to detain, inspect, disinfect had to be the product of general agreement between states east and west, north and south. Which, of course, was never quite attainable, especially when there were profound disagreements on how the disease arose in the first place and whether it was humanly transmissible. When the British delegate at Constantinople insisted that cholera 'was not contagious at all', he was echoing what had become a truism in the British public health establishment. John Simon, the formidable chief medical officer, first of the City of London and then of the whole of Britain, referred to the idea of any contagiousness as a 'peculiar doctrine' against which 'almost insuperable arguments have been stated'. Sanitary measures to correct the conditions from which the disease arose, he insisted, were necessarily and specifically local. The distinct impression given by British representatives at this and all successive Sanitary Conferences was that they had come only to pre-empt quarantines of suspected shipping lest they interrupt the flow of imperial trade, especially that from Hong Kong and India, to Britain. One of the British delegates objected to the assumption that cholera was transmissible and declared forthrightly that his task was to oppose any measures interrupting trade unless their absolute necessity could be irrefutably demonstrated.[19]

The proceedings at Constantinople, like all other international gatherings in the nineteenth century, were conducted in French, the lingua franca of most modernising projects in the Near East and Maghreb. But it was also the prominence of French doctors and scientists in Constantinople that made it clear that France believed the Levant to be a field of social as well as clinical experiment. Though the introduction of French 'modern medicine' into the Near East was not, in and of itself, an exercise in medical imperialism, it is true that alongside its linguistic and educational exports, the export of French science was intended to succeed where military conquest, courtesy of Napoleon Bonaparte, had so catastrophically failed. The

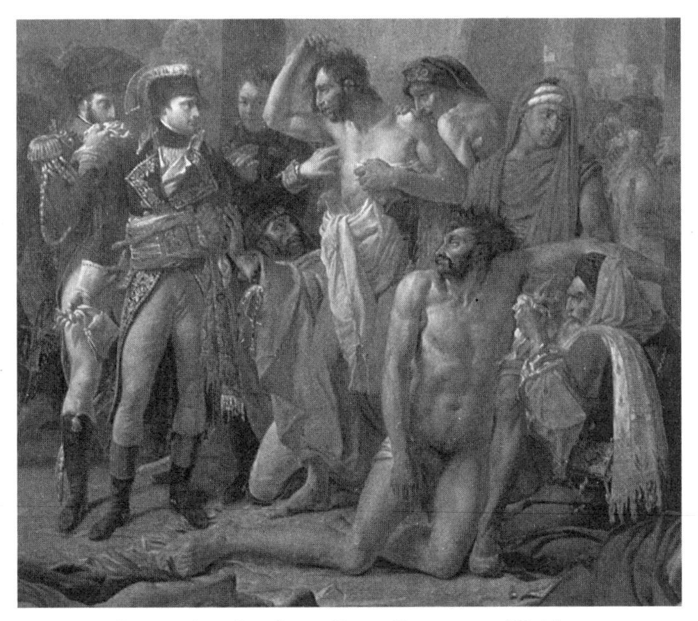

Baron Antoine-Jean Gros, *Bonaparte Visiting the Plague-Stricken in Jaffa*, 1804.

hero-general might reappear as engineer, linguist, scientist to 'regenerate' (a favourite word) the slumbering, torpid Levant. That power of revitalisation had already been immortalised in Baron Gros's painting of Napoleon visiting the plague house at Jaffa.

Prominently displayed in the Salon Carré of the Louvre, it is unlikely that Adrien Proust could have missed it. There stands the Hero as healer, compassionately and fearlessly touching the afflicted while his entourage shrinks back in horror, one with masking handkerchief (ineffectively) clutched to his face. But Napoleon (or his gifted visual hagiographer) marshals two sources of healing power: that of the Saviour, of course, but also that of the Merovingian-Frankish kings to whom unction had been mystically delivered and who by 'touching for the king's evil' cured sufferers of their scrofulous goitres. The truth was more macabre. Napoleon's actual – and characteristically cold-blooded – prescription for the unfortunate soldiers who had contracted bubonic plague in Palestine was that they should be put out of their misery by terminal overdoses of opium. Otherwise they would be a dead weight on his already exhausted and depleted army. Above all, the reputation of the Hero at home in France could not be compromised by any suggestion that his campaign for an Egypt liberated from barbaric superstitions and decrepit economic neglect was no match for nature's way with mortality (much less the British Royal Navy, whose victory at Aboukir Bay had left his army stranded). Gros, who had himself been in Egypt, was a reliable icon-maker and the painting went on display in the Louvre in 1804, between the announcement of Napoleon's elevation to emperor and his coronation in Notre-Dame. Thus it was that a shocking debacle was given an altarpiece of healing redemption.[20]

After the first, but not the last, desertion of an evaporating army, Napoleon never returned to the Middle East. But the cultural presence of the French was never forgotten by the post-Waterloo generations, either at home or in the Levant itself. 'Regenerators' arrived in Damascus, Cairo, Tripoli and Alexandria at regular intervals, many of them invited by the Egyptian khedive Mehmet Ali, for whom French-style modernisation was another weapon to consolidate his autonomy from the Ottoman sultan. The most famous of those medical regenerators was Antoine Barthélemy Clot, known in Alexandria, where he had instituted an exemplary public health regime, as 'Clot Bey'. But while Clot always had something to say and do about medicine in the Middle East, the figure who commanded most attention and respect in Constantinople in 1866, and who was, in effect, the reason why all the frock coats had gathered there, was Sulpice Antoine Fauvel.

Trained in Paris as a cardiologist, Fauvel had then switched to 'social medicine', specialising in typhoid and scurvy, becoming like Proust in the next generation a precocious *chef de clinique* at the next-door hospital of Hôtel-Dieu. But Fauvel, like Proust after him, came to believe nothing could be more urgent than epidemiology and in 1847 he left Paris to accept a post in the newly created Institute of Public Health in Constantinople.[21] One year later, during the next brutal surge of cholera – a reminder,

Dr Sulpice Antoine Fauvel

if ever one was needed, how fundamental public health was to the empire known to states in the west as 'the sick man of Europe' – Fauvel was given a seat on the Imperial Council and in the following year, 1849, as Turkey counted its bodies, he became professor of pathology.

Fauvel was the first to conceive of a permanent office for international health, even as he witnessed the diplomats and doctors of the assembled nations bickering among themselves about the legitimacy or illegitimacy of quarantines, the impossibility of *cordons sanitaires*; whether the interruption of business or the uninterrupted sweep of disease was ultimately more damaging; whether it was justified or prudent to impose restrictions on the hajj. The irreconcilability of conflicting interests, as ever, between British business-as-usual and French paternalism, on show for the long seven months of the Constantinople conference, reinforced Fauvel's belief in the ultimate necessity of an international health agency. His persevering efforts were noticed by the next generation of epidemiologists in France, and among them, the most passionately convinced was Dr Adrien Proust.

Which is why, three years later, in 1869, Proust found himself in Tehran, waiting in the shady throne room of the Golestan Palace to be received by the Qajar shah, Nasr al-Din, Zell'allah (Shadow of God on Earth), Qebleh-ye 'alam (Pivot of the Universe), Islampaneh (Refuge of Islam).[22] How far he had come from Illiers and the shelves stocked with beeswax and honey! His father, staring at the church opposite the shop, knowing his son was being schooled at Chartres, had set his sights on Adrien becoming a priest. But the youth's devotions had always been scientific. Much of his life would be spent in a flight from the parochial. He seemed always to have long-distance vision, measured not just in kilometres, impressive though that tally was: thousands of them tracked by the rumbling Paris–St Petersburg train, via Cologne, Berlin, Warsaw and Vilna, stops to refuel locomotive and passengers – '*Messieurs, mesdames, s'il vous plaît, vingt minutes arrêt de buffet!*' – and other stops for bodily relief (for there were as yet no on-board lavatories, which, given Proust's understanding of how cholera travelled, might not have been a bad thing). In St Petersburg, assisted by letters of introduction given him by the Ministry of Commerce, Agriculture and Public Health, which had commissioned the expedition, Proust exchanged niceties with officials from the Russian departments of Foreign Affairs and Public Health (a recent innovation), their guarded manner shadowed by lingering suspicions of interference persisting after the catastrophe of the Crimean War. Then, onwards to Moscow in Russian trains, surprisingly luxurious: blue plush velvet or flowered moquette to stretch out on; samovars on trolleys rolling along the corridor; silver cloches raised high by waiters in the dining car as if magicians, though the revelation was seldom more than greying carp surrounded by an entourage of boiled carrots and potatoes. Come evening, the candle-lighting in the compartment sconces became too dim for reading and Proust was reduced to drafting memoranda in his head, the train swaying and clattering on its rails as it made a turn to the north-east. At Nizhny-Novgorod, Proust gratefully alighted at the broad, sweeping bend of the Volga. Then south by high-chimneyed, coal-fired passenger steamer on the river, its surface changing from silvery to leaden depending on the heaviness of the clouds. Birch forests which further north came almost to the water's edge retreated as the boat sailed south, thinning out until, finally, somewhere between Samara and Saratov, the woodland disappeared entirely on the eastern shore, replaced by the dun emptiness of the steppe. As he stood on deck staring at tents set back from the shore, their Kalmyk occupants smoking long pipes in front of the flap-doors, Proust, the geographer of the pandemic, understood he had at some point crossed, somewhere or other, into Asia. Days stretched out like yawning eternity chugging smokily along upon the dead flat of the turbid water, before reaching Astrakhan where the trim hats and coat lapels worn by first-class passengers were made by wrenching unborn Karakul lambs from the wombs of the ewes, the tight-coiled glossy fleece peeled from the bloody foetus. Travelling downstream into the Volga delta, the shallower draught required smaller Caspian paddle-steamers, which on numerous stops filled with goats, Tatars squatting cross-legged on their packs while their horses clicked their shoes on the deck, Cossacks and Armenians in brightly embroidered sheepskin coats. By the shore, caravans appeared, the dromedaries kneeling to drink from the stream, or pushing slowly along, heavy bags hanging off their sides while their turbaned riders, high on the hump, shouted and switched the animals in futile attempts to get them to put on a burst of speed.

Disembarking at the Russian–Persian border port of Astara, Proust hardly had time to get his ground legs back before swinging them over the saddle on which, following ancient caravan routes, he would cover the remaining 500 kilometres to Tehran. He was supplied with the usual necessities: an additional horse for the baggage; a rubber mat and thickly woven blanket for sleeping; a steel cup and cafetière; riding boots, thick and tough; concentrated alkali to encircle his tent and keep snakes at bay; insecticide for the mosquitoes and biting sand flies; and, not least, a revolver.

The French legation in Tehran was housed in an urban palace, its architecture decorated with brilliant polychrome tiles and fluted columns the colours of marzipan twists. Heavy lengths of tapestry hung from the exterior of the windows, functioning both to shade room interiors on the second floor and, when they were sluggishly lifted by a breeze from the high plateau, to fan some air into the stifling chambers. Persians in tall, soft, conical hats were seated on cushions beside the reflecting pool of the paradise garden, or on the low terrace wall, one of them accompanied by a hooded falcon attached to a perch held in its keeper's right hand. Terraces bordering the length of the reflecting pool were thickly planted with rose and jasmine, but during the day they fought a losing battle against aromas invading from the streets and bazaars beyond the legation: the dung of mules, camels, horses and goats; raw sewage running in the gutters; but mercifully, less noisome odours wafted in from pyramids of drying limes and mounds of fenugreek leaves. It was late summer but by noon, uncovered faces roasted.

He was greeted cordially by the shah but also, as he had been warned at the legation, with polite professions of regrettable difficulty. Pivot of the Universe he might be, but Nasr had been unable to lever this exalted position into keeping his kingdom free from every manner of misery: swift and bitter defeats at the hands of Russia (which had stolen Persia's northern territories) and Britain (which had taken punishing exception to what was claimed to be Persia's infringement on its zone of interest in Afghanistan). Humiliation was compounded by periodic, horrifying outbreaks of bubonic plague, but more regularly, waves of cholera culminating in the devastating epidemic of 1865–6, along with droughts so severe they brought on bitter famine in 1869, the very year of Proust's arrival. He had seen black-draped bodies by the Caspian loaded into carts.

Nasr thought of himself as a modern ruler, which was not altogether delusional.[23] Enthroned at sixteen, he had looked for immediate guidance to the kind of counsellor who would enact the reforms that were meant to have turned neighbouring Ottoman Turkey into a modern state. Secular schools! Laboratories! Railways! Uniforms! Steam engines! Paddle-boats! Banks! Printing presses!

Hospitals! Amir Kabir, his chief minister, went through the recommended list and more. A modern polytechnic, the Dar al-Fanum, had been founded to bring western learning to Persia. Short on instructors, scouting recruiters had been sent to Vienna to hire medics, engineers and mathematicians to seed the new institutions. Alerted to these new opportunities in western Asia, the Europeans arrived in Tehran as they had gone to Alexandria and Constantinople. But there were never enough of them. Joseph Désiré Tholozan, former head of medicine at the French army hospital of Val-de-Grâce and a medical veteran of the Crimean War, in which more personnel died from cholera than the enemy's guns, became the shah's personal physician and persuaded him, by citing what had been done in Constantinople, to establish an official Council for Public Health. A telegraph service had been created, along with an official state newspaper, so that up-to-date scientific information might be communicated to provincial cities: to Isfahan, Khorasan and Qom. But Qom wasn't listening. It was, as ever, paying more attention to the Islamic ulema, which didn't hold with the new ways, especially the new ways of medicine. Cholera and plague continued to roll over the country and still the guardians of the faithful clung to ancient Galenic medieval texts, which held that sickness was the result of a disequilibrium of the four humours or else a punishment brought by Allah to scourge a people steeped in transgression.

The objections of the ulema, recited in Friday prayers, were forceful enough to incline the young shah to pause in his transformations. The reformer Kabir was made the subject of a demonising campaign and dismissed, the decision finalised in the traditional way by having the minister murdered. But the shah who greeted Proust so cordially, and who showered welcomes and exquisite carpets on him, looked disconcertingly not unlike himself: the same age, mid-thirties, the same carefully waxed moustache, bright bespectacled eyes, western-style military uniform crisply tailored. Between the exchanges of diplomatic smiles, Shah Nasr was happy to inform Dr Proust that earlier in the year he had convened the first *majlis* or assembly for sanitation. He promised, too, that he would send Persian delegates to any future International Sanitary Conferences. What he didn't say was what Proust and Tholozan already knew: that the ancient system of subterranean *qanat* shafts and tunnels, which brought water from their source near the northern town of Shemiran at the foot of the Elburz mountains, was an invitation to water-borne infection. The water in the shady pools in the shah's palace, and the drinking water for the court and the wealthy, were conveyed in conduits that never saw the light or dirt of day until stone slabs were safely removed in the interiors; in effect, a sealed system.[24] There was in addition an office of the *saqaa*, carriers who brought water directly from the sites where it first emerged to the surface, before it entered the city where it risked contamination. But for most of Tehran's people, *qanat* water surfaced through sluices in the city streets where it flowed freely into gutters in which clothes, including the linens of the dead, were washed, food was rinsed, and drinking containers filled. It was not uncommon to see children defecating directly into those running gutters.[25] The Farsi for the place of evacuation, after all, was *kinar-i-ab* or 'the water's edge'. Tholozan and the sanitation *majlis* might have pestered the shah on this, but with little success beyond ceremonial nodding of heads. Pressed by the Russians, who believed Persia to be the source of all the major epidemics that had passed through a porous border into their country, to institute a 'mixed' sanitation regime, with troops empowered to interrupt the passage of travellers coming from known infected regions, there was much shrugging of shoulders and raising of eyebrows. The shah knew very well that importing western medical, as well as economic, expertise did not come cost free. His neighbour the Ottoman sultan had paid it by in effect forfeiting full sovereignty over foreign policy and much else. So when Proust suggested that Persia might like to consider a comparable arrangement, whereby a regulatory health administration would be set up in which European epidemiologists would share duties with members of the Persian government, it was pointed out that, unlike with Turkey, the European powers eager to introduce their supervision into the shah's kingdom had not committed themselves to its defence against relentless Russian expansionism.

And besides, the royal government in Tehran had no power to dictate what provincial governors could or could not do in their cities and towns where *Vibrio cholerae* paid frequent visits. Nor would the shah stop thousands of the Shi'ite faithful from making pilgrimage every year to holy cities like Najaf and Karbala in Mesopotamia where cholera was endemic. Though Tholozan was distressed by the use of shallow temporary graves, pending reinterment close to the revered holy men at Najaf and Karbala, the shah wouldn't hear of forbidding the practice, much less stopping the hajj to Mecca, even though he had been told participants were likely to catch the disease from Asian pilgrims and then carry the infection back home. The shah knew that a significant body of opinion among European epidemiologists, especially the British, resisted any talk of the personal contagiousness of cholera. So if the disease could not be shown to be transmitted from the sick to the healthy, why court trouble by interfering with the pilgrimage? Besides, he had neither the right nor the power. Nasr Shah was, as the mullahs never tired of reminding him, merely the Shadow of God, not the substance. Proust resigned himself; took copious notes and accepted the graceful compliments rolled up in the shah's gift of rugs.[26]

Departing Tehran, Proust retraced his steps north and across the frontier to the Russian-Azerbaijani port of Baku. Sited as it was in the base of a bowl formed by surrounding hills, he could see that the city was twice blown: on the mountain side, the winds from the high country blew grit into the eyes of the inhabitants, who suffered from ophthalmic infections and a whole host of dermatological disfigurements; and on the Caspian side, damp winds and boggy ground bred anopheles mosquitoes that fed on human hosts and infected them, not just with malaria but with the dengue fever which Proust, along with a few other European physicians, reported on for the first time. The local *chef de service de santé*,

Dr Rustamov, inadequately paid by the Russian government, greeted Proust as if he were a veteran in the field of pandemics, walking him through the town and riding out with him into the surrounding country, all the time holding forth about the shortcomings of sanitary arrangements in the city, the web of provincial bureaucratic obstacles, and his frustrations at getting much in the way of improvements, with the result that the fatal surge of cholera, so horrifying in 1848–9 and again just three years ago in 1866, could be expected to return and men of goodwill and medical learning in the Bureau of Public Health at Baku were bound to get the blame for the entry of the infection into Russia proper.[27] What could he do (shrug) when it was known the disease had arisen over the border? The provision of guards was a joke. In St Petersburg, Baku was just another oriental backwater, not a way station to mass death. And some in the Russian medical establishment, like their counterparts in Britain, continued to assert – contrary to all evidence – that cholera was not, in fact, contagious. Faced with this fatal combination of indolence, obstruction and ignorance, there was nothing to be done to halt the caravans and boats crossing the frontier, not without cooperation between the two governments which, since Russia had annexed Persia's northern provinces, was unlikely to be forthcoming.

From Baku, Proust travelled west. An old-fashioned rattletrap *diligence* shook him to Tiflis and then, on horseback again, over the border to Turkish Trabzon and thence to Batum on the Black Sea. Having spent the first months of his epidemiological odyssey tracking the 'northern' route of the disease – and by some miracle never feeling a day's sickness himself – he now followed the secondary route, from the Black Sea to Constantinople and then, should infected pilgrims have returned to the Ottoman capital, the illness would follow out into the Mediterranean. In Constantinople, he was received with elaborate respect, admitted as a fellow of the medical society of the city and empire, and had the Order of the Medjidie bestowed by the grand vizier. After all this – more than 10,000 kilometres since he'd left the Gare du

Nord – a relatively leisurely sea voyage awaited on board a steamer of the Messageries Maritimes, stopping at Piraeus, Messina, Naples and Livorno before arriving at Marseille, where the traumatic experience of bubonic plague in 1720 had created its own vigilant system of inspection and quarantine. In between writing up the notes he had made in the Russian Caucasus and in Persia, Proust might have noticed flotilla traffic going in the opposite direction towards Alexandria and Port Said where, on 19 November, an event that would change the world was due to take place: the opening of the Suez Canal.

It was staged as the apogee of Napoleonic oriental imperialism; not a vainglorious conquest but the gift of French genius to the modern world. Vicomte Ferdinand de Lesseps had taken the first Napoleon's fixation with realising Caesar's dream of a canal from the Red Sea to the Mediterranean and made it happen. A chain, then, of three quasi-Caesars: Julius, Napoleon and, as Victor Hugo witheringly called him, Little Napoleon. To hammer home the message, the Austro-Hungarian *Kaiser-und-König*, Franz Joseph, sailed in aboard an imperial ship, as did Friedrich, *Kronprinz* of Prussia, and Khedive Isma'il, the ruler of Egypt, on their respectively pennant-heavy vessels. But the honours belonged to Lesseps's cousin Empress Eugénie, who rode a camel to the ceremonies, wore an haute couture outfit straight out of the opera house wardrobe for *Aïda*, complete with Pharaonic-Cleopatran head-dress, and who, aboard the imperial yacht *Aigle*, led the slow procession of boats up the canal from Port Said. On the second day, a French vessel, the *Péluse*, grounded just north of Lake Timsah, swinging helplessly about to block the entrance to the lake. To the imperial party's irritation, only the British HMS *Newport* had a draught shallow enough to manoeuvre itself past the obstruction, after which it did a little cruise of nautical schadenfreude around the lake before sailing on to Ismailiya.

The embarrassment would have been forgotten, or at least carefully avoided, ten months later when Eugénie conferred the imperial order of *chevalier* of the Légion d'Honneur on the shopkeeper's son Adrien Proust, at the Tuileries Palace. Nine months after the ceremony, the palace was burned down by Communards in the last days of their uprising. There was nothing left of the empire that had employed and honoured Proust for 'rendering exceptional services' to the cause of international public health. Eugénie had done the honours as regent, standing in for her husband, who was in command of the French army facing war with Prussia. Less than a month later, on 1 September 1870, that army found itself encircled and abruptly capitulated, the emperor taken prisoner along with more than 100,000 French troops.

Not a good moment for a wedding, then. But the day after the surrender of the Armée de Chalons, Adrien married Jeanne Weil, whom he must have met not long after returning from his mission to the east. He was the epidemiologist of the hour; she was the daughter of an Alsace Jewish dynasty that had made a fortune in the porcelain industry and, in the year of the first Napoleon's *coup d'état*, 1799, had staged its own coup by buying the porcelain factory of Fontainebleau. Jeanne was a catch, but then in his way, especially after his report was published in the *Journal Officiel*, so was Adrien: handsome, learned, held in high regard in government circles and the public health establishment, emphatically a man of the future.

None of which, for the moment, mattered. The day after their wedding, on 4 September 1870, the Third Republic was declared. Two months later, Jeanne was pregnant. En route to besieging Paris, Prussian troops occupied Chartres and Illiers. Adrien became anxious about the fate of his elderly mother, cut off from the rest of France by the encircling Germans. A bitter winter closed in on the capital. Food and fuel ran out. Famously, in desperation, when supplies of horses, dogs, cats and rats ran out, the animals of the zoo in the Jardin des Plantes were butchered and consumed, including its twin elephants, Castor and Pollux. Helpful recipes were published for dishes featuring kangaroo, antelope and wolf. As mortar shells from the Prussian guns continued to fall, Adrien took Jeanne and himself off to her uncle Louis's house in the leafy suburb of Auteuil, but that too came within artillery range and was badly hit. Through all the ordeal Adrien continued to go to work at the Charité Hospital, taking the Auteuil–Madeleine omnibus every day. He was *chef de clinique*; there were still patients to attend to. At the end of January, the provisional government of the Republic capitulated to the Prussians. But that was not the end of the ordeal. On 18 March, refusing to acccpt the terms of the armistice, soldiers of the National Guard declared Paris a commune. A second siege began, this time by the national army of the provisional government, led by Adolphe Thiers, marching on insurrectionary Paris. As this next round of fighting between Frenchman and Frenchman got under way, the heavily pregnant Jeanne begged her husband to stay away from the city. It was only when a stray sniper bullet grazed his head that Adrien saw she might have a point. In the last week of May, hawthorne blossoms unfurled, the Tuileries burned down and thousands were slaughtered in combat and executions.

On 10 July 1871, the baby boy was born in Auteuil. A little underweight, he was baptised Marcel. Not long afterwards, he started coughing and his father began to refer to the baby as '*mon pauvre Marcel*'.

Two months later in September, the government's National Sanitation Office issued advice on how to avoid contracting cholera. The instructions amounted to a curiosity shop of obsolete social and moral assumptions. Drunkenness, it was said, would especially make you vulnerable, while sobriety maximised your chances of going unscathed. Other filthy but self-evident vices which attracted infections of their own would compound the likelihood of contracting cholera.

Proust responded to these obstinate absurdities with his mighty *Essay on International Hygiene*. The title was a misnomer since the work was a 400-page survey of the history and geography of epidemics and pandemics from antiquity to the present time, from China to Guyana.[28] Aside from addressing at length the origins and transmission of cholera and yellow fever, a thread running through the volume was the startling persistence of bubonic plague well into the industrial era. Most Europeans thought of the Black Death as banished by modernity. But Marseille's episode of plague in 1720–2 had killed 50,000 of its 90,000 inhabitants in the first year. Before it receded, another 50,000 in the surrounding hinterland of Provence and the Vaucluse had died in addition. Mass graves were dug in Aix and Apt as well as Marseille. Such was the sudden shrinkage of the population that survivors had to be rounded up to collect corpses that had been decomposing for weeks on docksides and streets, and shovel them into mass graves. A 15-foot-high *mur de peste* was raised to close off the infected zone and the death penalty was imposed for anyone violating the *cordon sanitaire* ringing the city. When it was all over, the received wisdom, at least north of Provence, was that the Marseille epidemic had been an anomaly, an accidental import from the orient. It had arrived, after all, on a vessel trading Levantine cargos, the *Grand Saint Antoine*, which had sailed from Smyrna and called in at Sidon on the Lebanese–Syrian coast, Tripoli and Cyprus, all places where plague had never become extinct. In all likelihood, many said, it would stay there. For all his personal immersion in central Asia and the Levant, Proust's mind-set was still that of an imperial epidemiologist seeking to 'defend' western Europe. He knew that the Suez Canal was likely to increase the closeness of east and west and that it was impossible to have, as the British imagined, both untrammelled commerce and safety from infection.

He was not, however, a crude imperialist. Not least because he understood that if the epidemiological map of 'Europe' was extended north and east it would immediately reveal the stubbornness of plague in places far from the oriental conditions which, it was said, generated it. Early in the eighteenth century, there had been plague in southern Sweden; in the 1770s, a ferocious visitation in Russia had killed possibly a quarter of a million. Plague had stowed away on warships and in the Mediterranean theatre of the Napoleonic wars, stopping off at islands like Malta, Gozo and Cyprus. In regions of Turkey and much of Persia, it had never really gone away.

And as the European empires pushed further into north Africa, the Levant and south Asia, shortening distances with rail and steam, all kinds of infectious diseases would necessarily become Europe's business and its moral, as well as epidemiological, responsibility.

It was through the movement of humans, Proust wrote unambiguously, that pandemics spread their fatal net. Steam and rail were multiplying and accelerating those contacts, whether in the form of mass Islamic pilgrimages or the to-and-fro traffic of the colonial empires, expanding every year. This all now seems self-evident, but in the 1870s and 1880s it was anything but. Efforts to establish an internationally agreed system of inspection and quarantine, by Proust and his French colleagues at the International Sanitary Conferences in Vienna in 1874 and Rome in 1885, had been thwarted by the British and British-Indian delegates, hostile to any measures that might interrupt the imperial trade on which the domestic economy increasingly depended. Sir Joseph Fayrer, president of the Medical Board in the India Office and expert on snakebite venom, attacked the 'evil theory of contagion' for promoting 'futile' quarantine measures which would cause irreparable economic harm to imperial commerce and all those who depended on it. It was as though the mere notion that cholera could arise in India and then be exported west was an attack on the honour of British governance in the sub-continent and its habitual self-congratulation on bringing the blessings of modern civilisation to the benighted natives. When Alexandria suffered a severe outbreak of cholera in 1883, British delegates argued vigorously that the disease was actually endemic to Egypt, rather than India. It was, therefore, Mediterranean (for which read French) shipping that needed the most vigilant inspection, while British imperial ships, sailing directly from the mouth of the Suez Canal to the home country, should be able to do so without delay and obstruction.

Even Robert Koch's isolation of the cholera vibrio (the bacillus Filippo Pacini had seen thirty years before) in the intestines of a cholera victim in Calcutta in 1883 failed to convince the

Anglo-Indian medical and sanitary establishment that the bacillus was, in fact, the active agent of infection. Such was the inflexibility of the orthodoxy that even when the trio of Emanuel Klein, Alfred Lingard and Heneage Gibbes were sent out from Britain as an official 'Cholera Commission' and duly found the comma-shaped bacillus in the same Calcutta water tank from which Koch's victims had drunk, they nonetheless resisted the assumption that it was the exclusive agent of infection. The microbes, it was said, could be present as the result, not the cause, of the disease. D. D. Cunningham, Professor of Physiology at the Medical College in Calcutta, agreed with them, adding that he disliked what he called 'the German view' of microbial pathogenesis and transmission. Aetiology had become hopelessly political. The praise heaped on Koch in Berlin (by no means universal in Germany), and his elevation to the status of a patriotic hero, stung British imperial *amour-propre*. It was as if Bismarck was weaponising the science to embarrass the arrogant British, showing them just who had unlocked the fatal secret of cholera.

Imperial Britain was not yet ready to concede this prize of knowledge. Rebuttals were mobilised. The status of laboratory science as a guide to government action was itself questioned compared to the experience accumulated over generations by medical men and sanitary officers grappling with cholera in India since the first severe outbreak in 1817.[29] Compared with that body of knowledge, Koch, as J. D. Isaacs has put it, was discounted as no better than a scientific tourist.[30] The sanitary commissioner to the Indian government in Calcutta, J. M. Cunningham, who had issued his own report on a devastating outbreak in 1872, bluntly declared, in the preface of it of his *Cholera: What can the State do to prevent it?*, published in 1884, that 'the policy of the government of India is to reject all theories as a basis of practical sanitary work.'[31] A paper in the *Quarterly Journal of Microscopial Science* in 1886 was titled 'The Official Refutation of Dr Robert Koch's Theory of Cholera and Commas'. An alternative German, Max von Pettenkofer, long revered internationally as the dean of epidemiology, was coopted to contradict Koch. Pettenkofer conceded that the bacillus was a real presence, but continued to insist that it required locally receptive environmental conditions in the soil or the air, as well as susceptibility in the infected, before cholera could take lethal effect. Grudgingly, some officials in the India Office and in Calcutta acknowledged the recognition of vibrio might be helpful, but only in distinguishing between common diarrhea and cholera.[32]

Sitting at the 1885 conference in Rome, listening to some of these anti-contagionist arguments for rejecting quarantine barriers, Adrien Proust, now a convert to Koch's aetiology, alternated between aggravation and exhaustion. The arrival of the fifth cholera pandemic in 1881, beginning in Alexandria and quickly spreading across the Mediterranean to Italy, southern France and Spain, had made the need for an agreed sanitary policy, to halt the disease as close as possible to its place of origin, extremely urgent. In 1883, Proust had gone to Toulon and Marseille where the infection was raging, studied its progress from site to site, and predicted, accurately, that it was only a matter of time before it reached central and northern France. He may have been amused by the Turkish delegate turning the tables on Europeans who characterised the Ottoman empire as a nursery of infection. Possibly, now that the core of the disease was located much further to the west, Turkey should think of creating a *cordon sanitaire* to keep the 'European' microbe away? To the British sally that the lazaret quarantine stations in Turkey were notoriously unhygienic, Zerouas Pasha shot back that those in India were much worse.

Proust tired of the bickering, the reduction of what ought to be internationally agreed measures of public health to national and imperial games. He thought himself above all that. He was, now in his early fifties, a great authority on epidemics. After Fauvel's death, he had succeeded him as inspector-general of public health in France. The string of titles, honours and responsibilities got longer – secretary of the Academy of Medicine, professor of public health and hygiene in the Faculty of Medicine and so on. And in 1885, presumably on commission, Adrien was painted by a fashionable portraitist, Jean-Jules-Antoine Lecomte de Nouÿ, whose very name conjured up a Proustian salon fantasy. Adrien, fifty one, in prime middle age, was portrayed in Titian-Venetian style as a High Renaissance doctor in the original sense of that word, as a learned man, one not lost in airy abstraction but rather the personification of the Ciceronian ideal, a man in whom thought and action are indivisible. Hence the pose is that of the interrupted scholar, the gaze imperiously commanding as if to make it clear that he has no time to waste on idle chatter or the dreams of fools. His gown speaks of ancient authority sustained into modern academia. His beard is half white from experience, half black for energy. He wears his learning beefily. Time, says the half-empty upper chamber of the hourglass, is of the essence; it summons urgency. What, one wonders, would Marcel have made of *that*?

Pierre Lecomte du Noüy, portrait of Adrien Proust, 20 November 1886.

In December 1891, Proust made another journey into the hot zone of the pandemic. There was to be yet another Sanitary Conference in Venice in the new year. The British, together with representatives of the Austro-Hungarian dual monarchy, had made a pre-emptive strike to hold off any further attempts to impose quarantine measures that would damage or cut their trade with India. The substance of their proposal was that, at the entrance to the Suez Canal, ships' captains should make a formal declaration of bill of health on their vessel, and if no active cases were present, and the ships were going to sail directly to Britain, they should be able to pass through the canal without more ado. As France's senior delegate to the Venice meeting, Proust was not about to concede that, but nor did he want the stalemate of earlier conferences. In Cairo, he met the controller-general of Egypt, Evelyn Baring, and persuaded him of the urgency of practical measures, not least when it came to the hajj. More pilgrims than ever, many of them from Dutch Indonesia and British India, were being crammed on to steamboats in conditions that were ripe for a cholera outbreak. The event which had brought the British into direct control (after bombarding Alexandria) had been the revolt of the nationalist army officer Urabi Pasha against Khedive Tewfik's excessive compliance with European demands. But even after its suppression, Baring was aware of the beginnings of an Islamist movement that would break out in full force in Sudan in the following decade. On the one hand, he was wary of interference in the hajj; on the other, he bought into Proust's concern to bring an end to the cholera outbreaks which, in one year, had killed more than a third of the 200,000 pilgrims going to Mecca.

Looking over the British plan, Proust made another objection. Allowing vessels declared to be sailing directly to Britain through the canal – as long as they accepted two 'sanitary guards' on board to ensure no contact with land – overlooked the refuelling stop at Port Said. It was there that ships took on provisions for the remainder of the voyage. The merchants and vendors of all kinds coming on board might easily carry and spread the disease.

In beautiful, wintry Venice, afternoon fog rolling in from the lagoon, Proust reiterated those points and presented an alternative plan that stopped short of imposing the draconian quarantines on British Indian shipping that would be bound to frustrate any action. Dividing shipping into its different types from troop ships and passenger vessels to cargo ships and pilgrim boats, he prepared different sanitary regimes for each of the types. Ships that were declared to have an absolutely clean bill of health could proceed; others that came from infected ports of departure or had any active cases would be subject to a quarantine of five days, with passengers – European ones naturally – along with crew accommodated at a specially customised reception centre at the Wells of Moses, east of the canal, while goods and baggage were treated by the new steam engines of disinfection. Proust spoke about the poor conditions of crowding and shortage of water at the holding camp at Jebel Tor in the Sinai desert at the head of the Red Sea, but even as he did so he knew that the imperial sensibilities of sahibs and memsahibs would be mortally affronted by any suggestion that they should share quarters or even a site with Muslim pilgrims barely off the backs, as they imagined, of camels and donkeys.

The conference lasted three weeks – a mere twinkling of an eye by International Sanitary Conference standards – and this was because it came round to Proust's view.[33] Surprisingly, the British accepted the basis of his proposals (with a slight reduction of quarantine time). The delegates signed a Convention of Agreement. With one exception. The British, of course, withheld their signature, but promised to fall in line 'after due consultation'. Once they had extracted a few more concessions during an additional meeting in Paris in the spring, they kept that promise.

Adrien Proust was in a good mood, sampling the unusual harmony that had settled over the meeting like the pearly light of a Venetian January. For once he had managed to have international cooperation prevail over national self-interest, or rather had persuaded Scandinavians, Russians, Turks, Italians, Portuguese, Dutch, Persians, Americans, British from Britain and British from

India, and all the rest, that saving lives, forestalling pandemic catastrophe, was in everyone's interest; that they were all in the same global lifeboat together.

He went for a walk in the Piazza San Marco. The fog had lifted. Why not be a tourist? He bought a bag of corn kernels from a vendor, posed, half-smiling for a photographer, and fed the pigeons.

V

Sans frontières

Goodness, but he is beautiful, the Vaccinator. Or so Sarah Evangelina Acland – Angie as she liked to be called – has made him in her photograph.[1] Face on, Haffkine's countenance is broad, open, affably responsive. But in June 1899, when the picture is taken, he has become magnificent: the inoculator of Indians in their millions; their protector against lethal contagion. So Angie Acland poses him in profile, as chiselled as possible, heroically lit against a dark background. His look, now, is fine and foxy, the nose strong, the wide-eyed gaze projecting keen intelligence beyond the frame. Look a little more closely, though, and something else seems to register in the portrait: self-consciousness trapped in fine tailoring. The soft trench-coat is opened to reveal double-breasted lapels. The white winged collar belongs to the wardrobe of a casually superior gentleman. But although, two years ago, he was made a companion of the Order of the Indian Empire (CIE) in Queen Victoria's Diamond Jubilee Honours, Waldemar Mordechai Wolff Haffkine, though impeccably dressed, is not a casually superior gentleman.[2] He is a hard-working experimental bacteriologist, the creator and deliverer of vaccines, not just against cholera, but also against bubonic plague, for the Black Death has returned to prowl through the modern universe of Edisonian electricity and automobiles. Haffkine is also, and unapologetically, a Jew from Odessa, and this matters. Fourteenth-century Jew-haters had accused them of poisoning wells, of being demonic instigators of mass death. But here was a Jew who, given the chance, would inoculate the world against it.

So it was now put about in high circles of British science and government that Waldemar Haffkine was a Good Jew, and what was more, an admirable man, a saintly scientist; the very first to make an effective vaccine for humans against lethal bacterial infections. (Smallpox was viral.) A Jew to be trusted, moreover, since from the start he had demonstrated that he would recommend nothing not already tested on his own person. Accordingly, Haffkine had been the first to be inoculated in 1892 with the cholera vaccine he had developed in the Institut Pasteur. Discomfort had been minimal: local soreness at the site of inoculation in his left flank; a slight, temporary fever; nothing of serious concern. Russian friends living in Paris, one of them a doctor, then offered themselves to the needle. None of them were any the worse for it. In India, Haffkine insisted that only volunteers would be vaccinated. This complicated and slowed down the campaign. But strict adherence to the voluntary rule did not preclude exercises in persuasion, especially on captive populations most likely to contract contagious diseases: prisoners in Gaya jail in Calcutta and Byculla jail in Bombay; 'coolie' labourers and pickers in the tea gardens of Assam; native troops in the cantonments. Less confined people had also volunteered, often the poorest of the poor in the bustee slums of Calcutta and the tenement chawls of Bombay. It was the indigent he had most wanted to persuade since since their overcrowded lodgings were nests of contagion. Eventually, the results had been successful enough for senior figures in the government of the Raj – viceroys and governors – to use the vaccination campaigns as an advertisement for the benevolence of British medical imperialism. More locally, though, opinion was far from unanimous. It was not hard to find veterans of the Indian Medical Service grumbling about the 'Russian', without so much as a medical degree, who presumed to tell them that their obsession with sanitation was beside the point. What if microbes had been revealed by microscopes? Was not the lesson of cholera that disease bred in muck, and muck elimination, be it by sluicing with carbolic and limewash, the burning of possessions or even the demolition of the lodgings of the infected, was the only sure way to root it out? As for informing themselves about the new science of bacteriology, that was a foreign body of knowledge, cooked up in the laboratories of Paris and Berlin. The sick were packing the plague wards; the great entrepots of imperial commerce were shut down; there was no time to waste on airy speculations nor, for that matter, much of an inclination to learn.

Haffkine was not deaf to the muttering. His work in India had always been an uphill battle, especially to find funds and adequate space for research and vaccine production. Government wished him well, authorised his inoculations and permitted local administrations to lend him support. Somehow, though, the well-wishing was never accompanied by adequate sums of rupees; certainly not on the scale that would match his ambitions to create a 'bacteriological laboratory in India for the training of young medical officers in that science' or mobile inoculation facilities ready to go wherever need was most urgent.[3] Those who made decisions in Calcutta regarded Haffkine's apparent interest in mass vaccination, rather than reserving treatment for the Anglo officers and native troops on whom the security of the Raj depended, as utopian.

But by 1899, after six years of negligible official funding, his work made possible only by the enlightened patronage of wealthy patrons from minority communities – Parsis, Ismaili Muslims and Jews – Haffkine seemed, at last, to have secured official recognition. A plague research laboratory was to be established at Parel, in south Bombay, with him as director-in-chief. Should an all-India national institute for the study of dangerous infectious diseases get beyond the proposal stage, his name would be advanced as the most likely head. Plaudits rolled in, even from the traditional ranks of the Indian Medical Service. Colonel-Surgeon W. G. King wrote to him from Madras that, in honour of the invaluable service Haffkine had given, he had recommended that the vaccine ought henceforth be called 'Haffkinine'.[4]

But still, and not without reason, Haffkine suspected lingering opposition. For all the public praise, his position seemed insecure. Unlike Colonel King, there were veterans of the Indian Medical Service who shared the view of Leonard Rogers, an IMS expert in sea snake venom, that Haffkine was 'a foreign bacteriologist without registrable medical qualifications'.[5] His 'suitability' as director of any sort of institute was routinely questioned, even by some of those working for him. An official commission, launched in 1898, to inquire into the origins of the Indian plague pandemic and weigh the measures taken to contain and treat its victims was said to harbour doubts about the evidence Haffkine had produced to support the efficacy of his vaccine.[6] The six-month leave that had brought him back to Britain was partly to repair his damaged health, but also to mobilise support from the Victorian grandees of science and medicine.

In Britain, he had famous champions. Joseph Lister sang his praises; there was a meeting with the enthusiastic Florence Nightingale in her late seventies, even though, as Haffkine noticed, she was seriously unwell.[7] Four years earlier, in December 1895,

Angie's father, Sir Henry Wentworth Acland, lately Regius Professor of Medicine and Physiology at Oxford, had been astonished, listening to Haffkine, billed as 'an assistant to Dr Pasteur', lecture in the examination hall of the joint Royal Colleges of Physicians and of Surgeons on the demonstrated effectiveness of his cholera vaccine.[8] No less than Robert Koch had given it a glowing endorsement. If only such a life-saver had been available when the disease had torn through the university town in 1854! At that time Sir Henry had also been senior physician to the Radcliffe Infirmary and had taken upon himself to marshal the resources of town and gown against the spread of the disease. He had established a 'camp' for the infected on a field in Jericho (the poorer part of town, of course), while making sure, since he was not a jailer, that the quarantined would be decently supplied with food and shelter. As a result, so he believed, the outbreak had been contained. Two years later, Acland had published his account of that sanitary regime, along with a detailed map of infected Oxford.

When Angie's younger brother, Theodore, the only physician of the seven Acland children, was sent out to Egypt in 1883 to deal with an outbreak of cholera, he exported his father's inspection and quarantine regime to the Nile valley, in the face of official hostility to any measures that might halt Anglo-Indian traffic through the Suez Canal. A vast encampment of the infected and suspected was set up. This was not Oxford's Jericho. Military guards patrolled the desert camp with orders to shoot escapees. There were strict instructions about regular, intensive washing with carbolic acid solution. But the severity of the sanitary regime risked provoking a native revolt. Theodore Acland's surveillance, segregation and disinfection camps were unlikely to reconcile the empire's latest Arab subjects to the blessings of imperial rule. A decade later in the 1890s, Theodore served as principal medical officer to the Anglo-Sudanese army, attempting to keep the peace in the Upper Nile region, dangerously inflamed by jihadi preaching and on the brink of a rebellion led by the messianic Mahdi.

Confinement camps, or prohibitions on the hajj to Mecca, were asking for trouble. On the other hand, should it be allowed to run unchecked, cholera could cut a swathe through the Anglo-Sudanese army before it ever got to face its enemy in the field. If a vaccination campaign could obviate the need for the kind of measures that alienated native populations, so much the better for everyone.

There were, then, all kinds of reasons for the Aclands, father, son and daughter, to befriend the shy Waldemar Haffkine. Much taken by the younger scientist's modesty and eloquence, Sir Henry responded in the accustomed way of British hospitality: an invitation to stay at his house on Broad Street in Oxford. It was a famous place. Eminent Victorians – Gladstone, Lord Salisbury, Cardinal Newman – had all sipped from Acland cups and laid their heads on Acland pillows. Haffkine, whose impression of the British had, until this time, been formed by the starchiness of those who had held him at some distance in Calcutta society, was disarmed by the warmth of the Aclands' admiration. 'Your presence at the Examination Hall on the occasion of my lecture,' Haffkine wrote back, in boyish delight, 'I consider the greatest compliment . . . I have never seen Oxford before.' He would 'accept your kind invitation with great pleasure' but for the fact that, first, he had to honour previous engagements in London. When he eventually got to the Aclands, Haffkine found himself amid a tea party of guests who, though inevitably donnish, had been thoughtfully selected by his host. There were old India hands, including Sir William Hunter, the social statistician who had unsparingly documented the horrific Indian famines of the 1860s, and the entomologist Frederick Dixey, who had been taking Haffkine round the colleges and libraries. As happens on such occasions, once the introductions had been completed, the company chatted so brightly among themselves as the seed cake circulated that they failed to notice their guest of honour collapsing to the floor. Angie helped Haffkine to his feet. Despite regular medication with quinine and arsenic, a spasm of malaria, contracted while inoculating labourers against cholera in the tea gardens of Assam, had come upon him in the Oxford parlour. Painfully embarrassed, he was put to bed in one of the Acland rooms and ordered to stay there until feeling better. The following week, Haffkine wrote to Sir Henry apologising for having become such 'a burden' to his host.

Acland hospitality now verged on adoption. Conscious that Haffkine's stay in Oxford had been less than ideal, Sir Henry invited him to join the family in mid January at Killerton, the Devon estate of his elder brother. Acland must have held out the promise of riding (Haffkine could do that), country walks and lively conversations over dinner since the guest replied, 'I am delighted with the prospect of seeing that English life you describe [for] which I have already (from books) a great liking without having seen it for myself.'[9] On no account, though, would he hear of being met at Exeter station by Sir Henry – 'it will grieve me very much to see you tired'. Nor could there be 'any question of paying for any part of the ticket' (as the Aclands had offered). The two-day stay at Killerton was evidently a great success. Introduced to the high-minded as well as the horsey, Haffkine passed the acceptability test, winning, according to Sir Henry, 'golden opinions' from everyone – or at least everyone who mattered – in the county. Waldemar was, he wrote to his daughter, 'active in more good works, science, art and sound duties than any man I have ever known'. And it went without saying that he played both the violin and the piano.

In the clubs and mess rooms of British India, Haffkine's undisguised identity as a Jew continued to feed the reservations of the IMS old hands. A fish out of water; not really pukka; too clever by half. But in England itself, a strain of philosemitism circulated among the good and the great of science as well as literary culture. George Eliot's Daniel Deronda was just one of the incarnations of philosemitic romance. Haffkine, so handsome, so altruistic, so brilliant, perfectly fitted the idealised portrait of a modern Jewish hero. Racial biology had already lodged itself at the poisonous core of modern antisemitism and it had become a commonplace to claim, once again, that Jews were themselves walking hosts of infectious disease, especially typhus. Jewish dirt, the antisemites said, killed. But when a Jew appeared as a proven conqueror of contagion, he could be celebrated, not least by Gentile admirers, as the living rebuttal of that defamation.

Haffkine recovered enough to return to India in time to witness the outbreak, in the autumn of 1896, deep in the *chawls* of Bombay, of an epidemic every bit as lethal as cholera, if not more: bubonic plague. On the strength of his success with cholera, there were expectations that Haffkine could produce a comparable vaccine. And so he did. By the time of his second leave in the spring of 1899, almost half a million Indians had been inoculated against the plague. But the disease was outrunning the campaign. Government support was needed if the worst was to be avoided and Haffkine was not at all sure he had it.

A campaign of persuasion by lecture was organised: the Royal Society, London Hospital Medical School, a speech to the annual dinner of the 'Old Boys' of London Medical College and, crucially, the military medical college at Netley. In the midst of this non-stop schedule, Haffkine made sure to go and visit the Aclands in Oxford, where Sir Henry was now a frail invalid with only a year more of life left to him. Angie had moved her studio, along with her ailing father, out from the house on Broad Street to a quieter property on Boars Hill, north west of the city. Half-timbered pseudo-manor houses were going up in the fashionable suburb, including one belonging to the archaeologist Sir Arthur Evans, then in the process of a fantastically-imagined, high-coloured restoration of Minoan ruins at Knossos. That summer of 1899, the breezes blowing in from Tommy's Heath were uncomfortably cool, hence, perhaps, Haffkine's weighty trench coat, although the outfit also squared with Angie's vision of him as the picture of disinterested scientific wisdom. In a second plate, pairing Haffkine with Sir Henry, the old boy is similarly wrapped up against the mid-June chill. Seated in a wicker armchair, a velvet cap on his head, Acland smiles paternally at Waldemar who looks directly out at Angie's camera holding a book and some papers as if the two of them had just finished (or perhaps were about to begin) a lively discussion on how reliable vaccines might come to the aid of a world besieged by pandemics.

In June 1899, none among the scientific community championed Haffkine more passionately than Joseph, now Lord, Lister. Seventy two, retired from his chair in clinical surgery at King's College Hospital, Lister had been following Haffkine's career with intense interest ever since he had met him in late 1892, just before the young man's departure for India. He had been at the famous lecture on cholera in 1895 and, three years later, as president of the Royal Society, he had written to Haffkine in Bombay that it would make all the difference to its persuasiveness if he would come in person to read his paper on the plague vaccine. On the strength of the invitation, Haffkine requested, and secured, a longer stay of leave. His reception at the Royal Society vindicated

Sarah Evangelina Acland, photograph of Waldemar Haffkine with Sir Henry Wentworth Acland, summer 1899.

its president's advocacy. So when the eminent chemist Raphael Meldola asked Lister to propose the toast to Haffkine at a dinner to be given in his honour by the Maccabaeans in mid-June, Lister readily agreed. Founded just a few years before, the Maccabaeans was a 'friendly society' committed to helping, both philanthropically and educationally, the incoming tide of poor Jews fleeing persecution and destitution in the Russian Pale. Its membership was middle-class and professional: lawyers like Herbert Bentwich, Israel Zangwill the novelist and playwright, and not least, scientists. Many of its members were early Zionists.

Haffkine had not come to St James's Restaurant to speak about Zionism (about which he had reservations). Nor was he going to rehearse in much detail his long bacteriological road to vaccine-making. That he left to a generous summary account given by Lister in his introduction. But both Lister and Haffkine set their remarks in the context of bad things that were being said about Jews and bad things that were happening to Jews. Lister went out of his way to characterise Haffkine as 'an honourable man' for making himself the first human subject of the cholera vaccine, and 'putting his life in jeopardy' when working against cholera and plague. All that, as well as his untiring efforts at persuasion with Indians suspicious or hostile to vaccination, was the mark of a hero. But then Lister changed tack in a way that made the Maccabaeans, drowsy with port, sit right up. 'There are some who begrudge his ability and success because he belongs to the Jewish race. I am happy to say that in this country such ignoble sentiments do not exist.' This was wishful thinking on the old boy's part, but it drew a burst of appreciative applause from the diners, some them hammering their hands on the tables. The French, on the other hand (of course), were mired in unconscionable prejudice. 'We sympathise with Dreyfus,' said Lister, 'and rejoice with you at the prospect of his speedy deliverance . . . we remember we had a prime minister [Disraeli] who was of purely Jewish blood . . . of all contemptible things there is nothing more contemptible than hatred of the Jewish race. Yours is the noblest race on earth . . .'

And so on. More ringing applause and cheers broke out before Haffkine got to his feet.

Those who heard him speak in English – his third language, after Russian and French – were frequently taken aback by Haffkine's eloquence, though he was also said to speak slowly and meditatively, as if translating as he went along. But what he had to say to the Maccabaeans was dramatic, in the rhetorical manner of a cumulatively tragic performance. Pleased as he was at the honour done to him, and despite the fact he had not been back to Russia for many years, 'recollections of the vicissitudes' of the Jews there 'and the conditions of anxiety and uncertainty' in which they lived, waiting for the next eruption of violence, had never left him. Even as they were dining in St James's, 'millions of Jews in Poland and west Russia are now being subjected to a well-calculated, well-planned and systematically, unremittingly carried out, policy of extermination'. He was concerned that, on the part of the Jews themselves, there appeared to be no 'pre-concerted plan of resistance' to the physical harm that would certainly come their way.

Haffkine moved from dark prophecy to personal history, describing the immense difficulties he had experienced, especially in the first year of the cholera vaccine in India in 1893. Only once he was there did he realise how elusive measurable subjects, in meaningful numbers, might be, not least because cholera had receded that year. He had needed not only sizeable numbers of volunteers, but equal numbers from identical living conditions who would decline inoculation and thus be used as a necessary control. But even when he did manage to find such comparable groups, the poorest people, those most likely to suffer infection, were also those least likely to stay in one place, making follow-up reports of results difficult. It dawned on him that perhaps 'the task seemed impossible', even though he also knew that it was also impossible to walk away from the project. Trapped by the dilemma, his mind succumbed to tormented indecision, so bitterly painful that it brought him to the edge of suicide. 'In the fatigue and despair of ever seeing the end of those efforts becoming effective, the image of death appeared to me as a welcome liberation; the whole creation around me seemed to have exhausted all that which normally supports a man in his activity.' He continued to vaccinate when and where he could, but loneliness and a sense of being alien, both to the British and to the Indians, bore down on him with crushing force. Those he was treating were separated from him by language and cultural habits, 'unable to greet me with a single gesture of friendship or approval'. So his heart and mind returned to whence he came. Paradoxically, what sustained him at those moments of Indian melancholy were visions of Jewish Ukraine, 'the distressed of the plains of the Dnieper, the Vistula, the Niemen and the Danube', and 'the prospect of one day being able to alleviate *their* grief by a single ray of hope'. At this moment in the speech, Haffkine turned directly to old Lister sitting beside him and asked him conversationally whether, when they first met and many times after, 'in the inexhaustible spirit of helpfulness' extended to him, Lister had remembered 'the race I belong to'. 'Yes!' said Lister loudly. More applause. Again, from Lister, still more emphatically: 'YES.' 'Then,' Haffkine resumed, full of emotion, 'you bestow on me a reward which is the sweetest, the proudest, a man can ever get. At the same time,' he went on, looking at Lister, 'you foreshadow, perhaps unconsciously, the possibility of better times for our people. For it seems to me that in pursuing our own ideals, endeavouring to do things which seem right to us which answer to the most intimate demands of our nature, it so happens that a feeling of approval is generated in those with whom we come into contact, in the nations who watch us; it seems to me that a time of conciliation, of mutual affection and brotherhood between our people and other nations becomes not only a possibility but a realisation.'

A wave of cheering broke from the listening Jews in their evening dress. But before the astonished acclaim ebbed away, one of the non-Jewish guests, the Reverend Charles Voysey, abruptly stood up and said that, on hearing he was to dine with 'that dear Haffkine', his daughter, just back from Karachi, had given him orders to tell Haffkine 'that he saved my husband's life'.

All this – the apocalyptic prophecy; the drama of despair and near suicide; the visions of defenceless Jews in Ukraine wondering when the hammer of hatred would beat down on them again – must have been unexpected by the Maccabaeans. But that was only because, as yet, they knew little of the true history of Waldemar Mordechai Wolff Haffkine.

He had grown up in the shadow of violence, just eleven years old when a brutal pogrom struck his own community in Odessa in 1871. Jewish Odessa, supposedly a model of cosmopolitan co-existence, had been put to the torch that Easter. Six Jews died in the mob assault; twenty one were savagely beaten and wounded. No fewer than 183 houses and 550 shops and businesses were destroyed. The experience must have been traumatic for the boy who was motherless after Rosalie Haffkine's death, four years earlier, at the age of forty.

The pogrom took place in Holy Week, ignited by fantastic, immemorial slanders of alleged Jewish desecration of images and ritual objects in Greek Orthodox churches. There had been other attacks earlier in the century driven by religious paranoia. But in 1871 the violence was also coloured by a streak of economic resentment. Greek rioters were joined by Russians and Ukrainians objecting to Jewish penetration of the trade and craft businesses of the port city; in other words, fury at the only modern Jewish success story of the Russian empire. In 1860, the year of Haffkine's birth, one in four Odessans was Jewish. Opportunities for Jews were unlike anywhere else in imperial Russia. They could get a secular education, including the sciences and mathematics; they could learn Russian and other European languages. The unlearned could take manual jobs in the docks; the somewhat learned could open shops or small businesses; and, at the top end, Jews could enter university and qualify for entry to the professions. This was the only place in Russia where Jews could live a life comparable to that led by their co-religionists in the German-speaking world, England, France or America. In Odessa, you could be a Jewish lawyer, journalist, teacher, physician. And you could also be a businessman working in the international grain trade on which Odessa's fortunes were built.[10] Haffkine's father, Aaron Wolff Pavlovich, was a commercial factor for the most ambitious and successful of these new family firms, the Gunzburgs, and a classic example of an Odessa *maskil*: a believer in and practitioner of Jewish enlightenment and optimistic integration into Russian culture and society.[11] His second wife, Waldemar's mother, Rosalie Landsberg, also came from this milieu in which it was unproblematic to be both Jewish and a modern Russian. At least one of Odessa's synagogues campaigned to replace Hebrew in its services with Russian prayers. No one in the immediate family circle of the Khavkins/Haffkines spoke Yiddish. The names Aaron and Rosalie gave their son embodied this optimistic cultural mix: Waldemar was a Germanised version of Vladimir or Wolodymyr; Mordechai the imprint of an unapologetic Jewish identity; Wolff a family favourite. His early childhood was spent amid a crowded nest of siblings: three sisters – Henrietta, Maria and Rebecca – and two older half-brothers from Aaron's first marriage: Alexander (like the tsar) and Salomon (like the king). Even domestic disaster was coloured by this mixed identity. The seven-year-old might know no Yiddish, but coached by his maternal grandfather David Landsberg, a teacher at the Jewish College, he could, poignantly, recite by heart the Aramaic *kaddish* at his mother's graveside.[12]

Without abandoning his Judaism, the precociously clever Waldemar would get his education not from a religious *cheder*, but from a government school where his classmates would include non-Jews. This would happen at Berdyansk, some 500 kilometres to the east, another modernised Ukrainian port city at the head of the sea of Azov. The reasons for the widowed father to move his younger son and the girls are uncertain, but since Aaron's own fortunes shrank in the more provincial setting, the migration is likely to have been a shocked reaction to the pogrom of 1871, which exploded the myth that the more secular and modern the Jews became, the easier it would be for them to settle down to an unproblematic co-existence with their Gentile neighbours. Not that this darker view of the future altered the course of Waldemar's education. At the Berdyansk classical gymnasium, he learned Latin, Russian and French as well as the sciences. At fifteen, so a classmate remembered, he was already a schoolboy paragon: strikingly handsome, ferociously hard-working and temperamentally serious.

There was no university at Berdyansk. So, in 1879, the nineteen-year-old was admitted to the Imperial New Russia University in Odessa to read physics and mathematics. The decline in his father's fortunes meant that Waldemar would join the throng of hard-up students living in bad digs in the grimier side of town, wolf kept from the door only by the ten-rouble allowance sent each month by his step-brother, Alexander. Student Odessa – like the city itself – was a pot of simmering politics: socialism, peasant populism and Zionism, all finding their voice, sometimes in a prudently muffled underground, sometimes recklessly loud and open. The cafes, taverns and print shops buzzed with enough dangerous ideas to draw the attention of government spies who went undercover among the young men and women while keeping their eyes and ears open for an incoming freshman. The talky utopianism was infectious, also dangerous. In his first year, Haffkine, together with his schoolfriend Haim Mordechai Rabinowitz, was already drawn into protests against restrictions on student assemblies. He was vocal enough and visible enough for the police to open a file that, two years later, would cause him a world of trouble.

But there was someone within the academic, scientific world of the university who exercised a more powerful spell on Haffkine than any number of student zealots. It took almost no time at all after walking into the classroom lab where Ilya Mechnikov taught comparative anatomy and zoology for Waldemar's life to change.

Elie Mechnikov

Like his protégé, Mechnikov was a scientific vagrant, footloose in the dawning world of microbiology. His life had begun in the Ukrainian village of Ivanovka, near Kharkov, and it would end in the Institut Pasteur in Paris; in so far as he was ever at home anywhere, it was there. But for much of his life, Mechnikov belonged to the nomadic community of scientists strung out from Russia through Germany, Italy, France and Britain, connected only by the exhilarating experience of seeing a universe of micro-organisms: pathogenic microbes and 'wandering cells' as Mechnikov would call them, defending host bodies against toxic invasion.

Mechnikov was a mish-mash. His aristocratic father had been a Moldavian officer in the Imperial Guard who had Russified his family name from the Romanian, while preserving its vocational allusion to the life of the sword. Ilya senior drank away much of his inheritance and retreated to what was left of his estate near Kharkov to breed horses. For most of his son's childhood, he was a missing father: a big swiller of champagne while giving off a smell of stables. This made Emilia, Ilya's mother, the formative influence in Mechnikov's life. Hers was an altogether different pedigree: wordy not horsey, a Nevakhovich, from a family of famous boundary crossers. Her father (Ilya's grandfather), Lev Nevakhovich, was one of the first of the Russian *maskilim*, optimistic enough to believe that, sufficiently immersed in Slavic culture, Jews could be unproblematic subjects of the Tsarist empire. This was the theme of his *Lament of the Daughters of Israel*. He began by translating Hebrew works (as well as German and Swedish writing) into Russian, and made the usual notional conversion (to Lutheranism), allowing him to be employed at the Ministry of Finance in Warsaw and get his plays performed on stage at St Petersburg, on at least one occasion, before Tsar Alexander I. One of his sons became an imperial *ritmeister*, a master of horse; the other invented Russian comics, the first titled *Yeralash* – which also means mish-mash.

Emilia Nevakhovna Mechnikova guided the steps of her precocious son who delivered science lectures to his captive siblings. At the Kharkov Lycée, she encouraged his aptitude for botany and zoology and bought him his first serious microscope when he was fifteen. In short order, science replaced the Deity as an object of faith, a conviction Ilya repeated so incessantly that his nickname became 'God is Not'. When it came time to enter Kharkov University in 1863, Emilia persuaded Ilya (not least because of his delicate physical disposition, weak eyesight, nervy temper) to follow the life of a research scientist, rather than that of a physician. The mother had good reason to be watchful. Her elder boy, Lev, seven years Ilya's senior, had been expelled from the medical faculty at Kharkov, probably for embracing revolutionary politics. There was something about the brothers Mechnikov which made them, intellectually and politically, a handful. Not that this deterred Emilia, convinced as she was that her sons' generation was on the threshold of a whole new world of scientific knowledge and wishing them both to be among its pathfinders.

That new world was borderless in both time and space. Rejected by the medical profession, Lev Mechnikov became a master of ten languages including Arabic and Turkish and was disinclined to settle down anywhere, there being too much to be getting on with. Wounded fighting with Garibaldi's Redshirts, he became, in succession, a reporter-correspondent in Spain and a teacher in Japan, ending as an 'anarchist geographer', a collaborator with the perpetually unorthodox Elisée Reclus: Communard balloonaut, campaigner for the abolition of marriage, nudist, conservationist and militant vegetarian. Ilya's intellectual travels were tamer than those of his older brother, but still not those of the conventional Russian gentleman-scholar. Two of the books that formed him were Henry Buckle's *History of Civilization in England* (1858), with its paean to science as lighting the way to a future free of misery and disease, and, published just one year later, Charles Darwin's *On the Origin of Species,* initially introduced to Mechnikov via Fritz Müller's *Für Darwin.* When added to Ernst Haeckel's 'Biogenetic Law' (first outlined by Etienne Serres), holding that the embryonic development of animals 'recapitulates' the adult forms of evolutionary ancestors, Mechnikov had before him, pretty much for the rest of his life, the epochal, unbroken, anatomical chain connecting the simplest invertebrate organisms to every successive form of species including his own. To look closely at those primordial, unicellular organisms, then, was to follow Serres, Haeckel and Darwin in canvassing the entire history of life on earth.

So he did just that, wherever and whenever he could. In Heligoland he studied microscopic marine fauna. While still a student at Kharkov, he published a paper on the tissue structure of a protozoan freshwater ciliate, *Vorticella*, first seen under a lens by Antonie van Leeuwenhoek in 1676 and known to have a symbiotic relationship with bacteria. At Giessen University, working with a parasitologist, he saw intra-cellular digestion in flatworms – a revelation. Naively impassioned, just twenty years old, Ilya Mechnikov gave off an unworldly air, which made him an easy mark for the unscrupulous, including his own professor, who pirated and published the work on flatworms without giving Mechnikov any credit whatsoever. Impotently outraged, Ilya moved to Naples to work with a Russian zoologist, Alexander Kovalevsky, on the embryonic development of cuttlefish and the minute crustacean Nebalia. This was his Darwin moment and, together, the two Russians determined to vindicate the basic evolutionary principle of common ancestry by comparing the development of germ layers in the embryos of invertebrates with those of vertebrates. Before they could complete this work, the wave of cholera which triggered the great meeting of the International Sanitary Conference in Constantinople swept through Naples. Kovalevsky and Mechnikov departed north to St Petersburg, where Ilya earned his doctorate and, together with Kovalevsky, won a grand prize for the work on cuttlefish and Nebalia.

He was twenty two.

And ripe for a life scripted by Dostoevsky. The doctoral dissertation won him a position as a docent in the zoology department at Odessa University. Younger than all his pupils, Mechnikov raged against the ludicrous formalities and petty feuding of academic life. Possibly, he thought, things might be better in St Petersburg. They weren't. In the humming, ruthless capital city, he subsisted in dire poverty and misanthropic loneliness, his eyesight faulty and his mental stability unpredictable. To the rescue came Ludmila Fedorovich, consumptive but adoring. The two of them fell instantly and obsessively in love. But as if supplying material for grand opera, Ludmila was already so eaten up with tuberculosis that on their wedding day in 1869, pale as the bridal sheet, she had to be carried to the altar in a chair. As her illness accelerated in its fatal course, Ilya took her to Madeira where the warm Atlantic breezes were supposed to help. They did not. When Ludmila died in 1873, her distraught husband resolved to kill himself with an overdose of morphine. But he swallowed so much of it that a paroxysm of vomiting got rid of the drug involuntarily and expeditiously. Purged and gastrically scalded, Ilya left it at that.

Surfacing from his drowning pool of grief, Mechnikov moved back to Odessa, no more reconciled to conventional academia, but on fire every day with his research. In the apartment directly above his, he heard a high-school student, Olga Belokopitova, daughter of a family of landed gentry, stomping about in her schoolgirls' boots, chattering and laughing. She was sixteen, Ilya thirty. They courted, betrothed, married. But when, in 1880, Olga contracted typhus, her husband swore that it was his fate to be cursed in love, his premature contentment preyed on by hostile microbes. Surrendering to romantic despair, he had another go at killing himself, this time in a style that he hoped would be remembered, at least, as Useful to Science. The sacrifice took the form of injecting himself with bacteria from relapsing fever to see if the disease was transmissible by blood. Should the experiment succeed, he would be posthumously acclaimed. Immortality. Somehow, this made the prospect of extinction tolerable. But the second effort at suicidal consummation was no more decisive than the first. Not done yet, Ilya immersed himself in a hot bath, then went out into the freezing night in an attempt to catch a fatal chill. But moths fluttering around the light of a street lamp distracted him from his botched self-destruction. At that moment, biological curiosity overcame romantic egotism. Or so his posthumous biography, written by Olga, who survived her husband by many years, claimed. Ilya was prone to rhetorical suicide, recollecting that attacks on his work were sometimes so unsparingly ferocious 'I was ready to get rid of life'.

The year 1879, though, found Mechnikov very much alive; a professor, in his mid-thirties, at the Novorussiya Imperial University in Odessa, which today is named after him. Into his lab that year came a Jewish student fifteen years his junior, equally fascinated by the anatomy of single cell organisms. He had odd given names – Waldemar/Vladimir, also Mordechai, also Wolff. But Ilya Mechnikov, being one of Them himself, knew all about culturally hybrid prodigies. Under the skin they were a couple of clever Jews, the master with a full set of whiskers, the pupil with a wispy moustache. After winning the Nobel Prize in 1908, Mechnikov told a reporter from *The New York Times*, 'I ascribe my love of science to my descent from the Jewish race.'[13]

Would the Aclands, Joseph Lister, Florence Nightingale – the whole crowd of British admirers who thought of Haffkine as restrained, reticent, the epitome of the quietly reflective life – have recognised his younger Odessa persona: an armed agitator? As one of the leaders of protests against an official ban on student assemblies, Haffkine was conspicuous enough to be arrested. Once released, he was put under police surveillance. A dossier was opened, which, over the next eight years, would grow with recurring reports on Haffkine's flirtation with subversive politics. Did that extend, as the police claimed, to the breathtakingly dangerous step of joining the clandestine revolutionary terrorist organisation Narodnaya Volya, 'the People's Will'? There is no doubt that the organisation was strong in the student hothouse of Odessa University. On the strength of his conversations with Haffkine's eighty-five-year-old step-brother, Alexander, the Soviet journalist Mark Popovsky had no doubt that police suspicions were well-founded.[14] But Popopsky had an obvious interest in giving the man he calls in his biography 'my hero' a revolutionary as well as a scientific pedigree. On the other hand, once he had shaken off the harassment of the KGB and emigrated to New York, Popovsky made no changes to the picture of Haffkine, the student radical, he took from Alexander's memories. What is not in doubt is that a letter written to the governor-general of Odessa about Haffkine's politics and their effect on his standing as a student at the university certainly states that he belonged to the Peoples' Will.[15]

On 1 March 1881, a group of their comrades, including the daughter of the military governor of St Petersburg, bombed the carriage of Tsar Alexander II while he was returning to the Winter Palace. When, in the bloody panic, the tsar got out of his coach, a second bomb blew him to pieces. The second, decisive bomb-thrower, Ignacy Hryniwiecki, was said to be a Jew. In fact, he was not, but Hesya Helfman, the 26-year-old strikingly beautiful revolutionary at the heart of the assassination, certainly was. It hardly

Waldemar Haffkine aged twenty four, Odessa, 1884.

needed ingrained antisemitism to convince many Russian minds that the Jews had killed the tsar just as they had killed Christ. Rumours spread that this was just the beginning; the *Zhidy* were plotting death and terror.

What came next was predictable: a wave of pogroms: 187 separate attacks in cities, towns and villages, from Minsk to Kiev (as it then was). The assaults were especially brutal in Ukrainian 'New Russia' because economic resentments ran strongest there. As always, ancient madness played its part. The pogrom in Elizavetgrad in late March was triggered by yet another accusation of the blood libel: the allegation, originating in medieval England, that Jews required the blood of Christians, especially children, for the preparation of their Passover matzah. But the moment was seized by outsiders beyond Odessa, to push back against Jewish inclusion in commercial enterprises from which they had previously been barred. A number of pogromniks in Ukraine, particularly railway workers, were hired by Moscow merchants to rough up the Jews and attack their homes and property. Local government officials were taken aback by the pogroms, and in some cases gave orders for their suppression, even bringing in Don Cossacks to keep order. But the efforts were half-hearted. For some in the government, it was expedient to let matters take their course and certainly to refrain from any kind of systematic prosecution of those responsible. The incoming ultra-reactionary government of Alexander III was hostile to the economic and civic liberalisation that its predecessor had extended to Russian Jews. It was precisely the prospect of Jewish competition in trade and finance, not to mention their actual presence in Moscow and St Petersburg, as well as the entrepots of New Russia, that had so enraged the Moscow merchants. The aftermath of Alexander II's assassination, and the vague but potent attribution of Jewish responsibility for the crime, gave the merchants the opportunity to put Jews on notice that competition with Christian businesses and professions would put them in jeopardy.

The Ukraine pogroms of 1881 were what Haffkine was remembering when he made his surprising remark at the Maccabaean dinner about threatened Jews not organising resistance. But his listeners had paid attention to Theodor Herzl's tragic argument that while Jews had been targets of exterminating hatred when they lived separate lives from their host cultures, they became even bigger targets when attempting the opposite: integrating themselves, by language, education, social customs, and working professions, into the Gentile world. Odessa, the supreme example of how the liberalising of Jewish life had failed to disarm paranoid antisemitism, turned young Jews towards the only thing left to protect their community: armed self-defence.

In May 1881, with reliable information that a pogrom was being organised for later in the month, a group, about thirty strong (swelling quickly to eighty), of Jewish students and young professionals – including Haffkine as well as the physician Kostya Puritz and the lawyer Lyova Albert – met to organise the defence of Jewish Odessa. Their meeting place was the house of the sister of one of Haffkine's oldest friends, Haim Mordechai Rabinowitz, with whom he had shared lodgings during their gymnasium days in Berdyansk. Helpfully, Rabinowitz's brother-in-law, Hillel Yaffe, seemed to know where and how to acquire guns. With the likely pogrom only two days away, the group went through the motions of appealing for official help. They alerted the local military authorities to the imminent threat, though without much expectation of help from that quarter. Those low expectations were more than realised. What, the police shrugged, could be done about something that had not yet happened and perhaps never would? Instead of protective assistance, Haffkine and his friends got warnings from the police about not provoking potential assailants.

There was no time to lose. Discussions at the Rabinowitz–Yaffe house turned practical. Jewish Odessa was divided up into self-defence districts, each to be barred with metal barricades. Projectiles – bricks, paving stones, rocks – were to be piled up to be used against the thugs. Most crucially, Hillel Yaffe had managed to find guns, fifteen of them, pistols and shotguns, not an arsenal, but enough to give the enemy an unwelcome surprise. Haffkine himself packed a Webley revolver. All this was a startling turn, not just in Odessa, not just in Russia, but in Jewish history. Any sort of armed militia or gang was of course illegal in Tsarist Russia, so the risks the group was taking were extreme. But defencelessness, relying on the authorities to protect them, was a greater folly.

The violence was known to be planned for Sunday. The day before, on Shabbat, a recruiting visit was made to the synagogue favoured by Odessa's kosher butchers. Haffkine's lack of Yiddish prevented him from addressing the *shul*, but Rabinowitz did the talking. Suspicions of 'agitators' were overcome; the butchers in turn mobilised their craft brethren, the ritual slaughterers, *shochetim*, professionally handy with razor-sharp hardware. Word was put out to shopkeepers to close down their premises on Sunday. Barriers were erected.

The Jewish self-defence of Odessa was manned and armed. This had never happened in the long history of the European diaspora. In the person of the gun-toting Waldemar Haffkine were combined the two faces of how a modern Jew might engage with a hostile world: science and action.

When the first group of attackers appeared in Jewish districts on the Sunday evening, they were hit by a storm of bricks and stones, and beat a surprised retreat. A second sally was foiled in the same way. At five o'clock the next morning, Haffkine, Rabinowitz, Puritz and Albert rose early to make sure shops were shut and that no Jews were exposing themselves to danger. They found a solitary onion-seller who explained that he needed to feed his family. Before anything could be done, the five of them, none carrying weapons, were surrounded by a hostile crowd and only saved from harm by the sudden appearance of a detachment of police. As the sun rose over Odessa, there were more attacks on Jewish districts, almost all of them repelled by defenders, especially when they were butchers carrying the tools of their trade. Street fights continued through the day. At some point Haffkine, now carrying his revolver, ran into Hillel Yaffe, just arrested by the police. The pistol got Haffkine frogmarched to jail, from which, following a vain search of his lodgings for incriminating documents, and a few nights spent locked up, he was released. More than a hundred of the defenders were interrogated for their part in the resistance.

Haffkine was free but not out of trouble. In November, he was expelled from the university as one of three leaders of the unlawful armed self-defence group. No sooner had this happened than a petition for readmission was organised, one of the signatories being Professor Ilya Mechnikov. It succeeded but it was made clear that the troublesome student was on closely watched probation. He returned to physics, mathematics and the study of single cell invertebrates. But while the Odessa pogrom had abated, ugly moments recurred. Seeing a Jewish shopkeeper abused by army cadets, Haffkine impulsively threw himself on the assailants; he was badly wounded in the head and, to add insult to injury, arrested again. Mechnikov came to his rescue once more, testifying to his student's good character at the trial. But the Haffkine dossier was getting ominously thick.

In February 1882, he was arrested for a third time. The accusation that he belonged to a seditious terror organisation could hardly have been more serious. If convicted a Siberian prison camp awaited or still worse. Somehow, the case never got to trial. It's possible that Haffkine's father called on the powerful influence of the Gunzburgs in St Petersburg. But it was Mechnikov who, for a third time, pulled all the strings to extract his brilliant protégé from this dire situation and spoke out in his defence. In return, Mechnikov himself exacted a sacrifice from the renegade student. Henceforth Haffkine was sworn to devote himself entirely to science. Soon enough, though, Mechnikov eventually began to realise that it was he as much as his student who was the real target of the political enforcers. Sure enough, he was dismissed from his post in the faculty, triggering yet another suicide attempt. This time the shoe was on the other foot. A student petition for Mechnikov's reinstatement was organised with Haffkine among the signatories. The gesture of gratitude rebounded on him. His own expulsion from the university was renewed; professor and prize student were both cast out.

While Haffkine barely kept himself alive by giving private tutorials and holding down a menial job as guard at the Odessa Museum of Natural History, Olga's inheritance allowed Mechnikov to think of rebuilding his research laboratory somewhere warmer and free from politics. In the spring of 1882, they settled at Messina at the northern end of Sicily. The port city was notorious for the earthquakes and fires that periodically destroyed it. But Mechnikov was happy there. And it was in Messina, so he relates in his memoir, that he had his microbiological epiphany.

Mechnikov's *Souvenirs* were written thirty years later and in the afterglow of the Nobel Prize, when the temptation to weave a dramatic narrative fitting the magnitude of the honour was irresistible. Which is not to say the story of the starfish epiphany was pure fiction. But the moment did not arise out of a conceptual vacuum. Others like Ernest Haeckel had seen something similar and Ivan Turgenev had described Bazarov, the tortured hero of *Fathers and Sons,* observing 'infusoria' ingesting 'green specks' under his microscope lens. Nonetheless, the essence of the moment might well have happened the way Mechnikov described it.

According to the story, the rest of the family was at the circus watching trained monkeys go through their stunts. Alone in his private lab, his eye glued to the microscope, Mechnikov was looking at 'wandering cells' as they did their wandering in the completely transparent larva of a starfish. It occurred to him that those cells might defend an organism against the invading agent of infection. 'I got so excited that I started pacing around and . . . walked to the shore to gather my thoughts.' Supposing he was correct, those cells would surround and engulf something like a thorn 'similarly to what happens to a human finger with a splinter'. Close to a mandarin bush he had decorated for the children in lieu of a proper Christmas tree, Mechnikov found a rose thorn (this is Sicily after all) and inserted the micro-splinter into the glassy, bloodless larva. After a night sleepless with excited tension, he saw that the motile cells behaved exactly as he predicted, engulfing and digesting the invasive particle. In Vienna, when he reported the discovery a fellow scientist suggested the phenomenon be called 'phagocytosis': cells that ate.

What Mechnikov had observed went well beyond digestion. The eating cells, his 'macrophages', were also fighting cells. This put him at sharp odds with 'humoral' theory, which argued that it was elements in blood serum that were indispensably responsible for defence against infection. To which Mechnikov made the obvious retort that the vascular system of starfish pumped seawater, not blood. What was more, Mechnikov insisted, the response he had witnessed in the invertebrate was similar to benign inflammation in vertebrates. In 1883, he published the results of experiments on tadpoles and frogs in which it was clear that, in addition to attacking invaders, macrophages also scavenged dead and dying cells within the host. Most important for him was that all this seemed an incontrovertible vindication of the theory of Darwin (who had just died in April 1882) that the same processes of defence were at work from ancestrally simple forms all the way through to complex mammals. An immune response, as it came to be called, was, then, innate and universal.

How, though, could this revelation be applied to push back against pandemics? Any hope that the most savage of the nineteenth-century diseases – cholera – was on the wane was dispelled by the outbreak in Egypt in 1883 which sent Theodore Acland to take military charge of containment and quarantine. The epidemic was serious enough to trigger two missions from rival laboratories, in Paris and Berlin, to try to identify the pathogen and make accurate reports on its transmission. Louis Pasteur, already famous for producing a vaccine for chicken cholera based on attenuated doses of the bacillus – as well as cures for diseases of two of the commodities on which the pride as well as the profit of France depended, wine and silkworms – sent his colleague Emile Roux and a 27-year-old assistant, Louis Thuillier, to Alexandria. Thuillier confirmed the contagiousness of the disease in the worst possible way by contracting it himself while examining specimens of intestinal fluid from victims and dying from the infection. After the Pasteur mission returned to Paris in grieving dismay, Robert Koch arrived in the same port city and it was there that he first identified the 'comma vibrio' as the pathogen. There could be no doubt now that the elimination of such diseases would come from the new science of close examination of the microbial pathogens. Logic would dictate a pooling of resources and research across frontiers. But the French and the Germans differed even on the names they gave to their new discipline. Koch and the Germans called their work 'bacteriology'; Pasteur, Roux and their colleagues, on the other hand, were 'microbiologists'. Mechnikov, a zoologist, was drawn more closely to Pasteur. The lines were not hard and fast and professional allegiances were not necessarily mutually exclusive. Young scientists, especially the Japanese, were eager to work in both Berlin and Paris. As they became more interested in the field, the British were more likely to go to Koch than Pasteur, whom they suspected of French microbiological imperialism.

Back in Odessa, the continuously monitored, thrice-arrested Waldemar Haffkine had somehow managed to be readmitted to his studies; better yet, he had completed his doctoral dissertation on protozoa and had published five papers in two years. In 1886, he gave a lecture on a hitherto unknown microscopic organism of the kind known as rotifers or 'wheel animals' which Mechnikov thought should be named for his tireless student, *Pleurotrocha haffkini.* He was, evidently, more than well qualified to be appointed lecturer in chemistry and biology. But there was, naturally, a condition. The reactionary government of Alexander III had reinstated prohibitions on Jews entering certain professions, university instruction being one of them. It was made clear to Haffkine that conversion to the Russian Orthodox Church was the necessary condition of any academic position. Even the Mechnikovs had understood that. Why should the pupil not follow the teacher in an apostasy of convenience? But if, after his unhappy experiences in jail and his narrow escape from a treason trial, Haffkine was no longer an armed fighter, the pogroms had made him determined to be more Jewish, not less. Without more ado, he rejected, as had countless Jews before him when pressed hard, the blandishments of Christians.

Around Haffkine at the university, others were responding proactively to the opportunities opened up by this new world of bacteriologically driven public health. Close to him was another group of young Odessa-born researchers – Nikolai Gamaleya (one of the few hospital clinicians in the field) and the Jewish Yakov Bardakh – both inspired by Pasteur's success in developing vaccines for anthrax and rabies. In 1886, Gamaleya won a competition funding a research trip to Paris where he not only got to work alongside his hero at the Ecole Normale Supérieure, but seeded the possibility of creating a 'bacteriological station' in Odessa that would treat patients, both prophylactically and therapeutically. Unsurprisingly, Pasteur was flattered and gave Gamaleya starter vials of rabies vaccines to take back home. There could be only one possible senior director for such an institution: Ilya Mechnikov, who to some people's surprise said yes, though by now everyone must have known that this would never be a guarantee of his staying put for very long. The Odessa Bacteriological Station – the first of its kind in the world for treating rabies victims which, since this was Russia, included wolf bites – opened its doors in the handsome building surmounted by a mansarded cupola and grandiose classical portico in July 1886, two years before the Institut Pasteur itself was inaugurated in Paris.

Odessa Bacteriological Station

But even with Mechnikov back in Odessa, Haffkine was stuck in his dead-end job as *kustos* of the Museum of Natural History: all the single cell slides he could need but none of the experimental excitement. Nor, as long as he refused conversion, was there any prospect of a paying teaching post. In 1887, on the suggestion of one of Mechnikov's colleagues, he moved to Geneva to work for Professor Moritz Schiff, then under attack from anti-vivisectionists for his removal of thyroid glands from dogs as a way of showing this could be fatal in both animals and humans.

Geneva was full of young Russian students including, of all people, Hillel Yaffe. No longer smuggling guns, Yaffe was among those organising anti-Tsarist meetings and protesting against the repression of free speech in Russian universities and their infiltration by networks of police spies. This hot-blooded exile politics was well and good but Haffkine's most passionate allegiance had now become microbiology. But he soon discovered that the crippling burden of pathology-teaching and basic lab assistance that Moritz Schiff laid on his back made any independent research all but impossible.[16] Before any time at all, he was itching to leave. An application for a post in Norway was rejected, but soon enough his old mentor came to his rescue yet again – though for the last time.

Delayed by its dependence on private funds, the Institut Pasteur finally opened its doors in 1888. One, and only one, invitation went out beyond France, to head its section on microbial research. With the Odessa Bacteriological Station operating under its own steam, Mechnikov, to no one's surprise, agreed, not least (and naively) because he believed that in Paris he could finally be free from the snares of politics. Almost immediately, he did what he could to find a position for his thwarted, wayward, politically incautious but gifted

Waldemar Haffkine, Geneva, 1889.

student. But no one at the Institut was helpful. Mechnikov had brought with him a number of young Russians and Ukrainians, and xenophobic and antisemitic papers accused the Institut of harbouring a nest of undesirable foreigners. Haffkine's presence only added to the suspicion. But there was a vacancy as assistant librarian to the Institut. It was yet another demeaning job, but Mechnikov promised Haffkine that were he to take it, there would, once more, be time and space for his research. Waldemar accepted with gratitude.

Twenty eight years old, no longer a promising student, but not a seasoned professor either, Haffkine arrived in Paris in early 1889. The moment was a cultural, as well as scientific, epiphany. Haffkine had long been a devotee of French literature, and not long after his arrival in Paris began to transcribe long passages from the novels of Honoré de Balzac in a journal which also included letters of Edward Jenner. An *Exposition Universelle* opened in May at the Trocadéro to celebrate the centennial of the first great Revolution. Haffkine made it his business to learn as much as possible about that epochal event, copying relevant pages from François Mignet's history.[17] Thrillingly, the technical and scientific future was on display as well as the romance of the insurrectionary past. Thomas Edison's electric lights illuminated one of the pavilions. Two legs of the completed Eiffel Tower formed the gateway to the exhibition park, although if Haffkine wanted to get to the top tier, he would have had to climb the steps since Otis's miraculous elevator – in evidence elsewhere at the show – was not yet operational at the tower. But were he of a mind to take his ease in the showgrounds, Haffkine could sample the Brazilian or the Japanese pavilions, a Wild West show starring Buffalo Bill, Javanese gamelans and '*danses nègres*'. Amid all this exotic excitement, the eyes of Russian government spies were still on him. Reports reached St Petersburg that Haffkine kept company with expatriates notorious for their radical and subversive views: Ivan Vilbushkevich, a Moscow engineer, and two Georgys, both physicians, Vaveyn from St Petersburg and Tamashev from Tiflis.

They were just one of the groups of young Russians breathing the heady freedom of Paris, enjoying the show and the drink that went with it. All serious imbibers at the Exposition gravitated towards one of the prize exhibits: a single, titanic oak barrel, bigger than most houses, allegedly holding the contents of 200,000 bottles of champagne. The monstrous object was there to boast of the supremacy of French viticulture. But Haffkine knew that it was his new *patron* who should be credited with saving that industry. Asked by Napoleon III to inquire into the contamination of wine, Pasteur had identified the primary toxin and recommended the high heat process – known to us now as pasteurisation – by which it could be killed. Likewise he had isolated and analysed the two principal diseases – hereditary and contagious – affecting French silkworms and had proposed measures for separating infected and healthy worms. But Pasteur often declared that, as important as his work on agricultural and zoological diseases was, his ultimate aim was to extend the revelations of microbiology to the epidemics visited on humanity. Whatever anxieties Haffkine might have had about being confined to his job in the library, he was exhilarated by the possibility of playing a part in that mission.

The Institute at 25, rue Dutot, where Louis Pasteur lived as well as worked, had only been open a few months when Haffkine arrived. Its very existence had been made possible by donations from admirers following Pasteur's successful vaccination in 1885 of a nine-year-old boy, Joseph Meister, who had been severely mauled by a neighbour's rabid dog. The boy's wounds were horribly severe, but Pasteur inoculated him before morbid symptoms began to appear. Nothing in microbiology happened *de novo*. The veterinarian professor Pierre-Victor Galtier had already demonstrated that sheep injected with saliva from rabid dogs survived the infection. Pasteur confirmed that the pathogen would maintain its virulence when transferred into the brains of dogs, but it was the experimental confirmation that it could be weakened (or 'attenuated' as the Pasteurians now coined it) when transferred to other species that opened thrilling possibilities. Serially passed through a succession of rabbits, that same strain of highly toxic rabies derived from dogs progressively lost its lethal strength. In keeping with his belief that oxygenation stalled the pathogen's capacity for growth, Pasteur then air-dried sections of infected spinal cord taken from the rabbits. The end product was then injected into fifty healthy dogs, all of which survived. That would be the vaccine injected, thirteen times over eleven days, into young Joseph Meister.

Unlike vaccines Pasteur had previously developed entirely for animals – poultry cholera in 1877 and, as shown in an 1881 demonstration before 200 observers at Pouilly le Fort, anthrax – the daring jump to humans set off passionate reactions in France. For all those who regarded the rabies vaccine as heralding a defeat of infections, there were more who were outraged and horrified by what they thought was a gamble. Not all the critics were reactionary stick-in-the-muds. Some of Pasteur's own colleagues, including Emile Roux, voiced reservations. But Meister survived and went on to become a caretaker at the Institut Pasteur until the Germans overran Paris in 1940, at which point he killed himself with a gas gun.

Contrary to legend, not every rabid bite is fatal, but the Meister case made Pasteur a medical-scientific hero in the Third Republic. In October 1885, he presented the case to the Académie des Sciences. The publicity brought some more rabies victims to the Institut and, in the following year, 350 were vaccinated – successfully in all cases but one. Pasteur's authority was assured; money followed the fame and the Institut on the rue Dutot opened its doors in November 1888. It might have been otherwise had it been known (as was recorded in Pasteur's private lab notebook) that before Meister, he had previously treated two cases, one of whom developed more severe rabies and died. Prophetically, Pasteur saw the moment as the threshold of a new age in which the infectious diseases that 'have repeatedly affected humanity and are a major burden' would be conquered.[18] Even though he was mistaken about depletion through oxygenation – it was in fact passing through dissimilar species that had that effect – Pasteur would turn out to be right about the historical magnitude of that moment.

Lowly assistant librarian though he might be, Haffkine owed his translation to the Pasteur Institute entirely to Elie Metchnikoff, as he now liked to be known. For a while in 1889, he continued the kind of zoological research he had learned under the older scientist's influence. A paper was published early in 1890 on infectious diseases of the protozoan organism Paramecium.[19] But as Haffkine settled in, he became drawn to the gravitational pull of the Pasteur's other powerful star, Emile Roux. Roux, who had worked with Pasteur on avian cholera, anthrax and rabies, was one of only two physicians in the Institut and a consummate technician of animal-modified bacterial cultures; he had become committed to a vision of the place as a home of clinical innovation alongside basic research. When Haffkine arrived, Roux's younger research collaborator and *préparateur* – teaching assistant – was the Swiss Alexandre Yersin. But Yersin's abrupt and unexpected departure to become chief physician on the Messageries Maritimes shipping line running routes between the Philippines, Vietnam and south China gave Haffkine the chance to step into his place. He had taken the Grand Cours on microbiology – the first such programme in the world – and, after Yersin's departure, became Roux's principal teaching assistant. For a time, Haffkine continued with work inspired by his teacher, who also taught sections of the Grand Cours de Microbie. Metchnikoff was observing the varying adaptability of bacteria, typhoid in particular, to different environments in animal anatomy and Haffkine's second Paris paper dealt with bacterial adaptability.[20] The two directions of research were not necessarily exclusive. But there is no doubt that Metchnikoff saw it that way, becoming cool and sometimes scornful of the work of the younger man. It may not have been the betrayal Metchnikoff imagined, but relations between the two of them were never quite the same.

When did Waldemar sleep? The job at the Institut library was not trivial, either in time or effort. To judge by his lecture notes, inscribed in many books, where his tiny, fastidious hand flits across the page, interspersed with drawings recording observations from experiments, his preparation for the teaching assignment Roux set was almost superhuman.[21] But at the same time, Haffkine was naturally drawn into Louis Pasteur's almost religious faith that vaccines could be created for any number of infectious diseases.

Roux himself was working on diphtheria. And as he spent more time working for and with Roux, Haffkine began to ponder the possibility that Pasteurian methodology of attenuation could be used against cholera. Eyebrows were raised; a quixotic quest, something a green novice would try on the assumption that what Pasteur had achieved with chicken cholera could somehow be translated into a viable vaccine against the human disease.

But along with tuberculosis and diphtheria, no vaccine could be more urgently needed. Annually, rabies killed hundreds, while the victims of cholera, in any severe year, numbered hundreds of thousands. Though cholera was ebbing in western Europe, the fifth pandemic of the nineteenth century, beginning in 1881, would be estimated as having killed 267,000 in Russia alone in 1892–3; 60,000 in Persia, 90,000 in Japan and 72,000 in the Spanish provinces of Valencia and Murcia. Paris was less badly struck than in earlier waves. But Emile Roux, the only physician at the Institut who had personally witnessed the death of Louis Thuillier in the midst of experimental research, never shook off a sense of tragic responsibility. With whatever pessimism that the work might not yield a vaccine, he nonetheless encouraged Haffkine to go ahead.

He was not the first to make the attempt. A cholera vaccine had already been tried the previous decade in Spain. Jaume Ferran i Clua, a self-educated Catalan bacteriologist and physician (also ophthalmologist and expert in electro-shock therapy) who insisted that it was he, not Koch, who had first isolated the cholera microbe (they were both taking credit for Pacini's discovery thirty years before), had long been following Pasteur's work. In 1884, Ferran had created an attenuated live cholera vibrio as the basis for a vaccine. After presenting the results to the Madrid Academy of Medicine, he embarked on a mass vaccination campaign but with results so mixed and methods so jealously kept secret that a Spanish commission concluded – after three days of cooling their heels while waiting for dependable information – that they had seen no scientific basis for the vaccine.[22] A second commission, this time international, comprising American as well as European biologists, declared, even more damningly, that the vaccine was more dangerous than reliable. That conclusion was not altogether unreasonable. When Ferran tried out the vaccine on eighty healthy nuns, thirty became ill and sixteen died following inoculation.[23] There were other reasons for scepticism. Aside from the most summary reports to Spanish academies, Ferran was never completely forthcoming about his experiments. A live vibrio with fixed virulence which could then be dependably attenuated eluded Ferran, so that the eventual vaccine was worryingly hit or miss. There was no thought of trials with a control population.

Attempting to succeed where Ferran had failed – at least in the eyes of the bacteriological community – Haffkine laboured without much sign that his experiments would ever bear fruit. For many months, he was dogged by one of the major problems that had compromised the Catalan's vaccine: the instability of the live vibrio's virulence. Nothing could possibly be accomplished unless and until he had managed to achieve a measurably fixed virulence which, in keeping with Pasteurian orthodoxy, would then be attenuated by exposure to heat aeration at 50 degrees Celsius. But all through 1890 and 1891, this will o' the wisp eluded him and it

Metchnikoff's lab staff in 1890. Haffkine is standing, far left; Olga Metchnikoff is seated, far left, with Elie Metchnikoff seated, second from right.

was depressingly obvious that both Metchnikoff and Roux had come to think it might never be attainable. Frustratingly, within the stomach, the comma vibrio became inert, and Haffkine was not yet aware of Richard Pfeiffer's new technique of injecting the organisms directly into the peritoneal cavity of guinea pigs.

There were other obstacles blocking his way: to begin with, a shortage of experimental time. Crushed by teaching obligations, Haffkine asked if his lecture load and work at the library could be lightened. When he was refused, he became convinced that it had been Roux himself who had told Pasteur to say no. If anything, Metchnikoff was even less helpful. Despite the fact that the pay-off of Haffkine's work would depend on the machinery of the immune system operating much as Metchnikoff had charted it, the patron now wrote off his erstwhile protégé's work as naive and utopian.[24] The rejection by his mentor must have been deeply painful. But whenever Haffkine was on the brink of complete demoralisation, something improbable would happen to lift his spirits. For there was one exception to the sceptics: Louis Pasteur himself, who was retired from experimental research but went out of his way to encourage the young Russian and remind him of the Pasteurian maxim that, 'It is only by subtle, well-planned and persistent experiments that one can force nature to reveal a secret.'

In early November 1891, Pasteur was showing Prince Damrouy, the brother of the king of Siam, around the Institut. When the prince mentioned cholera, he was brought to Haffkine, who described the encounter in his lab book:

> 'Here' – and he [Pasteur] indicated me – 'is the person who is now very near, nearest of all to this discovery. We even hope it will be this year. This gentleman is the nearest to the discovery.' I felt embarrassed and did not know how to behave. The prince told me through his interpreter 'If you find a remedy against the cholera, Siam will erect a statue of you.' Monsieur Pasteur added to the last words of the interpreter 'A golden statue'. . . . Well this is how things stand at the moment. Let's see what we are now capable of.[25]

Still suspicious of Roux's mixed messages, unsure how truly committed Pasteur himself was, Haffkine pressed on. The spring of 1892 saw a conceptual breakthrough. If Ferran's much-attenuated culture of the bacillus was too unstable or weak to trigger an adequate and predictable immune response, why not try the high-risk opposite: a more virulent concentration? *Exalté* – exalted – was the Pasteurian term, apt for the scientist who, unlike the atheist Roux, remained a devout Catholic. Haffkine's procedure, aiming for maximum virulence, followed the principal guidelines of Pasteur's work on rabies, but if anything was both more daring and more laborious, requiring, in the end, no less than thirty nine guinea pigs to achieve a satisfactory result. A culture of the cholera vibrio was now injected into the peritoneal cavity of the animal. What the science reports demurely describe as 'exudation' – faecal matter – was gathered very, very carefully by pipette, examined for the presence of viable microbes and then, while Guinea Pig 1 was expiring, injected into Guinea Pig 2. As the procedure was repeated seriatim, Haffkine saw – just as he had hoped – that with each successive passage, the vibrio grew more abundant and more virulent. Oddly, perhaps, the smaller the animal, the more the microbes multiplied. By the thirtieth guinea pig, the brew was rich in disease. And still more encouragingly, there was no loss in virulence.

Then came the critical test. A dose from the highly 'exalted' culture was injected *under the skin* of yet another animal as in vaccination. Thrillingly, this resulted in no ill effects other than the development of a crustily necrotic sore at the site of the injection. Still more sensationally, when that same animal was subsequently injected with a virulent dose of the cholera vibrio – anywhere in its body including the peritoneal cavity – its immune system resisted the fatal infection. On the other hand, this jolting of the animal's immunity was unlikely to find acceptance by humans if the result was a dramatically ulcerating wound. But, so Haffkine reckoned, if an attenuated dose was delivered, also subcutaneously, some days before the stronger concentration, the lesion would erupt but with much less effect and would obviate the possibility of more dangerous necrosis arising, or any other lethal complications following the second shot. So indeed it proved. There was some swelling, then the eruption of a crusty nodule which, however, quickly disappeared. All the animals given the two-dose inoculation survived.

There are moments of clinical breakthrough which in hindsight seem so inevitable as to be barely worth noticing as out of the ordinary. Compared with Pasteur, Roux, Metchnikoff or Koch, Haffkine was not a profoundly original bacteriologist. He could not have arrived at his vaccine without the precedent of Ferran's exploratory failure, Pfeiffer's technique of peritoneal cavity injection and Pasteur's own methodology of serial passaging through animals to enhance or deplete microbial virulence. But not only did Haffkine persevere in the face of demoralising scepticism, before being rewarded – after the thirty-ninth passage – with a vibrio of fixed virulence, he also intuited that a preparatory awakening of immune response from an attenuated dose, followed, some five or six days later, by an exalted vaccine could secure maximum protection with minimum peril.

Nonetheless, given the public suspicion over the rabies vaccine, as well as the continuing devastation wrought by cholera, Haffkine's proposal to deliver a virulent dose of it to humans was a breathtaking move, fraught with the possibility of disaster – for him and for the Institut Pasteur. But he was giddy with confidence. On 9 July 1892, he reported the results of the guinea pig experiments, along with an account of his working method and preliminary results with a control group, to a weekly meeting of the Paris Biological Society.[26] Injected with the same lethal dose of the vibrio, unvaccinated guinea pigs died; the vaccinated recovered. Jaws dropped. A week later, on 16 July, Haffkine returned to report that identical results had been observed in rabbits and pigeons.[27] Now, it seemed reasonable to try the double-dose procedure on humans. So he would begin, of course, with himself.

This was science as theatre: the self-administration of a pathogen which had killed millions. In 1885, Ferran had inoculated himself and a friend with his own version of Pasteur's anti-rabies vaccine.

Tragedy ensued. Three of his friend's children became seriously ill and one died. So the moment was terrifying, but it was also pure Haffkine: at one and the same time daring to the point of recklessness but, as his conscience prompted, morally essential. Because Haffkine had chosen a site on his left flank between the ribcage and the iliac bone, and perhaps because he wanted a doctor to administer the vaccination, Georgy Vaveyn did the honours, six days intervening between the attenuated and exalted doses. Side effects were minimal. After four days, swelling went down; after six, he had no pain at the inoculation site. On 30 July 1892, a momentous day, then, Haffkine could triumphantly report again to the Biological Society: 'Inoculation of the two anti-cholera vaccines whose protection was established experimentally in animals does not present the slightest hazard to health and can be administered to humans with the most perfect safety. At the same time I believe that six days after the second vaccination the human body will have acquired immunity against cholera.'[28] At a time when cholera cases in Paris were doubling every month, the news became a sensation, although not necessarily the kind the Institut courted. There were those at the Institut who remained stubbornly unpersuaded, including, most painfully, Metchnikoff. Thrillingly, though, Emile Roux was a convert to what his assistant had achieved. As for Waldemar himself, romantic sensibility overcoming scientific dispassion, he had entered a state of grace.

The timing, so excruciatingly drawn out, turned out to be serendipitous. In November 1892, Louis Pasteur, who had been following Haffkine's progress with increasing excitement, wrote to Jacques-Joseph Grancher,

> Did you know that there may be a great development in the making, concerning cholera? Pettenkoffer [*sic*] announces in a publication from Munich that he has swallowed a cubic centimetre of a pure culture of the virulent comma bacillus without becoming uncomfortable, perhaps just a little diarrhoea, that he has noted a very abundant culture of this bacillus in his intestine . . . Here,

> Metchnikoff was triumphant, or much encouraged. For the past few weeks indeed he carried out experiments with the conviction that the comma bacillus is not the determining cause of cholera, that when there is cholera the bacillus must be associated with another microbe which is responsible for fatal accidents . . . But the pure comma microbe causes the death of animals. All this is supported by many precise facts. Haffkine will become aware of this with a certain stupefaction.[29]

Pasteur's letter chuckles with undignified glee at the prospective embarrassment of his own colleague, Metchnikoff. But even in the face of decisive evidence to the contrary, Metchnikoff refused to abandon his theory of a second, as yet mysterious, triggering micro-organism. And the uptake of the Haffkine vaccine and the triumphant publicity surrounding it could only have made things worse for the survival of a good relationship between erstwhile mentor and protégé.

A young, studious English bacteriologist was closely observing the drama and taking copious and careful notes.[30] The son of a Hertfordshire parson, born six years after the appearance of *On the Origin of Species,* Ernest Hanbury Hankin was just the type to be looking for a more modern devotion: science. Initially torn between a medical calling and the excitement of basic research, studying at St Bartholomew's Hospital School cured him of ambitions to be a doctor. A double first degree in natural sciences at Cambridge made it clear that he was, as he liked to tell his contemporaries at St John's College, 'an experimental animal'. This may have explained why, as a graduate and young prize fellow, he kept rabbits in his rooms, injected with cultures of bacilli he had isolated from anthrax, to see if they conferred immunity.

Hankin wore his impatient cleverness lightly. To his friends he was both boffin and card: the man who rolled heated cannonballs over his stomach claiming they helped digest the steak pies and steamed puddings dished out by the college kitchen. But Hankin was in deadly earnest about microbiology and knew he would have

Ernest Hanbury Hankin

to leave Cambridge to learn it properly. A competitive bursary won him time in Robert Koch's laboratory in Berlin, but Hankin followed it, that momentous summer of 1892, by landing at the home of Koch's adversaries in the Institut Pasteur. (Metchnikoff had been especially stung by Koch's public, theatrical dismissal of phagocytes at a Berlin conference in 1890.) It may, in fact, have been in Koch's laboratory that Haffkine and Hankin first met, since Waldemar had also stopped off there on returning from a visit to Russia at the end of 1891.

In Paris, the Home Counties vicar's son and the Odessa Jew, Hankin and Haffkine, became close. They both knew they were on the verge of a breakthrough in the possibility of human immunisation. Hankin volunteered to be one of the seven human guinea pigs who would follow Haffkine in receiving the cholera vaccine. And it was Hankin who supplied the first news of what had happened to the *British Medical Journal* on 10 September 1892:

> August 16th. Dr Roux inoculated me with one-eighth of a 24 hour old agar culture of the attenuated virus suspended in 1 cubic centimetre of bouillon on my left side, about 2 inches above the crest of the ilium. 1 pm Slight pain locally. 4 pm Swelling noticeable and pain on movement. 8 pm Owing to pain on

> moving, found walking about difficult. 10 pm temperature began to rise; noticed a feeling of malaise. 12 midnight. Noticed temperature of 100 degrees. This was the highest point reached.

On the 17th, Hankin's glands were swollen. Like Haffkine, he went to work at the Institut as usual 'but had a bad appetite and a feeling of fatigue which once or twice impelled me to lie down for a short time . . . the swelling extended during the day to the crest of the pubes and to within an inch of the umbilicus . . . felt bilious. Took four capsules of castor oil.'

Thereafter the symptoms abated, so on the 21st, five days after the first inoculation, Hankin got the more dramatic '*exalté*' dose. The next day he couldn't get out of bed, slept a lot and took some liquorice powder to deal with constipation. By the 25th, all the swelling had gone down, leaving his skin yellow. 'At present the only trace of the inoculation is a small hard nodule less than half an inch in diameter at the seat of injection.' He pointed out that though the swelling had been 'startling' after the second inoculation, neither he nor Haffkine thought it would become dangerous. Indeed, after his own second inoculation, Haffkine worked for twenty four hours non-stop without eating anything. And the others? 'All the gentlemen who have been inoculated experienced only a trivial elevation of temperature and of discomfort after their second inoculation.'

Sixty people, women and men, from within the Institut and other medical and research institutions in Paris, were vaccinated, none of them suffering serious or prolonged side effects. A triumph, then? A new dawn for vaccination? Perhaps. Hankin conceded that, even with a test group of sixty, the experimental vaccinations, given to an atypical population and in a non-epidemic moment, were not the same as conditions in the real world, and that while the results with live bacilli seemed dramatic, it might be prudent to begin a vaccination programme for humans with vibrio bacilli sterilised by heat. He also conceded that some bacteriologists in England had commented that there were in fact several differing strains of cholera and that vaccine produced from one might not necessarily deliver immunity to subjects infected with another. Moreover, no one on the basis of the Paris tests knew yet how long immunity might last. Nor, crucially, had there been as yet a control group – one of the criticisms levelled at Jaume Ferran. Nonetheless, Hankin concludes emphatically that 'the evidence . . . shows that M. Haffkine's method of inoculation is not attended by any grave disturbance of health and that it can be practised on human beings with perfect safety'.

Just twenty seven, Ernest Hankin wrote with the authority of someone who was more than a mere visiting research assistant. As indeed he was, recruited to the medical empire of the British Raj. His job description was 'chemical analyst' (no one quite knew what that title meant) and bacteriologist (that, on the other hand, was thrilling) to an enormous swathe of upper India from Kashmir, the North-Western Provinces and Punjab down to the Ganges valley, with his station office in Agra. In the course of a long career, this would lead Hankin to wonder in print if the sacred river did not itself possess some pathogen-eating organisms that explained the relative rarity of cholera in the lands bordering its banks. Agra would also turn him into a serious student of patterning in the decoration of Mughal architecture. But there was also something much more serious than the latest edition of intellectual orientalism at work in his decision to go to India. The first issue – shared by Haffkine – was that cholera vaccination should go where it was most desperately needed: Asia. In fact, the disease had not disappeared from Europe. Even as Haffkine's vaccine was revealed to the medical world in the summer of 1892, a brutal epidemic of cholera broke out in Hamburg.[31] The same wave of the disease would kill more than a quarter of a million in Russia.

In December 1891, already convinced that he was embarked on a project that would change world history for the better, Haffkine envisioned an international commission (of the kind Adrien Proust would have recognised) verifying the results of his experiments. But in Haffkine's mind, that vision always had to be inter-continental if it was to fulfil its benevolent potential. Following official vindication by the 'commission', Haffkine grandly proposed that he would 'ask permission to transfer temporarily our laboratory to the banks of one of the great Indian rivers and attempt to immunize the population against cholera'.[32] This was not a conventional imperial utterance; if anything, it was the reverse: an empire of disinterested scientific knowledge. But it would only be when trials were successfully translated from a rarefied experimental lab in Paris to the impoverished Asian multitudes, who could supply a comparative control group, that the vaccine might reliably prove its value. Moreover, echoing Voltaire, Gatti and Holwell, Pasteur, who was deeply interested in the global history of inoculation, thought that long before anything had been created in experimental laboratories, folk cultures had ancestrally practised the treatment. In 1880, in a lecture to the Académie des Sciences on the poultry cholera vaccine, Pasteur noted that 'the practices of vaccination and variolation have been known in India for the longest time'.[33] Why then, would Haffkine and Hankin meet resistance when they took vaccination to south Asia?

Testing effectiveness and safety against a properly comparable control group is now, of course, standard procedure: the indispensable condition of viability. But Haffkine was the first scientist to make it integral to a visionary project beyond Europe. Just where he might be able to execute a programme of vaccination involving such control groups, though, was another matter. Despite Hankin's appointment at Agra and the Englishman's pivotal role in taking the vaccine out of Paris and into the real world, the first choice was not necessarily going to 'the banks of the Ganges'. No one at the Institut had forgotten the visit of the Siamese prince and once Haffkine thought he had succeeded in making a safe and powerful anti-cholera vaccine, one of the first things he did was to tell Louis Pasteur that he could now inform the prince that the promise had been fulfilled (leaving aside the golden statue). But the reason for thinking of south-east Asia as a promising place for vaccination trials – the active presence of French colonial ambitions in the region – was also a major factor in its rejection by the king of Siam and his government. Albert Calmette's establishment of an Indo-Chinese outpost of the Pasteurian empire in Saigon could never be regarded in Bangkok as a politically innocent event. However pure scientific and clinical intentions were (and they were often impure), vaccination and imperialism would never be disentangled.

If not Siam, then where? Could it be time to go home? While Haffkine was making his vaccine at the Institut, French foreign policy was laying the ground for what would be a formal Franco-Russian alliance, sealed in 1894. Despite Haffkine's compromised political history, he had actually returned to Russia in 1891, carrying specimens of cultured vibrio with him; though exactly what then transpired in his brief trip is unknown. But with a cholera epidemic exploding there, how could Haffkine not want to make Russia the place for large-scale trials? Just three days after Haffkine's last presentation to the Biological Society, three letters were sent from the Institut to Duke Alexander of Oldenburg, president of the Russian Academy of Science (and a physician), requesting permission to vaccinate in the tsar's realm. One of those letters was from Louis Pasteur, giving the work his official blessing and grandly offering to 'donate' the 'method' to Russia, another from Emile Roux certified that the work to produce the vaccine had been done under his rigorous supervision (not entirely true), and the third from Haffkine himself making the case. Unaware of the thickening spy dossier on his whereabouts and activities, Haffkine must have thought the prospects promising. In 1892, Russia was suffering the worst cholera outbreak in decades, with nearly a quarter of a million deaths by the year's end. The scale of the epidemic surely suggested the urgency of the need for a prophylactic vaccine and no problem at all finding a large and reliable control population. Needless to say, however, the first thing the grand duke did was to refer the request to the Russian Ministry of Police, which in turn made inquiries from the political police about Waldemar Haffkine of Odessa.[34] A history of multiple arrests and imprisonment, as well as documented membership of the terrorist organisation that had assassinated the late tsar, settled the matter. But Oldenburg's written response to the three letters coming from the Institut disingenuously confined itself to the scientific history of vaccination. The failure of Ferran's work in Spain, Oldenburg noted, made it clear that an anti-cholera vaccine was a pipe-dream. The fact that Haffkine's procedures were expressly designed to correct what he saw as the failings in Ferran's work cut no ice. The Russian door closed.

But an Indian door was opening. The last decades of the nineteenth century in the sub-continent were a turning point in the history of the Raj. British pretensions that imperial rule was justified by measurable improvements in the lives of its millions of subjects had been exposed as a cruel joke, given the scale of lethal horrors that had fallen on India: famines and epidemics taking devastating tolls in cities and countryside while painted elephants trundled along in the vainglorious ceremonies of the durbar. The record was not invariably callous. The creation of filtered water supply systems in the late 1860s had made a difference to the susceptibility of some major cities to cholera.[35] Nothing like the horrifying epidemic of 1817–21 was seen again until the 1890s. So there is good reason to believe that Ernest Hankin saw his impending work as part of a demonstration that British rule was not entirely hypocritical when it came to claims of ameliorating the material lives of Indians.

And in Paris in 1892, there was someone else who would have shared that sentiment, albeit from an uneasy mix of pride and guilt, based on direct, personal experience. Frederick Hamilton-Temple-Blackwood, Marquess of Dufferin and Ava (the last piece of the title added on, triumphally, like a garland about his shoulders for annexing a large chunk of northern Burma) was then occupying the enormous building on the rue Faubourg Saint-Honoré as British ambassador to the Third Republic. Dufferin and his wife, Hariot, both of them tall, forthright, good-looking and voluble, he with a strong nose, she with a strong jaw, were distant Irish cousins. He was from the sturdiest stock of the colonising Protestant 'ascendancy', but she had

United Irishmen nationalists in her family as well as loyalists. Freddie had made his way from that world of green pastures and stone houses via the usual nurseries for Victorian eminence: Eton and Oxford. He sailed into the Foreign Office after skippering a boat called *Foam* around Iceland and the far north Atlantic. A book about those exploits made money for him but, like many of his contemporaries, the dashing Freddie was attacked by a sudden strong sense of imperial mission. This he discharged smartly enough in Syria and Turkey, and rose and rose through the hierarchy to become governor-general of Canada when its constitutional balance was fragile. Dufferin had brains and easy charm, and he took the trouble to speak excellent French to the restive Québecois while living for a time in Quebec City. It helped that Hariot was liked on both sides of the cultural divide for her independence of speech and mind which often took precedence over protocol.

Then came India, in the immediate aftermath of a crisis stirred up by the unexpected liberalism of the then viceroy, the Marquess of Ripon, a favourite of Gladstone's, who had had the unmitigated gall not only to establish local councils to which Indians could stand for election but, even more outrageously, to make it possible for Indian judges to try English men and women in Indian courts.

Left: Frederick Hamilton-Temple-Blackwood, Marquess of Dufferin and Ava *c.* 1880; right: his wife Hariot, Marchioness of Dufferin and Ava, 1880s.

Obviously the viceroy had gone mad or native or both. In 1884, the rogue Viceroy, Ripon, was replaced by Dufferin. It was a back-handed tribute to his diplomatic reputation for being able to placate mutually hostile parties that both the Anglo-Indian establishment infuriated by Ripon's liberalism and budding nationalists inspired by it initially expected the best from the new Viceroy.[36] Dufferin's viceroyalty unfolded against the first meetings of the Indian National Congress, largely the work of westernised, liberal Indians in professions like the law and high journalism. Initially he did nothing to discourage Congress; rather the opposite, but relations quickly soured with its prime mover and organiser, the retired Indian civil servant A. O. Hume. In 1888, at the end of his viceroyalty, Dufferin delivered an uncharacteristically contemptuous speech, dismissing Congress as representing only 'an infinitesimal and partially qualified fraction of the people of India'.

Any residual liberalism on Dufferin's part seemed to have given way to full-on imperialist paternalism which fancied itself as aligning with 'real' India – rajas, landlords and peasants – rather than the westernised 'babu' journalists and lawyers of Congress. But in August 1887, Dufferin launched an inquiry into landlessness and poverty which, though never published in full, would have consequences well beyond its original remit. The 'Dufferin Report', the Bengal section of which was completed in May 1888, was not free of political strategy. In 1882, Dadabhai Naoroji, the second president of Congress and later a Liberal MP for Finsbury Central in the British Parliament, published *The Condition of India,* a frontal attack on the smug truism that at whatever cost to native sovereignty, the British empire in India had at least materially improved the lives of its millions of subjects. Assessing outflows of capital, the destruction of the indigenous cotton industry, Naoroji argued, on the contrary, that British rule had been a 'drain' on India's economy. But he argued from a cost-benefit analysis based on macro-aggregates of investment and taxation. Dufferin's inquiry was designed differently, a state-by-state, city-by-city and, in some areas, village-by-village investigation of actual social experience:

tenancies, wages, diet, shelter and the like. In some respects, this exercise in data collection was of a piece with the fashion in Britain itself for social investigation into the condition of 'The People', especially in the great cities and manufacturing towns. Such reports could be embraced both by Liberals, especially those on the radical wing, and, following Disraeli, by Conservatives who liked to point to improvements in material conditions and public health as an alternative to their opponents' obsession with political mobilisation and home rule.

None of this is to say that Dufferin and the Revenue and Agricultural officials delegated with the work of collecting evidence were entirely cynical in their project. The Viceroy's Irish roots might have made him sensitive to the horrifying scale of the great Indian famines of 1876–8 and their jarring contrast with the pomp of the Durbar. The completion of the first parts of the report – dealing with Bengal and Bihar – made it evident that landlessness, indigence, impoverished diet were becoming worse not better and that, as Naoroji had seen, the standard claim that Indian social conditions were constantly improving under the firm but benevolent rule of the British was an indecently hollow brag.[37] As more data flowed in, the report became a greater embarrassment; only a severely redacted part of it ever saw the light of day. And when, in 1902, there was a demand for it to be brought before the House of Commons, the Secretary for India, Lord George Hamilton, responded that it was, alas, 'very bulky . . . and fifteen years old'.

The Dufferins left India (without seeing any end to the inquiry), but India did not leave them. Frederick's next posting was as Ambassador to the Kingdom of Italy, where Pope Leo XIII had become committed to creating a Catholic clergy in India, Ceylon and Burma. Protestant missionaries in the Raj, however, had another powerful vocation: the saving of bodies as well as souls. Years earlier, before their departure for Calcutta, the Dufferins had been made aware of medical missionaries – especially the 'lady doctors' providing obstetric and midwifery help for high-caste women confined to the separate quarters known as *zenana* and thus deprived of male-administered medicine – by none other than Queen Victoria. In 1881, at Windsor Castle, the Queen had received one such medical missionary, Elizabeth Bielby, with a request from one of her patients, the Maharani of Poona, encased in a large silver locket 'praying me', the Queen wrote, 'to sanction female doctors being sent out to attend the ladies in the zenanas of India many of whom died for want of proper medical attendance, no man being allowed to go near them. Miss Bielby gave a melancholy account of these poor ladies . . . I expressed my deep interest and hope something might be done in this matter.'[38] Two years later, another of the women doctors, Mary Scharlieb, about to become a lecturer in midwifery at Madras Medical College, was introduced to the Queen by Sir Henry Acland and made a profound impression. In turn, Victoria made sure that the new Viceroy would give his support to the cause. But it was Hariot who in 1885, and with the help of wealthy Indians including a raft of maharajas, established the National Association for Supplying Female Medical Aid for the Women of India with the Queen-Empress as its patron.

Catering to high-caste and middle-class women secluded in the *zenana*, the Fund was self-evidently a conservative project, envisioning British 'lady doctors' coming to the rescue of what Elizabeth Bielby told the Queen was the harm done by ignorant (though, it was always added, well-meaning) native midwives.[39] But resistance to examinations by male doctors, especially during epidemics, was by no means confined to the well-off. Some of the most furious and violent opposition would come from poor Indians, so that the provision of women physicians, especially Indian, would indeed have a benevolent effect, very much as Hariot Dufferin hoped. In the event, the finance which had been forthcoming, especially when it was known that the Queen was personally involved, dried up after the Dufferins departed from India. But in the teeth of Anglo-Indian opposition to the admission of women to medical colleges in Madras and Calcutta, never mind paying for Indian women to be sent to England and the United States, the Association supported the education and training of eleven women doctors. Six of them, including Kadambini Bose Ganguly, Anandibai Gopal Joshi and Rukhmabai, constituted the founding generation of Indian female physicians. Conservative ideas can have radical consequences. Congress nationalists praised the Fund and one of them suggested that instead of the expected statue of Viceroy Dufferin set on the Calcutta Maidan, it would be a much better idea to have one of Lady Hariot.

Whatever the course of Dufferin's tacking between dilute liberalism and conservative paternalism, there is no doubt at all that he understood very well that the pretensions of the Raj to have materially bettered the lives of Indians was most severely tested by its response to epidemic. Landed in the Paris embassy in March 1892, both he and his wife were bound to be interested in the work being done at the Institut Pasteur, especially in the possibility of a vaccine that might correct the impression that India was, as one British newspaper put it, 'the home of cholera'. A year earlier, however fancifully, Haffkine had already imagined the translation of his laboratory to 'the banks of the great Indian rivers'. Now, Dufferin's invitation to take his work to Calcutta struck home. Discussing the venture with the India-bound Hankin strengthened his intuition that this would indeed be the decisive trial of the vaccine. At any rate he was sufficiently taken by the idea to go to London in December 1892 and, as arranged by the ambassador, to meet the Earl of Kimberly, the secretary of state for India. During that same visit he gave his first verbal report on the cholera vaccine to the teachers and cadets of the Army Medical School at the Royal Victoria Hospital, Netley. The hospital, much visited by the Queen after whom it had been named, was one of the great institutions of the medical empire, and as awed visitors often commented, the longest, if definitely not the handsomest, building in Britain, slung along a shore site near Southampton. Haffkine received a dutiful hearing, but that first encounter with the serried ranks of the medical Raj must have been daunting. There he was, the Russian Jew from France and not altogether what was expected of Those People: fair-haired, well dressed and eloquent, if heavily accented. And there they were, the rows of moustaches, many of them sceptical about the Frenchified science they were hearing. What did this man really know about India? Not even a proper doctor. Generations of their kind had been battling disease: smallpox, typhoid, malaria, cholera and plague along with the obstinate superstitions that got in the way of treatments and cures. Jenner's vaccine had run up against the Hindu reverence for the cow. Their cemeteries in Calcutta, Bombay and Madras were lined with young men and women who had themselves fallen victim to the diseases of the country, India's revenge on their presence. Well, he would learn soon enough.

Instructions went out to the Municipal Corporation of Calcutta to give assistance to Waldemar Haffkine, though it was as yet unclear whether this amounted to an actual position. February 1893 saw him on board a P&O liner sailing to India, via the epidemic bottleneck of Port Said. Knowing that refrigeration would be well-nigh impossible, he had taken a supply of 'devitalised' (sterile) cholera vaccines with him and wasted no time trying them out, persuading the young Surgeon-Captain C. C. Manifold, returning from a six-month leave, to undergo vaccination while still on board! Haffkine made sure to take his usual scrupulous notes. Manifold was vaccinated on 1 March, shortly after the ship had docked in Aden where the vaccinator had a biblical epiphany. Walking through the town, he encountered two old men, rags falling off their bony forms. At first, Haffkine assumed they must be poor Arabs. But then, looking again, something entirely unforeseen called to him, and brought forth from his lips the Jewish prayer the *Shema*, recited thrice daily but also in moments of great mortal peril: '*Shema yisroel, adonai elokeinu, adonai achad.*' 'Hear, O Israel, the Lord is God, the Lord is One.' At which point the Yemeni Jews joined Haffkine in completing the prayer. It was Haffkine's initiation into direct contact with Jews, seemingly worlds away, on the surface as unlike him as it was possible to be and yet, inseparably, also kin. He would never forget the moment.

Four days out from Aden (half the interval used in the Paris trials), Haffkine gave Manifold his second, stronger vaccination.

On 5 March, they disembarked at Bombay, Haffkine beginning his journey to Calcutta where he was to lecture to the Medical College.[40] There, he was able to report that the obliging Manifold had suffered no serious side effects from the procedure. Though he knew that there was resistance in the ranks of the Indian Medical Service to what was still a wholly new science, Haffkine thought his reception at least respectful. But – understandably, with no track record as yet – the Indian government in Calcutta warily treated his work as strictly and provisionally experimental. Officially, C. J. Lyall, Chief Secretary to the Government of India in Calcutta, 'suggested that a few selected Civil Surgeons in each province might be deputed to assist Dr. Haffkine while in their district'.[41] That was all.

In March 1894, a photograph was taken of Haffkine vaccinating a young Indian girl. If the British Raj had wished to stage a tableau enacting their claim to be interested in nothing more than the well-being of the millions of subjects of the Queen-Empress, this surely would be it. It is exactly the opposite from the horrifying photographs, taken by missionaries on their new box cameras, of the murderous famines that struck much of India in the same decade. In those brutally unsparing images, the sahibs are nowhere to be seen while Indians perish in their untold millions. Bodies are left to rot in the road where the remains are scavenged by carrion birds and feral dogs. This, on the other hand, is different. The image is composed like a Baroque history painting, a Caravaggio or a Rembrandt, crowded with attentive benevolence, a redemptive charge running through it from the hand of the ministering healer to the observing native apostles and helpers, to those on the receiving end of the saving touch; some enthralled, some indifferent, some outright suspicious. It is also an unusual document of medical collaboration between Europeans and Indians. Behind Haffkine is William Simpson, the staunchest of his helpers and champions, and defender against the legions of sceptical or downright hostile personnel of the Indian Medical Service.[42] To his left are the Medical Inspector of Calcutta, G. N. Mukerjee and next to him Dr Jagendra Dutt, analyst to the city health department. Haffkine always acknowledged that without Indian doctors, acting as translators, exemplars (who were themselves publicly inoculated as was Haffkine) and reassuring persuaders, his entire programme would have been a non-starter. Some of those medics, trained by Haffkine, then acted on their own initiative, expanding the range of inoculation well beyond an area he could personally visit. One of those Bengali pioneers not in the photograph but singled out by Haffkine in a letter of recommendation was a Dr Chowdry who offered his services 'at the time when no prospect of any promotion or any personal benefit for him would prompt him to exert himself'.[43] But Haffkine was not an arms-length worker; so there he is in the centre of the tableau, St Waldemar, pale-skinned saviour of the poor, protector of little children, vaccinating a young girl while a white-bearded grandfatherly figure tries to allay her fearfulness with a gentle hand laid on her shoulder.

The setting, with its mud and thatch walls, looks rural but is in fact one of the Calcutta *bustees* (or bastis) where Haffkine wrote that his little crew worked 'under a burning sun without any other shade than an umbrella [absent here] uninterrupted from the hours before dawn to mid morning; then back to the lab to prepare vaccine for the next inoculations and then in the evening around four or five pm till late at night'.[44] The *bustees* were (and are still) patches of land both on the outskirts and in the heart of the city, rented out to middlemen who then built as many cheap dwellings as they could to lodge migrant workers pouring into the city from the surrounding countryside.[45] The materials were rudimentary; the huts without windows or ventilation; sanitation non-existent. Individual houses were crowded with extended families or even groups of them. From Haffkine's own scrupulously detailed reports, we know, for example, that twenty five people lived in Gopal Das's house in the Shampukar Street bustee in Ward 1, and twenty one in the Chomilal Koormi house in the Joraghan bustee. And while the density of the population in individual houses and throughout the bustees generally made a perfect incubation for *Vibrio cholerae*, it also gave Haffkine ideal units for comparative trials, since he could divide a house into inoculated and uninoculated groups, knowing that both shared nearly identical living conditions. Nineteen people lived in Mungo Jamadar's house in the crowded bustee of Kathal Bagan; eleven were inoculated, eight not. Of the latter group, four suffered cholera attacks: three of them, a man and a woman, both forty, and a five-year-old girl, died, while a 25-year-old recovered.

A letter of thanks in April from Mohammed Ayub Khan to Haffkine for inoculating himself, his family and servants, and presenting a Kashmir work silver jug in gratitude makes it clear that Haffkine started work soon after he got to Calcutta in April 1893, at least among well-to-do Bengali society. But the overwhelming majority of the 22,703 vaccinated against cholera in Haffkine's first Indian year were inoculated in the first months of 1894. There were many reasons for the delay: a host of independent variables which Haffkine later admitted he had not anticipated, patterns of migration being among them. Unlike in Europe, cholera was most rife among the rural poor, but as the Indian economy became more interconnected and rail networks shortened distances, infectious diseases came with the migrants to the crowded shanty settlements of the cities. Then, too, India was a subcontinent comprising, in effect, many differing kinds of country. In many of those regions, cholera was endemic; in others it broke out as epidemic. When countless people fled from the latter to the former, reliably comparative controls were very difficult to establish. As it happened, 1893 was one of those rare years when epidemic outbreaks were relatively rare in Bengal and the Ganges valley. But they came with a vengeance in the following year, 1894, providing Haffkine with a measurably large population of the infected.

Even so, there was resistance from sections of both the Indian and British communities in Bengal. Suspicion among Calcutta Muslims sometimes spilled into violence. At Serampore in West Bengal, where thousands of migrant workers in the jute mills were housed in the crammed slum colonies that were optimal breeding grounds for the cholera vibrio, rumours spread that the vaccine killed rather than protected and that 'inoculators were hovering round every corner to lay hold of people who were then forcibly inoculated'.[46] Tempers became ugly, especially among the Muslim community, where some mill managers subscribed to the scaremongering and others feared there would be riots by their workers. Threats to Haffkine himself became so heated and sinister that he was forced to beat a retreat.

But the most frustrating obstruction came not from hostile native communities but from dug-in opposition in the military and medical establishment of the Raj. The physicians and surgeons trained at Netley, whose first responsibilities were to ensure that the army, both European and Indian troops, was not struck down by disease, mostly clung to the traditional conviction that cholera was spread atmospherically or arose locally from putrefying organic matter. In his 1895 report to the Indian government on the effectiveness of the vaccine, Haffkine recalled the 'incredible difficulties' he had encountered, by which he meant institutional resistance; the refusal of state funding; the reluctance of the official medical service to give any kind of official recommendation for the vaccine. Surgeon-General W. Rice was typical in his view that Haffkine's vaccine lacked 'the century of experience which justifies Government in making the Jennerist [smallpox] discovery compulsory'.[47] Rice, who evidently had not read the scrupulous reports published by both Hankin and Haffkine, claimed that the vaccine was not as harmless as the latter claimed and was strongly against government 'taking any active share in propagating this system'. And if Haffkine wondered why, notwithstanding a glowing letter of introduction from Dufferin to his viceregal successor, the Marquess of Lansdowne, the medical Raj wanted no part of his campaign, he needed to look no further than the viceroy's honorary physician, surgeon-major of the Bengal Medical Service and Professor of Pathology at Calcutta Medical College, David Douglas Cunningham.

Unlike Sir Joseph Fayrer or J. M. Cunningham (no relation), David Cunningham, who had entered the Indian Medical Service in his twenties, was not a scientific reactionary. Although his opposition to Koch's insistence that the comma bacillus was the exclusive cause of cholera grew more furious in the 1880s and '90s, Cunningham, who had his own laboratory in the grounds of Calcutta General Hospital, recognised the reality of the microbes, but remained steadfast in believing that Pettenkofer was right to insist that local conditions of soil and climate ultimately determined whether the bacillus would become epidemically toxic. That the comma bacillus could not possibly be the exclusive cause of lethal infection he thought was proven by the fact that he had, for a few rupees, persuaded one of the sweepers in the grounds of the Hospital to swallow bacillus-rich samples with 'no ill effect'. Such were the opportunities offered by imperial science. Further futile experiments were conducted to see if colonies of bacilli could be cultivated in specific types of soil, in cow dung or deposits of human ordure. He also set great store by the arguments of the Swiss mycologist Ernst Hallier that cholera, along with measles, typhoid and syphilis, were the result of fungal spores entering the human body. Believing that what had been seen in the excreta of cholera victims were those fungal 'zoospores' or 'micrococci', Cunningham devised what he called an 'aeroconiscope' to trap them. This was a contraption in which glycerine-covered slides were set in a cone, facing the wind to collect the airborne pathogens. When the fungus theory failed to stick, he took other tacks. Koch's bacillus was not one organism but polymorphous, some forms of which were infectious and others not.[48] Often Cunningham's real attention was elsewhere: as curator of the Calcutta Royal Zoological Garden and official naturalist to the Sikkim Expedition of 1886.

None of which is to say that Haffkine would necessarily have expected indifference, much less opposition, to his vaccine from the likes of Cunningham when he arrived in Calcutta in 1893. Perhaps the contrary, since, in 1892, Cunningham had actually sent him specimens of highly virulent vibrio, when he was in the midst of his work at the Institut Pasteur. It would have been a bitter disappointment then to learn of Cunningham's belief that the vaccine 'would not, in any way prevent the occurrence of cases of the disease but . . . will modify their character in a fashion similar to that which hygienic improvements in a locality may provide'.[49] Making the best of it, Haffkine hoped that this crumb of recognition might at least offer the possibility of support alongside traditional localist campaigns at exterminating 'filth'.

Fortunately, Haffkine had some more enthusiastic local allies. There was, to begin with, Simpson, the chief medical officer in Calcutta, who went out on a limb to support him, sometimes accompanying him on his long vaccination campaigns.[50] It was Simpson's report on the inoculation programme which, eventually, persuaded the Indian government to give Haffkine 'every facility' for his work and to have a local board of two medical officers and two magistrates report the results. Not many did. At the beginning, when he was still deemed a rank outsider, the only substantial help came from Hankin, now settled in his laboratory at Agra where he commanded staff trained in the preparation and safe storage of vaccines. Hankin's territory was also a region – with an eye on the possibility of Russian military movements – of heavy troop concentrations. Once a new wave of cholera began to move through those cantonments, Hankin succeeded in persuading military medical officers to offer voluntary vaccinations to both Indian and native soldiers. At Lucknow, for example, the 640 soldiers who remained unvaccinated had 120 cases of cholera, of whom 79 died, while among the 133 who were vaccinated, there were just 18 cases, of which 13 proved fatal. (One of the conclusions of these trials was that the vaccine was effective in preventing infection but less so in stopping the infected from dying.) When news got out concerning the military vaccinations, questions were raised in Parliament about the wisdom and the legality of what seemed unauthorised trials, not least because it was thought military vaccination risked disaffection or even mutiny in the barracks, especially if it was believed that animal matter was involved. But this was based more on memories of the great sepoy uprising of 1857 than contemporary reality. Total unfamiliarity with the rudiments of bacteriology did not prevent one letter-writer to the *Pall Mall Gazette* from arguing that it would be 'insane to tolerate an experiment of this nature' on the bodies of Gurkha soldiers or any higher-caste troops, since how could the army explain 'the repulsive manner in which they have been inoculated with the "excreta" of men of no caste, lepers and the like'?[51] In contrast to the usual race assumptions about native 'backwardness' and 'superstition', Indian troops were sometimes more willing than British soldiers to get inoculated. At Dinapur, where the 2nd Battalion Manchester Regiment was stationed, there had been thirteen severe cases of cholera in the summer of 1894, including nine deaths. But only 193 British soldiers came forward for vaccination, while 729 declined. After another three men died in short order, panic set in and another 387 volunteered for the jab.

This greater willingness on the part of some native troops, Sikhs in particular, to accept vaccination was less surprising to anyone familiar with Indian medical traditions of the kind described by John

Zephaniah Holwell a century earlier. In the spring of 1896, when Haffkine inoculated volunteer prisoners at Darbhanga jail in Bihar, the superintendent (and surgeon) E. Harold Brown reported that eagerness to be vaccinated was largely due to benign associations with *tika*, the smallpox inoculation they had known in their villages.[52] But while Brahmin inoculation may have helped ease Indian fears about what, on the face of it, was a deeply counter-intuitive procedure, Brahmin inoculators would not have been overjoyed at the prospect of being replaced by Haffkine's vaccinators.

Other considerations slowed the campaign. When Hankin and Haffkine had first talked about vaccinating in Asia, they had wondered if, especially in the midst of a raging pandemic, it would be safer to use sterile 'devitalised' doses. Initially, Haffkine was convinced that the safety of the vaccine required a live but attenuated initial dose, kick-starting the body's immune response without bringing on severely necrotic side effects. But as time went on, he wondered if two inoculations were, in fact, indispensable, especially given the logistical difficulties in getting patients to return for their follow-up dose. In his 1895 report to the government of India, Haffkine commented that perhaps the weaker starter dose might after all be unnecessary. As always, he never ventured a change without first trying it on himself. Once he had been inoculated with the 'exalted' dose alone, and without ill effects, he proceeded to do the same to his thousands of subject patients.

Though Haffkine always emphasised that he would only carry out trials on volunteers, a number of his experimental groups seemed, on the face of it, to some degree, captive. In July 1894, at Gaya jail in Bihar, one of the most chronically impoverished and famine-prone provinces in northern India, there had been a sudden, violent outbreak of cholera, the worst in a decade, so Surgeon-Major Macrae reported when he requested Haffkine's vaccination programme. Although originally built in the time of the East India Company, Gaya prison was not the reeking hell-hole of the condemned seen elsewhere. The original structure had been replaced by a new building boasting model sanitary conditions of light and air. Its water supply came from a well rather than an open tank. But pilgrims returning from the Kumbh Mela festival at Allahabad, where millions gathered for the ritual dip in the Ganges, had brought the disease south-east to Gaya. The previous year had seen 768 deaths from cholera in the district, but in 1894, 6,005 had died in July alone. In the second week of the month, five prisoners of six who had suffered an attack died. 'Many of the prisoners, on being told about preventive inoculation, wished to be inoculated,' so Macrae reported to the *Indian Medical Gazette*, 'and M. Haffkine . . . whose zeal and enthusiasm in the cause he so well advocates are beyond praise, arrived here on the 18th July and in the presence of Surgeon-Colonel Harvey who kindly assisted and myself, inoculated 147 prisoners and on the 19th, 68, making a total of 215 out of 433 at present in the jail.'[53] Haffkine himself described the procedure, designed to be rigorous about the random control group drawn from identically comparable living conditions. 'The prisoners had been told to seat themselves on the ground in rows and every second man or woman as they happened to place themselves, was inoculated.' At the beginning of the exercise, comparative results were mixed, not least because during the five days before the first dose took full effect, prisoners had been unprotected from infection. In the end, Haffkine concluded that the unvaccinated were five times as likely to go down with cholera as the vaccinated. 'I henceforth accepted these conclusions confidently as guidance for future work.'[54]

Between February and July 1895, Haffkine visited forty five tea-growing plantations in Assam, Sylhet and Chittagong – the 'gardens' of Burnis Braes, de Gubber, Chargola, Kalain, Cachar, Pallarbund, Lungla and Kalaincharra – and it was at one of them that he caught the malaria that ended his first Indian mission. Initially there had been an attempt to use local indigenous people – the Naga especially – to clear rain forest as well as cultivate the crops. But they had resisted the brutal work regime either by fleeing the tea gardens or by succumbing to disease. So the tea industry became dependent on large numbers of imported peasant labourers culturally familiar with what was needed to raise and harvest tea.

Housed in crowded huts, living in near-slave conditions, the 'coolies', mostly Indian migrants, were nonetheless not expendable even to the most ruthlessly exploitative tea estate-owners, hence their grasping at the vaccine's possibility of reducing losses to their labour force. To try to pre-empt the infection, Haffkine went to centres of coolie recruitment – Purulia in west Bengal and Bilaspur in the Arpa valley – inoculating thousands of prospective labourers.[55] On the plantations themselves, Haffkine vaccinated long lines of workers, returning to ensure that a second dose had been properly administered. Comparative results were conclusive. At Cachar garden, only 4 deaths occurred among the 2,381 inoculated, while there were 61 deaths from 2,976 who refused the vaccine.[56] Overall, in Assam it was estimated that the vaccinated were seventeen times less likely to die of cholera than the unvaccinated.

By any standards, this was a hard-earned success and it had been accomplished not at the usual distance of the imperial rulers but by deep immersion in the social realities of India. By the time his malarial shakes and sickness compelled a return to Britain, Haffkine could hardly be accused of aloofness, of seeing India with the eyes of a white sahib. His odyssey, in companionship with just one or two doctors, Indian as well as British, had been heroic, almost as mythic as the gods and stories he encountered on his way. In the Himalayan foothills he had sought out pilgrims (a primary chain of contagion) on the Haridwar trail to the festival which, in 1872, had killed 100,000 people in a terrible cholera outbreak; at Mussoorie; and at Nainital where the lake was named after the eye of Sati, one of the fifty two parts of her body sliced up by Vishnu, this one dropped in its limpid waters, making it supremely holy. He had ventured into villages where only Kumaoni was spoken. In keeping with British fearfulness that its power in India could be weakened by epidemics sweeping through its native troops, Haffkine had inoculated soldiers in 64 regiments, the vast majority of whom were Indian. But he had also treated multitudes of Indians whose welfare was not a priority for the governors of the Raj: lower castes shunned by Brahmins, prisoners, beggars, slum dwellers and coolies. In the Brahmaputra valley, the Himalayas glowing on the horizon, he had inoculated 294 British officers, 3,206 native troops, 869 civil servants – but 31,056 Indian villagers and townspeople.[57] The odyssey had taken him to Oudh, Punjab, the North-Western Provinces, the coalfields of Jharia and the jute mills of West Bengal; to some of the most destitute and desperately famished villages of Bihar and Orissa. In 1893–4, 42,000 were vaccinated; another 30,000 in 1895. One journey alone, from Serampur to Haridwar, took a thousand miles of travel.

Haffkine's meticulously entered expenses books record payments for every kind of transport: trains (both first class and third); boats and boatmen; *gharri* horse carriages and, for his bags, bullock carts, mules and ponies.[58] *Chokidar* watchmen, ferrymen, sweepers, innumerable coolies for menial tasks and *punkah wallahs* to keep him tolerably cool, all needed paying. Heaven forbid he forget to keep grain coming to the Calcutta lab to feed the guinea pigs. Should his travel be delayed, as it often was, there were payments for last-minute

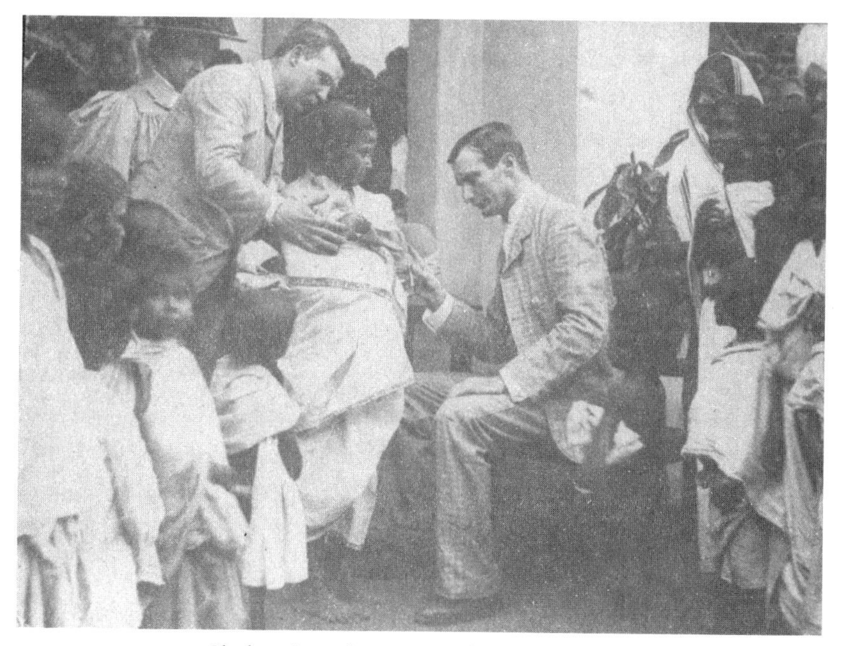

Cholera inoculations at Chaibasa, Bengal.

overnight bungalow accommodation. En route, supplies of workaday things ran short, so orders were put in for sealing wax, stamps and stationery; ink and blotting paper; laces and braces; soap and string. Not least, there were the needful items to keep Waldemar Haffkine in good form: black silk socks, evening shoes, 'sleeping suits' (especially for long train journeys); gallons of lemonade and bottles of wine spirit; and (at Srinagar) a serious craving for chocolate and walnuts.

Haffkine prepared his vaccines wherever he could: in railway carriages, station waiting rooms, improvised sheds. Nothing quite like it – the organisation and improvisation – sustained over immense distances had ever been seen in the history of public health or medicine, much less in the heart of European empires in Asia or Africa. Waldemar had sat half-naked children on his lap, soothed the fears of old women, reassured imprisoned thieves and corporals with rotgut liquor on their breath. Up in Lahore (he had been there too), Rudyard Kipling had heard of the travelling Jew, the hypodermic missionary of modern medicine. You would suppose he might have been perfect for one of Kipling's tough and wild short stories. But for some reason, neither Haffkine nor anyone remotely like him ever featured in his pages.

PART THREE

Power and pestilence: plague

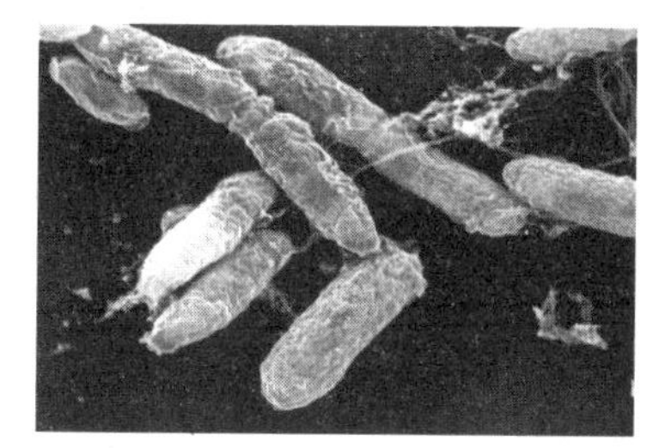

vi

THE DEATH OF RATS

The imperial bond turned out to be the imperial bind. The difficulty was this: the moustaches in the Colonial Office (and for that matter the Quai d'Orsay), the India Office and the Viceroy's House in Calcutta; the brigadiers-general in their cantonments; the tea-planters in their vast 'gardens' peering at rows of toiling coolies; the magistrates enthroned beneath courthouse portraits of *Kaiser-i-Hind* Victoria wearing her face of unrisen dough; the customs inspectors of the Canton godowns prodding the chests and bales; the chiefs of police, I-see-you–you-don't-see-me peaked caps pulled down over their eyes, and their subordinates with their boiled cheeks and swagger sticks tucked under their arms; the hatted memsahibs tending herbaceous borders in gabled Ooty and half-timbered Simla, or bending over the bowling green; all those morning-bright schoolmasters declaiming Shakespeare – THE QUALITY OF MERCY IS NOT STRAINED! – while the punkah rose and fell on downcast ranks of boys; the polo riders nursing a long G&T as they ha-ha-ha-ho-ho-ed in the club; *all* believed, or at any rate said they believed, in the disinterested mutuality of their rule, the ultimate benevolence of their authority. They conceded, to be sure, that the empire had been made in their interest. Of course! What bloody fool or bloody hypocrite would deny it? But (they *all* believed, or at any rate said they believed) what was good for the white rulers – their commerce, their laws, their troops – was self-evidently good for the ruled as well. P&O lifted all ships, did it not? Was there not an Indian *munshi* standing to attention by the Queen's bedchamber door? Funds banked in Glasgow and Liverpool, London and Birmingham found their way back east as investment. Well, then, the toiling cultivator, the rickshaw runner, the ticket inspector in his starched khaki and fiercely belted shorts, the driver of buffalo, the cutter of jute and the logger of teak would, sooner or later, all be beneficiaries of this stupendous system of interlocking parts, the globally unparalleled distributor of peace and prosperity, the guarantor of shared well-being. Think of it this way (they did). The empire was an invitation to become modern. Why would anyone, of any colour or confession, decline that?

So something as inconveniently medieval as the Black Death had a cheek showing up at the very places which were busy being modern: Canton, Hong Kong, Singapore, Bombay – especially when everyone was so looking forward to celebrating the Queen's Diamond Jubilee. What a way to spoil a party! Still, it gave one pause. There were, out there, undeniably, the starving, the sick, bags of bones blackening or bleaching in the roasting sun; heaps of them rotting on fields of battle (never sought, of course, by the imperial masters, but never shrunk from either), crumpled up or kicked by cattle, lying in fields and roads, dumped in rivers and woods in years of catastrophic dearth; and then when, as had just happened, plague came calling, ferried from the hospital boat anchored in the harbour to a remote mass grave or carted to the ghats, the pyres, burning day and night, giving off the sickly smoke that collected as a dark mantle hanging above the town.

For as Dr Proust had endeavoured to explain, in Vienna and Dresden but especially in Venice in 1897 where the International Sanitary Conference agenda was entirely devoted to the plague pandemic, the very means used to bind the parts of empires more closely – shortening distances, abbreviating shipping schedules, reducing costs, optimising profits, doing things the modern way – had themselves become the flowing conduits of disease and death.

*

The modern instalment of the Black Death arrived in Hong Kong in the spring of 1894 the modern way: by steamship, coming from Pakhoi, 300 miles to the west. Along with tea, silk and cotton, their cargoes included Yunnan opium, packed in hemp bales, amid which lodged the usual stowaways: *Rattus flavipectus*, long-haired yellow-chested rats.[1] If the hold also shared space with rice or grain, so much better for them. But the rodents were, in their turn, dining opportunities for their fellow passengers, *Xenopsylla cheopsis*, the biting flea. Should the flea have taken a meal from a previously infected rat, for instance in Yunnan province, where plague was endemic, the newly infected animal would die and the flea would hop off for an alternative host, animal or human.[2] In 1898, another of the students in Elie Metchnikoff's lab at the Institut Pasteur, Paul-Louis Simond, who had himself survived yellow fever in French Guiana, would discover the plague bacillus in fleas departing from dead rats, establishing the role of insect bites in transmission. But even without benefit of microbiology, the Chinese, especially in Yunnan and the south, treated the sudden appearance of a host of dead rats as the invariable herald of plague. In 1792, the sight of them moved a young poet, Shi Daonan, to write 'The Death of Rats':

Rats die in the east
Rats die in the west
People look upon dying rats
As if they were . . .
A few days following the death of rats
People die like city walls
Do not ask how many people die
The dim sun is covered by gloomy clouds
Three people take less than ten steps together
When two die falling across the road
People die in the night
But no one dares weep
The ghost of plague blows

> The light turns green
> Suddenly the wind rises and the light is blown out
> Leaving man, ghost, corpse and coffin in the same
> dark room . . .[3]

Just a few months later, Shi was himself infected and rode the fiery dragon into the hereafter.

The decades when plague swept through south and east Asia saw predictable arguments about its source and trajectory, each nation attempting to identify the origin as somewhere other than its own backyard. Chinese observers believed that Yunnan had become a plague province owing to its border with Burma, where they claimed it had been endemic for much longer. Others pointed to northern India, ruled by the sanctimonious British, as the place of original infection and noted that there were traditional words in Indian languages – *madhmari*, for instance, or *poli* – that supposedly suggested long familiarity. But when plague came to Bombay with a vengeance in 1896–7, it was unarguable it had come from further east, on vessels sailing from Hong Kong and Singapore. As far as most at-home nineteenth-century Europeans and Americans were concerned, the dominant modern contagions were cholera and yellow fever. The horrors of the Black Death, on the other hand, graphically described in the pages of Boccaccio and Defoe, were thought by most northern Europeans to have been left behind with wigs and quills. But this was a delusion. Rather than a long break between that second Great Pandemic (the first having been judged to have taken place in the sixth century CE) and the third in the 1890s, bubonic plague never actually went away. Between 1720 and 1722, more than half the population of Marseille, 50,000 in all, perished with another 50,000 dying in the Provençal hinterland. There were further epidemics in Scandinavia, Russia, the Baltic and the Balkans (Transylvania, Hungary, Moldavia) throughout the eighteenth century. In 1799 and 1800, it had erupted in Syria where it had caused Napoleon's plans lethal inconvenience. In Constantinople in 1812, plague killed the enormous number of

320,000 and from there it billowed out west, to Bosnia and Dalmatia, once again to Egypt, Syria, Palestine, Mesopotamia and, to the consternation of the Royal Navy, all the way to Malta and Gozo. In Anatolia, Persia and the Caucasus, on the Tibetan slopes of the Himalayas and on the Caspian shores, it had obstinately persisted through much of the first half of the nineteenth century. In 1878, nine years after Adrien Proust visited the city, Astrakhan was hit by a savage bout of the sickness. But then why should plague have ever disappeared? We now know that bacilli and viruses remain circulating in animal populations until they have opportunities to colonise susceptible populations including humans. Plague in Manchuria was confined to Siberian tarbagan marmots until their fur was discovered to be a profitably down-market substitute for luxury marten and sable. Trappers handled the skins, wrapped warmly about their bodies as they slept. Fleas were happy. More recently, Madagascar saw a major outbreak of plague in the latter half of 2017, and the Oxford laboratory that developed the AstraZeneca vaccine against COVID-19 is currently in trials for an advanced vaccine against *Yersinia pestis*.[4] This is just as well, given that *Live Science* reported on 3 August 2021 that chipmunks, those cheery little visitors to the picnic sites and beaches around Lake Tahoe where my children romped three decades ago, have been diagnosed with plague.[5]

In Yunnan, plague was familiar enough to be given colloquial names: *shih-i,* seasonal fever (the prime season being spring); *shu-i*, the rat plague; or in grim homage to the size and shape of erupting buboes, *luan-tzi*, the egg plague. Still, it was possible to imagine that, deep in its country of river-cut gorges and alpine peaks, one might be safe from the threat. The formidably tough missionary-botanist Père Jean-Marie Delavay, walking alone through the highlands of the Tapintze mountains in north-west Yunnan, about as far away from urban China as you could possibly get, nonetheless caught plague in 1888 while collecting specimens of wood peony fritillaries and the magnolia that endures for centuries in Yunnan temple courtyards, blooming only at night. Though the fatality rate was usually 90 per cent, Delavay survived to recuperate, albeit partially, in France and to catalogue his hundreds of thousands of specimens for Adrien Franchet at the Muséum National d'Histoire Naturelle. Broken in health but not in spirit, Delavay returned to Yunnan to continue hunting for yet more hitherto unknown species to add to the 800 he had already discovered. He died, still at work in Yunnan, in 1895, a year after the streets and alleys of Canton had been covered in dead rats.

In all likelihood, they and their infected fleas got there from Pakhoi, for the return of the very old thing that was plague had been enabled by the making of the new thing that was rapidly urbanising south-west China. (Pakhoi and Kunming, the capital of Yunnan, continue to this day to be some of the fastest-expanding conurbations in Asia.) This was not necessarily the result of gunboat diplomacy, the enforced opening of China to western commercial penetration. Carol Benedict has demonstrated that, well before the European imperial onslaught, the Qing empire's improvement of communications in the south-west moved disease faster and more frequently along rivers and roads. Wave after wave of disasters – famine, war and disease – sent millions on the move. The Taiping rebellion – by any measure one of the great catastrophes of the modern era (though often missing from nineteenth-century history textbooks) – killed 20 million but uprooted another 30 million. Yunnan, at once a place of rural desolation and urban transformation, endured more than its fair share of calamity. Demographic pressure further north and a sense of frontier opportunity had drawn a flood of migrants from the Han ethnic majority to a province where, though still a minority, they proceeded to treat the locals as primitive inferiors. Roughneck antipathies in the copper and silver mining camps became ugly. Sporadic fights between Hui and Han miners blew up into a full-scale insurrection in 1856 after the Qing governor invited Han people to cleanse Kunming of the entirety of its Hui population. Thousands of Hui were slaughtered in a three-day butchery, including the elderly and small children. The reaction against this horrifying massacre grew and with imperial armies elsewhere engaged, not least against Europeans, the revolt managed to establish an independent Islamic sultanate ruled by an imam-sovereign known to the Hui as Suleiman (and Du Wenxiu to the Han), with the city of Dali as its capital. The cultural genocide of the Uyghurs of Xinjiang today draws on (though is not justified by) the inherited memory of that astonishing nineteenth-century rebellion. Mandarin was replaced by Arabic as the official language of the sultanate. Five new mosques were built, and a madrasa opened. Though the leadership was Hui, the sultanate consciously tolerated other religions and embraced the many ethnicities of the region (including Han), especially border people like the Burmese Karen and the Bai and Yi, who lived on both sides of the blurred frontier with Vietnam. It took another sixteen years before the Qing armies inflicted a second genocidal massacre at Dali, murdering at least 4,000 women and children. Du Wenxiu took a suicidal overdose of opium before allowing his body to be surrendered to the Qing general. The corpse was decapitated and its head, preserved in honey, sent as tribute to the emperor in Beijing.[6] Tens of thousands of Hui fled into Burma, travelling through and settling in a country where plague was endemic. One of the major livelihoods of the Hui was cotton trading, transported on mule caravans. Where beasts of burden plodded, fleas went with them. When some of the refugees returned to pacified Yunnan in the mid-1870s, they may well have brought plague in their train.

So the arrival of Europeans in south China was not the initial awakener of plague from its dormancy. But newer commercial openings, as always, made the outbreak and progress of infectious disease a lot easier and a lot speedier. River and coastal steamboats shortened shipping times, increased cargo capacity and provided hospitality for the travelling bacillus. As ever, when cases arose, the British were averse to imposing quarantines, believing them epidemiologically futile and economically disastrous. Lost in the late nineteenth-century fever dream in which imperialism was the rough medicine for ailing traditional empires

(Ottoman, Persian, Chinese), Europeans were incapable of imagining that what they took to be the cure for social and institutional ills – the magic of modernisation – might actually kill the patient. Suggest to a map-poking functionary, or a cigar-smoking warehouse master in the godowns, that business would have an unexpected partner – sickness – and you would have been met with incredulous stares.

Some of them, though, had an inkling. Robert Hart – the son of a Belfast distillery worker, fluent in Chinese languages, father of three children by a Chinese woman and elevated to the office of inspector-general of the Imperial Maritime Customs Service (IMCS) collecting revenues on behalf of the Qing empire – was the first to commission maps of the routes of disease, especially plague. Medical men joined businessmen and the military as an integral part of the European establishment in China. The pretence, at least, was that this was not just a matter of European self-preservation but a concern for the welfare of the native population, especially in a culture in which both nosology and therapy were routinely dismissed as 'absurd'. But for at least some physicians and bacteriologists in China, devotion to bringing modern medicine to bear on endemic diseases was genuinely disinterested and often profoundly consequential. After a chequered teenage career which included apprenticeship to an Aberdeen ironmaster, serving as medical officer to Durham Lunatic Asylum, and forensic dissection of psychiatric patients, Patrick Manson, the son of a Scottish bank manager, followed his physician brother out to China where he worked for eighteen years for the IMCS, beginning in Formosa and then at the treaty port of Amoy on the mainland coast. Manson learned Mandarin and Cantonese, treated Chinese patients, and discovered the role of filaria worms in generating elephantiasis. In Hong Kong, working at the Baptist Missionary Hospital, he was the first to make a causal connection between mosquito bites and malaria. He would, among other roles, become the founder of the London School of Tropical Medicine.

History often works in paradoxical ways. Ruthless imperialist profiteering did not preclude breakthrough biological research. There was, to be sure, a great deal of nose-holding hypocrisy in the rulers' professing as much concern for the physical welfare of the ruled as for themselves.[7] Europeans made sure to reside in districts separated from densely populated streets where infection ploughed through tenements, even as they needed coolie labour, domestic servants, gardeners and sweepers to make their tropical lives tolerable. But the aim of bringing scientifically based medicine to vulnerable populations was not always spurious and self-serving. And even when it was, the fixation with scrupulous observation made imperial medical services amassers of data, without which public health measures were impossible. In 1879, Hart announced that his office, ostensibly a revenue-collecting operation, would act as a clearing house for medical reports from all over China, and that up-to-date information on the spread and severity of disease would be printed in official publications as well as in the Anglophone press. Emile Rocher's report on plague in Yunnan in the aftermath of the suppression of the Hui Islamic kingdom was published in 1879 and was followed by another specifically dealing with plague in Pakhoi, written up in 1882 by J. H. Lowry, another official working for Hart's IMCS. Every so often there were reports of local outbreaks in fast-growing Yunnan cities like Kunming and neighbouring provinces Guangxi and Canton, increasingly wired to the buzzing trading world of the south-west. As in Pakhoi and central Yunnan, plague became almost expected. What was not expected was what then unfolded in Canton and Hong Kong.

On 16 January 1894, an American missionary, Mary Niles, one of only two women doctors in Canton, went, as requested by a high-ranking Qing general, to take a look at his daughter-in-law. It was out of the question that the young woman could be examined by a male physician since the location of the troubling 'boil' was her groin. The inguinal swelling, combined with a fever of 104.8 degrees, were ominous signs of bubonic plague. But unlike 90 per cent of plague victims, the woman recovered and Niles gave no definitive diagnosis, believing the illness to be typhus. It was only a few weeks later, when thirteen dead rats were counted around a house that had been visited by a girl who took sick and swiftly died, that Niles knew for sure what had come to Canton. By the end of March, the bodies of rats were carpeting the alleys of the most crowded districts in a city of nearly two million. A Scottish doctor, Alexander Rennie, who in 1895 wrote the first report on what unfolded in Canton, described the terror of residents in hitherto healthy districts on suddenly seeing rats, 'the heralds of evil . . . come out of their holes in broad daylight, run and tumble about in a dazed condition and die'.[8] Though the danger was obvious, money was offered by the city authorities to anyone prepared to collect the rodent remains. The uptake was astonishing. One official received 2,000 dead rats in a single day and 35,000 in a month. At the west gate, 22,000 were carted from the city and buried in a pit (though not, of course, before their fleas had gone in search of alternative meal sources). Inspection of the dead rodents showed distended stomachs, congested lungs, enlarged livers and lymph glands: some, but not all, of the signs of plague.

It came soon enough to humans, most speedily to the poorest districts of Canton. As had been the case with every known visitation of the Black Death from the fourteenth century on, plague divided rich and poor, the living and the dead. Those who could afford to do so fled to the surrounding countryside in their thousands, many taking the disease with them to local villages. The division of domestic space was another separator between those at more risk and those at less. Rennie observed that people living in the higher storeys of crowded houses, elevated above the rat population, along with boat people and (needless to say) the European colony, had the best chance of escaping infection. Conversely, even among the poor, women and children, especially young girls, were disproportionately victims as they were more likely to be confined indoors during the rainy season of May and June when plague was at its height.

The progress of the disease in an infected body was shockingly rapid. It usually began with fevers swiftly climbing to 105 and sometimes even 107, accompanied often by vomiting and diarrhoea. Within a day or two, painful, hard bubonic swellings and nodes appeared on the neck, groin or armpits or all three. Following their appearance, death could occur within forty eight to seventy two hours. If the infected managed to survive for five days, the chances of recovery were good, but that only occurred in a small minority of cases. The non-appearance in some cases of buboes was not necessarily good news. Pneumonic plague, transmitted in droplets, was almost invariably fatal and could occur just twenty four hours after a fever set in.[9] Overall fatality rates in Canton were, as they would be in Hong Kong and Bombay, near 90 per cent.

In panic and desperation, Canton's people understandably turned to traditional Chinese medicine: ointments made from wild chrysanthemum pasted over the buboes, potions made from ground mace mixed with dandelion, rhubarb, crushed betel nuts or khan-ts'ao grass. When possible, buboes were incised to drain, a measure which would, in fact, have offered a scant hope of survival. Noticing that animals other than rats had been attacked by plague, including poultry and cattle, the slaughter of pigs – the principal supplier of meat – was forbidden. In hopes that the war god Gung Wa might be summoned to do battle with the disease demons, the streets of Canton filled up with thunderous noise: drumming, clashing cymbals, shrieking brass and firecrackers. Dragon boats, normally sunk in the river bed following the celebration of dragon festivals, were dredged up from the mud for celestial intercession. When all else failed, a second New Year was arbitrarily declared so that the terrible evil might be banished along with the old year.

Nothing worked. Rennie estimated that 40,000 lives were lost, but later reports put the total far higher, possibly as many as 100,000 out of a population of 1½ million. Witheringly superior as he was about the folly of traditional Chinese medicine, Rennie's own grasp of the source and transmission of infectious disease was more or less oblivious to the bacteriological revelations coming from Paris and Berlin. Even after the plague bacillus was isolated in Hong Kong in the summer of 1894 and the news quickly published in medical journals like *The Lancet*, much of the sanitary establishment of the British empire was reluctant to abandon decomposing organic matter and the miasma rising from it as the source of the disease. In 1882, Osbert Chadwick, the son of the patriarch of British sanitary reform Edwin Chadwick, published a report exhaustively detailing the state of public latrines (25), the number of seats (565), the number of users per day, and the flow, or lack of it, of open sewage. The shocking fact that the 106,000 residents of Hong Kong occupied a mere 6,402 houses, many of them shared with animals like pigs, was, for Chadwick, enough to mark out the city as a prime target for devastating contagion.[10] In keeping with his father's fixation and the received truisms of Victorian public health, it was assumed that a good clean-up of these gruesomely insanitary conditions would put an end to the scourge of infection. While as yet ignorant of pathogenesis (Robert Koch's findings would be published in 1883), the standard sanitary regime would not necessarily have done harm and in respect of thinning overcrowding would, of course, have made the tenements' lower

Tai Ping Shan before the plague.

storeys less of a fleas' paradise. But just how that was to be accomplished, and how much it would cost to demolish and relocate, was not anything at the top of the Hong Kong government's agenda. So nothing was done until terror struck.

One swarming zone which proved to be the nursery of infection was Tai Ping Shan in west Hong Kong, the district to which the Chinese had been moved in the mid-nineteenth century when the British merchant and governing class decided they needed to rebuild the central area of the city for their own needs and according to their own expansive, expensive, tastes. Tai Ping Shan – 'Peace Hill' – had been named for the surrender of the pirate Chang Po Tsai to the Qing viceroy in 1810. According to the writer, translator and westernising reformer Wang Tao, who came there on the run from the Qing government in 1862, the district was packed with gambling dens and brothels – 'gaudy houses with doors painted in bright colours and windows with fancy curtains'.[11] It was also a place for cheap lodging by any Chinese who toiled at the docks, pulled a rickshaw or cooked for the Europeans. Shopkeepers kept goods and often themselves on the ground floor, in unventilated, windowless spaces. Upper storeys were divided by wooden partitions into cubicle-like compartments where whole families might sleep. This was the place where, on 7 May 1894, James Lowson, the acting superintendent of the Civil Government Hospital, identified the first unmistakeable case of bubonic plague in Hong Kong, immediately isolating the patient.

Much of what was done, for good or ill, in Hong Kong during the devastating spring and summer of plague turned on the young Scottish physician. Rake-thin, pale of face, and hot of temper, Lowson had already faced and survived almost certain death, though not from disease. In 1892, he had been the star all-rounder of the Hong Kong Cricket Club, both opening batsman and demon fast bowler, especially decisive in the needle matches with its arch-rival in the Inter-Port series, the Shanghai Cricket Club. Having won earlier in the season, HKCC were tipped to be victorious again in the return game in Shanghai. But not least because Lowson was

James Lowson

badly injured by his opposite number, C. A. Barff, after having taken eight wickets for sixty six runs in the first innings, HKCC went down to ignominious defeat. At the end of Shanghai Cricket Week, the P&O steamer *Bokhara* taking the team back to Hong Kong was festively seen off at the docks. Lowson, fully recovered, was seen 'blowing his horn' (literally not figuratively, a full blast on the bugle) as the ship steamed off from the quay. On 9 October, close to Sand Island in the Pescadores, the *Bokhara* ran into a ferocious typhoon, struck a reef, and sank within a few minutes of a great gash being torn in its starboard side. Of 148 passengers and crew, 123 perished including the entire cricket team other than Lowson and a Lieutenant Markham, who was so badly hurt that he subsequently needed a lung removed. The handful of survivors were washed up on the island where they were initially attacked by locals brandishing axes and knives. Salvage from the wreck seemed to pre-empt further violence and the survivors were eventually rescued by the SS *Thales*.

Ripping yarns aside, it was clear, especially to himself, that Lowson had what it took – including physical courage – to confront peril and act expeditiously to contain it. On 4 May 1894 (a year which would see further devastating typhoons), he took the night boat to Canton to witness the horrifying extent of the epidemic, returning three days later to Hong Kong, deeply sobered by the chaos and mass death. With plague about to descend on Hong Kong, Lowson was aware that, aside from medical staff at the hospitals, there were just three full-time public health officials in the colony, one of whom, as things looked ominous, had gone 'on leave' back home. That left Lowson and the veteran Phineas Ayres, who had already served twenty four years as colonel-surgeon and, though not actively obscurantist, was not about to immerse himself in all that foreign microbiology.

Two days after his initial diagnosis, Lowson visited Tung Wah Hospital where he identified, alarmingly, another twenty plague cases. Tung Wah had been founded in 1872, expressly as an institution which would treat the sick, especially the poor, with traditional Chinese medicine. But it was also a charitable institution managing burials and looking after communal welfare. Because it was funded partly by wealthy Chinese donors and partly by government money, it was, at least for those in the British administration who favoured it, a symbol of the purported cultural cooperation by which they claimed to govern Hong Kong. For the short-fused Lowson, though, this was a ruinous delusion which harmed those it claimed to help. Like Rennie in Canton, he dismissed traditional Chinese medicine as a bad joke and the hospital as a spreader of infection. 'I cannot denounce this bed of medical and sanitary vice in sufficiently strong terms,' he wrote to the Sanitary Board. 'I venture to say that if the question of allowing this to remain was to be submitted to the public health authorities at home they would order its immediate abolition.'[12] On the alarmed and indecisive governing authorities Lowson thundered down with the full force of his Calvinist zeal. And with the governor of Hong Kong, William Robinson, away in Japan, the young doctor, still only an acting superintendent of the Civil Government Hospital, but whose mettle had been tested by his ordeal two years earlier, became the moving force behind measures to contain and fight the epidemic.

Soldiers of the King's Shropshire Light Infantry disinfecting houses in the Chinese quarter during the bubonic plague in Hong Kong, 1894.

On 10 May, the day that Lowson brought the bad news from Tung Wah to the Sanitary Board, Hong Kong was officially declared an 'infected port'. An ordinance was enacted giving the Sanitary Board emergency powers over the lives and livelihoods of residents in affected districts, which meant, of course, Tai Ping Shan and other poor and crowded districts – Kau Yu Fong, Sin Hing Lee, Nga Choi Hong and Mei Lun Lee. As deemed necessary, houses could be entered without permit or warrant, searched, belongings removed and burned, the premises disinfected and limewashed, or entirely demolished. Nothing was said about what the occupants were supposed to do or where they were supposed to go. Infected patients from Tung Wah were to be removed to the Civil Government Hospital, quickly replaced by the isolation ship *Hygeia*, a redundant hulk that had been used for cholera and smallpox victims, brought from its anchorage at the coast into the harbour.

By the third week of May, 300-plus plague victims were dying every day in Hong Kong. On the 23rd, 300 volunteers from the garrison of the Shropshire Light Infantry, serving under eight officers, began daily patrols and searches. It's not hard to imagine what the sound of military boots rounding a corner followed by the sight of pith helmets triggered in those terrified residents of Tai Ping Shan or Mei Lun Lee. Robert Peckham has pointed perceptively to the British obsession with what they called the 'indescribable rubbish' of Chinese household objects, within and without, kept in no particular place or order, cooking utensils and upholstered furniture in dismaying proximity.[13] Many of those pieces, passed on through the generations as family heirlooms – say, a screen – were now seen as reservoirs of infection. The wooden partitions dividing upper-storey spaces were often the first to be taken out into the street and burned. Morning bonfires became a daily spectacle in Tai Ping Shan.

The search for humans was even more terrifying. The fact that the British soldiers were accompanied by gangs of Chinese police, acting as both interpreters and enforcers, made no difference. Sick family members were hidden in closets and chests; others practised working as normal to try and hide any sign of illness. Family separations were especially traumatic, not least when patients were taken from Tung Wah, where families had had visiting rights, to the *Hygeia*, rumoured to be a 'ship of death' from which 'not one in ten emerges alive'. This was in fact true. Attempts were made by frantic Chinese relatives to seize family members from the ferries taking them to the ship and pull them into sampans. Rumours spread that people, children in particular, had been abducted for hideous lethal experiments: the harvesting of organs and other body parts including, strangely, eyebrows. Girls and women were said to have been abducted for sexual abuse by the British, especially the soldiers and sailors. As protests became more frequent, the warship *Tweed* was moored opposite the west, Chinese end of town, its guns pointed towards the shore.

As the *Hygeia* filled up, Kennedy Town police station was repurposed as an isolation hospital and when that filled to overflowing, a disused glassworks nearby became the next place of confinement. This last Kennedy Town hospital became so crowded and so drenched in misery and despair that it came to be seen as no more than a waiting room for the dead. Rows of houses found to be infected were not just disinfected or altogether demolished but walled off to prevent access. Whole neighbourhoods were simply destroyed or turned into no-go zones.

It's easy to depict the gaunt, beetle-browed Lowson as a pitiless monster, grimly presiding over this human catastrophe. It's an understatement to say that he never suffered fools gladly. In his own mind, what he saw as the idle complacency and genteel amateurishness of the Sanitary Board was at least as harmful to the safety of the city as the nonsense (as he dismissed it) of traditional Chinese pharmacopoeia. He was much given to upbraiding figures like Henry May, the superintendent of police, and James Stewart Lockhart, the registrar-general (who in fact was one of the few in the city government to be fluent in Cantonese), both in his journal and to their faces as incompetent 'idiots' and 'obnoxious bloody fools'.[14] Draconian as his measures were, the disinfections, burnings and demolitions in areas where plague had hit hard were meant to annihilate the rat habitat even without any serious understanding of pathogenesis. The same was true for Lowson's obsession with ensuring that plague corpses were buried feet, rather than inches, deep, as he discovered was often the case. Of course this concern could as easily have been attributed to an obstinate belief in miasma, but deep burial was another way in which the flea-borne infection might have been limited. Whether or not Lowson could properly take credit for the difference (and he himself never to my knowledge made the comparison), infection and mortality rates in Hong Kong ended up being substantially lower than in Canton. For that matter, Lowson was not oblivious to bacteriology. On 12 May, in the same week as he made his appalled diagnosis in Tung Wah Hospital, he began injecting rabbits with fluid taken from the buboes of plague victims. But immediate urgencies and his constant conviction that the only capable and physically fearless person (he was contemptuous of the 'cowardice' of colleagues on the Sanitary Board who steered clear of the hospitals) capable of managing the public health crisis was James Lowson, overwhelmed any thought of sustained, systematic experimental research. 'The question', he wrote in his diary, 'as to whether one will find bacterial or other causes for pathological change will be in abeyance for some time.'[15] That same day, twenty six more people died of plague in Hong Kong.

Enlightenment, however, was just round the corner. In the first instance it came in the person of two Japanese bacteriologists – Kitasato Shibasaburo and Aoyama Tanemichi, who arrived together in Hong Kong on the steamboat from Yokohama on 12 June. What were Japanese scientists doing there and why? It was, inevitably, as much a concern of state as of science. A war with China over Korea was likely if not inevitable, and should there be an invasion, the Japanese authorities wanted to know more about a microbial enemy that could be more lethal than the Banner Regiments of the Qing army. Notoriously in east Asia, disease moved with masses of troops. It was perhaps significant that William Robinson was in Japan at that doubly critical moment in the late spring of 1894, ostensibly on holiday. But given strong British interest in east Asian politics and Foreign Office inclination towards understanding with the new, muscular empire, it would not be surprising if there was a back-and-forth between Tokyo and London that triggered the bacteriological mission.

A cable from the Japanese consul in Hong Kong to the Bureau of Public Health in Tokyo about the infection triggered an alert. A meeting was arranged between the Tokyo Institute and the government (from which it was notionally independent) at which an immediate mission to the plague-stricken port city was decided and mobilised.[16] It had the blessing and, more to the point, some funding from Kitasato's most powerful and influential patron, Fukuzawa Yukichi, the most famous embodiment and practitioner of the Japanese paradox: that western 'learning' would be the most potent weapon in the arsenal of purely Japanese national 'self-strengthening'. Fukuzawa, the son of an impoverished samurai, was a Confucian scholar turned translator, first of Dutch, and then English, precisely at the moment in the late 1850s when the Tokugawa shogunate was beginning to open up from centuries of enforced seclusion. The opening was thrust on Japan by the thinly veiled threats of gunboat diplomacy, but Fukuzawa was the personification of pragmatic westernisation. On the first embassy to the United States in 1860, the indispensable translator had himself pictured in San Francisco together with the daughter of the photographer. Back in Japan he wrote *All the Countries of the World for Children in Verse* and a ten-volume guide to *Things Western: An Outline for a Theory of Civilisation.* Still more important, in 1882 he became the editor of *Current Events*, the gazette by which his countrymen and women were meant to familiarise themselves with the strange new world of parliaments and politics, shaped in the selectively Japanese manner more by the faux-liberalism of imperial Germany than the alternative pole of emulation, Great Britain.

If there was one guiding principle to Fukuzawa's mission to persuade his readers that western means could be safely adopted to serve purely Japanese ends, it was that knowledge was the most important weapon in the arsenal of national rebuilding, more formative and potent than any battleship or heavy artillery. At the core of weaponised knowledge was, of course, science and technology. Even during its long centuries of ostensible seclusion, there had been intense, if selectively policed, interest in 'Dutch learning',

William Shew, photograph of Fukuzawa Yukichi with the photographer's daughter, San Francisco, 1860.

especially optical lenses, while confining the Europeans to a strictly regulated village compound on the island of Dejima. Once the doors opened a little, the British in particular flattered themselves that they could act as tutors to Japan's curious students. Minds would be opened along with ports. In 1863, five years before the overthrow of the Tokugawa government, five young men from samurai families of the *han* of Choshu in south-west Japan were smuggled out on a ship belonging to the opium trafficking company Jardine Matheson. At Yokohama they were hidden in the more usual contraband cargo of opium and split into two groups. Three sailed to Britain as passengers on the *Whiteadder*. The remaining pair, boarding the *Pegasus*, were assumed (since anything else was hard to imagine) to be looking for work as deckhands and spent the 130-day voyage trapped in gruelling routine. By way of compensation, once they all reached London, Robert Matheson introduced them to Alexander Williamson, professor of chemistry at University College. Not only were they enrolled in his courses, but three were lodged in the Williamson family house in Camden; the remaining two stayed in Gower Street, a short walk from the college. The course of studies was cut short for some by news of naval conflict between Japan and the Anglo powers in the Straits of Shimonoseki, but for a while they were the children of samurai, scholars and villagers, turned Bloomsbury science students, as if they had jumped from the late medieval to the industrial world without much of an interval between remote past and bustling present. One of them, Ito Hirobumi, would become Japan's first prime minister.

Kitasato Shibasaburo the bacteriologist was the product of the astounding phenomenon that was the Meiji Restoration of 1868. Operating in the name of traditional, even reactionary, ideals – a romance of samurai honour – and against flirtation with the west, those who deposed the shogun and made the emperor once more the focus of national devotion almost immediately embraced the very modernisation against which they had rebelled. Needs must. The means morphed into the ends. In short order the 'Restoration' developed into a radical revolution. The volte-face did not go unchallenged.

In 1876–7, a ferocious civil war took place in Satsuma between samurai leaders like Saigo Takamori, who believed that the original ideals of the Meiji Restoration had been betrayed, and modernising pragmatists whose model of realpolitik was Bismarck. Needless to say, the nostalgics were crushed by the heavy guns of the new Japan. But their spirit was undefeated and mutated before very long into terrorism. The patron of Kitasato's colleague on the Hong Kong mission, Aoyama, was Japan's financial wizard Okuma Shigenobu, also a negotiator with western powers. Demonised for conciliating Europeans and Americans, Okuma had his right leg blown off by a terrorist bomb. In fact, the Meiji oligarchs were never uncritical consumers of what the west had to offer. On the contrary, they were clear-eyed about the pragmatic road to national self-affirmation. Japan would take from the west only what it needed to ensure that it would never suffer the wretched fate of broken imperial China; would never become dominated by European and American instructors. They were the temporary hired tutors. Western learning was the means; the rejuvenation of Japanese imperial power was the end.

Science and technology were always integral to this national reawakening. Predictably, there was from the start a heavy emphasis on engineering, both civil and military. In the 1870s, yet another Scot, Henry Dyer, established a six-year foundational programme at the brand new Imperial College of Engineering which combined basic courses in physics and chemistry along with mining and civil engineering that could be applied at the enormous Akabane Works he helped to found. But biology, including microbiology, quickly became a state priority, especially since neither the old policy of isolation or the new one of regulated openness spared Japan from waves of cholera. Its visitations reinforced the view of one of the most formidable leaders of the Meiji governments, Okubo Toshimichi, that the mastery of contagious diseases had to be as much a major concern of state as munitions and mines. Before he was assassinated in 1878, Okubo saw to the establishment of the Tokyo Hygiene Laboratory and the Bureau of Public Health.

Kitasato Shibasaburo

Though the British, never much concerned with fine distinctions between one Asian and another, failed to notice it, Aoyama and Kitasato had different origins and career paths as well as those crucial patrons. Kitasato was the son of a provincial mayor from south-west Kyushu, the region closest to the Korean peninsula, so that all through his time in Hong Kong he must have been intensely aware of the military crisis. Aoyama was a metropolitan from Edo, renamed Tokyo, and worked as an assistant pathologist in the university while getting his medical degree. The two of them overlapped as students in Robert Koch's institute at Berlin University so that it seems inconceivable, given the rarity of Japanese in Germany, that they would not have formed some sort of mutual bond. But Kitasato's close collaborator could hardly have been more different. The provincial scientist from Kumamoto shared a bench with the son of a Prussian village schoolmaster, Emile von Behring, in another demonstration of how microbiology and the mission of understanding and mastering infectious diseases was, from the beginning, a work that jumped cultural and national frontiers, even as they were ominously hardening in almost every other aspect of modern life. Working together, Behring and Kitasato managed to grow tetanus bacillus in culture and use it to make anti-toxic serum. But they also found that graduated amounts of the toxin drawn from the blood of infected animals and then injected into healthy ones would deliver immunity to the latter. It was on the strength of that work, together with its application to diphtheria, that Kitasato was able to found the Tokyo Institute in 1891. By the time of his mission to plague-struck Hong Kong three years later, Kitasato was at the centre of a whole network of microbiological research in Japan that connected him with the levers of government without ever being subject to their authority. It was widely assumed that the Bureau of Public Health took action on his opinions and instructions.

Once in Hong Kong, things moved fast.[17] Perhaps tipped off by Governor Robinson in Tokyo, James Lowson, the effective head of the Sanitary Board, was obliging. The two Japanese guests were lodged at the grand Hong Kong Hotel. Much more importantly, they were given improvised working space at the Kennedy Town hospital grounds, where most of the patients were being accommodated and from which, appallingly, a regular supply of plague corpses could be provided. Nominally there was supposed to be a division of labour: Aoyama assigned to more forensic dissection, Kitasato to taking blood samples and, as he had done with tetanus in Berlin, injecting them into rodents to see if they developed plague symptoms and died, both of which duly happened. On the second day of work, 15 June, Kitasato told Lowson, via an interpreter, punctuated with occasional German, that he had seen a dense abundance of microbes under his microscope and that he had as a result isolated the plague bacillus. Electrified, Lowson cabled *The Lancet* with the revelatory news. Guinea pigs and mice continued to die for bacteriology. Two weeks later, while dissecting a body, Aoyama cut a finger and contracted the sickness, though unlike 90 per cent of its victims, he managed to survive. But for weeks afterwards, he was effectively disabled as a collaborator. A week earlier, war had broken out between Japan and China over Korea. Kitasato abruptly departed from Hong Kong but was already famous in Japan, hailed, in the overheated atmosphere of the Korean war, as a national hero. Honouring him, the imperial

Prince Konoue declared that Kitasato's discovery 'reflects credit on Japan's medical science and makes our civilization shine to the heavens. Such achievements can only raise the level of our nation to universal acclaim.' Not long afterwards, Kitasato married into the *daimyo* aristocracy and would end up being a baron himself. Aoyama, who remained in Hong Kong through August 1894, evidently resented his colleague's monopoly of credit for the discovery. His own report on the mission barely mentioned Kitasato at all.

Still less were either of the Japanese going to give any praise to the European scientist who arrived just three days after them, and to whom they were introduced that same day, 15 June, when the bacilli made their momentous appearance under the microscope. From Lowson to Kitasato to most of the members of the Sanitary Board, there was a general response of 'Who asked you?' to the arrival from Saigon of the Swiss Alexandre Yersin. 'The Frenchman arrived' was the terse entry in Lowson's journal for that day. And though Yersin had grown up in Aubonne in the canton of Vaud, facing Lake Geneva and the Alps on the horizon, it was understandable that he was taken to be French. He was known to have worked at Louis Pasteur's institute in Paris, and it was no secret that he was involved – though

Alexandre Yersin in the doctor's uniform for the Maritime Postal Fleet, Nha Trang, Vietnam, 1890.

unclear just how – in creating a branch of the Institut in Indo-China. He was, in short, a suspicious figure, and, most damningly, spoke no English at all. Yersin was greeted with ostentatious unhelpfulness. Once the grudging introductions were made, he was on his own. No cadavers were provided; no dedicated space was set aside for his work and Yersin refused to share Kitisato's matshed.

They could not have been more different. Alexandre Yersin – short, slight, deceptively delicate in appearance for someone who had slogged through impenetrable rain forests and over daunting Vietnamese mountain ranges – seemed in every conceivable way the opposite from the studious, buttoned-up, unswervingly disciplined Kitasato. Both in their different ways, and for good or ill, were empire-builders. And they were also recognisable personifications of two late nineteenth-century types: one, a believer in and practitioner of institutional hierarchy: scientific, pedagogic, professional, political; the other, an escape artist, lone wolf, tropical adventurer, who ended up doing his science in a thatched hut in a Vietnamese village by the sea (now a huge beach resort) where houses perched on stilts.

It is possible, of course, to make too much of these picturesque contrasts. Yersin went to plague-struck Hong Kong not out of pure bacteriological opportunism but because he had been told to go there by Théophile Delcassé, the ardently imperialising French minister for the colonies. So he was as much an instrument of imperial science as Kitasato and Aoyama. A telegram from Paris had arrived on 9 June on the desk of the governor of Indo-China, Jean-Marie de Lanessan, who took it to Albert Calmette, the chief of the newly created Institut Pasteur in Saigon. And as Aro Velmet has perceptively noticed, the Institut, which had trained both Calmette and Yersin, was busy building its own scientific empire.[18] Its bacteriological mission, moreover, was inseparable from research into tropical diseases. Calmette knew all about Hong Kong and the British medical service there because in the 1880s he had collaborated with Patrick Manson on his work on filaria. But in Paris he had been taught by Emile Roux, and Roux in turn was a long-time friend of Delcassé. In Saigon, Calmette brought the famous specialisations of the Institut Pasteur – rabies, diphtheria and tuberculosis – to his lab, but his own speciality was the distinctively Asian need for an antidote to snake venom, especially cobra toxin. All this criss-crossing added up to another chapter in the French mission – going back to Bonaparte in Egypt, Fauvel in Constantinople and Tholozan in Tehran – to make their presence felt in the non-European world through the engineering of public health as well as roads, bridges and harbours. So Yersin's sudden dispatch to Hong Kong was a sortie in this great game of competitive medical imperialism. In all kinds of ways, it was the scientific projection of the ferocious Anglo-French rivalry at work all over Asia and Africa: the hunting of microbes was akin to the race between those powers to locate and possess the source of the Nile, a competition which very nearly brought them to war four years later.

None of this, though, much qualifies Yersin's unpredictable character: a maverick in a world of budding scientific institutionalism, averse to hierarchies of all sorts, especially those pickled in imperial protocol and racial prejudice. Like Waldemar Haffkine in India, he insisted on doing things his way and paid a price for it since he made pink-faced men uneasy by much preferring the company of Asians with whom he worked, ate, travelled and often lodged. Asia returned the compliment. There is still a Haffkine Institute in India; Yersin's house at Nha Trang is now the Yersin Museum, where an inscription says he is revered by the people of Vietnam.

He got to know them very fast and very well. Working as ship's doctor on the Saigon–Manila and Saigon–Haiphong routes of the Messageries Maritimes was an education in local languages and the world of coastal Vietnam. But Yersin aimed to be at home not just in the port cities but in the village life of the many ethnicities who made up the peoples of Cambodia and Laos as well as Vietnam.[19] It's likely that he was inspired or at least triggered by the famous and ongoing expeditions of the 'Mission Pavie', a staggering enterprise of ethnographic, geographic and political exploration led by the ex-soldier Auguste Pavie and covering, in the end, 260,000 miles of Indo-China, from the Siamese border to the coast and down to the gulf of Tonkin. Yersin met Pavie in Saigon in December 1893, but always refused invitations to join those expeditions, suspicious of becoming just another in Pavie's team of forty-odd topographers, entomologists, biologists and itinerant diplomats. Always he wanted to remain his own man. Aiming to reach the remote Moi people, he set off with a few native guides in March 1892 from his base at Nha Trang to cross the Annamite cordillera, and hack his way down to the Mekong valley. In M'sao, he paid studious respects to an old, formidably obese Moi chief, surrounded by teams of wives, while he worried about tigers padding out from surrounding forest at night. At Kheung, rice and rice wine were laid out on a mat for him but only, he learned, on the assumption that he would settle an intra-village tax dispute, always a tricky business. By May, he had reached the Mekong where his expedition's two canoes carefully avoided close contact with crocodiles capable of capsizing the boats. On land they swatted away whirling clouds of mosquitoes which Yersin knew, even before Ronald Ross demonstrated it was so, were the carriers of infectious diseases, including malaria. Yersin would be the one who would bring Peruvian cinchona trees to Vietnam to fight off the shaking infection from which he was himself somehow spared. Every so often he would come upon isolated 'consulates' of some quasi-official Frenchman-cum-merchant-cum-smuggler, sometimes deposited there by Équipe Pavie, sweatily established in clearings close to the sluggish Mekong. In June, he made it all the way to Phnom Penh, then a small lively town already boasting the architectural brag of the '*mission civilisatrice*', usually in the form of Beaux Arts or neo-Baroque post offices. The most grandiose construction in the city, already almost completed, was the enormous pointed-arch bridge thrown across the Mekong, its masonry rising on both banks into operatic towers, named then and now for the local supremo, Albert-Louis Huyn de Vernéville, who received Yersin with perfectly calibrated aristocratic condescension. But then Vernéville's birthplace, Metz, had been ceded to the Prussians after the debacle of 1870. Louis-Napoleon's empire had unconscionably betrayed him and his country on the Moselle, so the least the Republic could do by way of recompense was to make another empire and give him a small piece of it to govern.

In October 1892, running short of funds, unable to persuade Lanessan to cough up any, yet still refusing to throw in his lot with Pavie, Yersin sailed back to Paris. Though he had departed abruptly and inexplicably from the rue Dutot, and was listed in its records merely as 'former preparer' of experiments, the Pasteurians embraced their prodigal. Old Louis himself, '*mon maître*' as Yersin respectfully called him in his letters, was, as he had hoped, generous, not least because an Institut Pasteur in Saigon, thanks to Calmette, seemed to be a reality, the first outpost in their empire of the new science and tropical medicine. Anyone with something of an Indo-Chinese story to tell could open doors. So Yersin topped up his funds with money from Henri, the Prince of Orleans, bought shotguns, electrometers and thermometers, and used the networks connecting the Institut with the Colonial Office to present himself as a quasi-state agent, scientifically and strategically important. Back in Indo-China in the winter and spring of 1894, once again refusing to join Pavie, Yersin travelled through Laos, picking leeches out of his legs every night, and crossing the Annamites again all the way down to Tourane by May from where he took ship back to Saigon. Hardly any time at all passed before he had to pack his microscope and his autoclave and sail to Hong Kong.

He spoke no English and no Cantonese. The cold shoulder from the bigwigs of the Sanitary Board, Lowson in particular, turned frigid. An Italian priest, Father Rigano, came to his aid as interpreter when he looked over the terrible scene at the Kennedy Town hospital. The place, Yersin thought, 'lacked all humanity . . . first and last there was a feeling of horrible haste and one of absolute discouragement, both from the doctors who no longer even attempt to cure their patients, and among the sick themselves.'[20] Yersin was also shocked by the contempt shown by Lowson and his colleagues for anything associated with traditional Chinese medicine; another aspect of what he judged to be the inhumanity of the British. Since western medicine was obviously incapable of doing anything about plague, what right had the British to rob people of their own customs and practice, adding to despair? Yersin even suggested that death rates in Canton, where Chinese medicine had been practised alongside western measures, had been better than in Hong Kong. He was wrong about that, but right about the complete demoralisation of the poorer Chinese in Hong Kong.

Frozen out from meaningful work, Yersin toured country villages outside the port where terrified Hong Kong people had fled, finding the same despair, fear and helplessness. Complaining to the Sanitary Board, and then to Robinson on the governor's return, about his treatment, he was at least not prevented from erecting a rudimentary matshed next to the Alice Memorial Hospital, living and sleeping in another straw hut built beside it.[21] But there had been instructions not to release bodies. With the Italian priest's help, Yersin paid sailors and dockers responsible for ferrying bodies from the *Hygeia* to mortuary pits to divert some of the corpses to his matshed. On his original visit to the Kennedy Town matshed, he had noticed Kitasato Shibasaburo working mostly on blood samples. Yersin, on the other hand, concentrated on buboes, which housed, he correctly believed, the greatest concentration of the bacilli. On 20 and 21 June, less than a week after Kitasato had made his discovery, Yersin saw the bacilli in copious numbers in pulpy matter drawn from the buboes and developed in culture. When the matter was injected into guinea pigs, mice and rats, the animals swiftly developed the symptoms of plague and died shortly thereafter.[22] Yersin also noticed that in culture the bacilli developed chains or 'colonies' and 'constellations' and, more promisingly, when injected into the lab animals, those newly created organisms seemed markedly less virulent. Pigeons injected with the 'colony' bacillus did not die at all. On the basis of Pasteur's work with rabies and cholera, this seemed to promise the possibility of a prophylactic plague vaccine.

Yersin's first-hand experience of the French colonial postal service via the Messageries Maritimes helped him get the word out to France. A 'note' from him about his discovery appeared in the bulletin of the Academy of Science in Paris as early as 30 July 1894, and a longer description, together with plates of the bacillus in humans and animals, followed early the next year. An immediate and bitter dispute broke out as to whether the Japanese or the 'French' bacteriologist had been the first discoverer. With the Sino-Japanese war raging in Korea, and national amour-propre at stake, this became more than a matter of pure science, hence the imperial reception and honours bestowed on Kitasato back in Tokyo. But the evidence presented by Yersin was decisive. The bacillus samples isolated by Kitasato had been contaminated by the presence of pneumococci, compromising their integrity. In 1884, the Danish microbiologist Christian Gram had created a method of distinguishing between two different types of bacteria. Those with a thicker cell wall which retained a crystal violet dye were termed 'gram positive', while bacteria with a thinner wall which degraded and lost the stain were 'gram negative', and by a mistake in identification Kitisato had described the bacilli as positive when in fact they were negative. In Japan, initially, these telling differences were brushed aside and treated as yet another instance of envious condescension on the part of Europeans, a refusal to recognise the fact that an Asian power was every bit as advanced as the pretensions of western states. But in 1896, another young Japanese bacteriologist, Ogata Masanori, working on plague in Formosa, had the temerity to argue that what Kitasato had discovered were in fact streptococci, causing second-stage septicaemia, and not the plague bacilli which Yersin had correctly identified. Ogata published the account in 1897 in German to a deafening silence in Japan.

Yersin returned to Paris, delivered lectures to the Institut and the Academy and was treated as a hero of bacteriology by Emile Roux, and as a patriotic scientist, the best that the Third Republic could produce, by the colonial and political establishment. But the hero was well aware that the identification of the plague bacillus was at best half the battle. Once infection had begun, it was already too late. What counted was the use made of the germ to develop a prophylactic. Together with Albert Calmette and Amédée Borrel, Yersin worked on a blood serum vaccine and believed that the results in animals showed it to be effective. But no one could know until real-life trials had been done where plague raged: in Asia. If it had receded a little in China, it had made an entry in Singapore and India. And in any case Yersin's tolerance for Paris seldom lasted more than a year or so. In 1895, he was back in the thatched long-house lab at Nha Trang. Though he knew that Calmette had established an Institut Pasteur in Saigon, Yersin had no qualms about invoking the patriarchal name for his own modest, but busy, outpost on the coast of Annam. But by late 1896, it was frustratingly apparent that there were never enough local cases of plague for a convincing trial of his serum vaccine. He might have to go to where the disease had got out of control. To India. To the British. Again.

vii

Calamity snaps

Little Jacky, 'A Negro boy' (as the caption calls him), is dead: one of ten thousand in Bombay who perished from bubonic plague between September 1896 and the spring of 1897 when this photograph was taken. Perhaps eight or nine years old, Jacky is laid out on a small table between a rudimentary shanty of the type in which many of his people, the Sidi, lived, and the freshly dug dirt grave into which he will shortly be interred. White socks cover his feet protruding from the table's edge. A white pillow cradles his head. His mother is half-mantled in a white cloth. For many Africans, white is the hue of death.

The Sidi, sometimes called 'Habshi' (or 'Abyssinian' from the Arabic *Habash*), were descendants of enslaved east Africans in the Swahili-speaking world taken, as early as the fifteenth century, by Omani Arab traffickers across the Indian Ocean to the west coast of India. There they had served Gujarati masters as domestic servants, bodyguards, soldiers and stable grooms. In some isolated enclaves in Gujarat and Konkan, Sidi soldiers and sailors had shaken off servitude to become independent local potentates. One of them had even established a ruling dynasty claiming descent from Ethiopian aristocracy and priesthood. During the reign of the Mughal emperor Aurangzeb in the late seventeenth century, a Sidi admiral commanded a fleet, while a fortified base on Janjira island held off attacks by the Hindu Maratha princes. In the following century, Janjira became an autonomous Muslim micro-kingdom. Like some medieval Moorish city state, Janjira was governed in the nineteenth century by a Jewish prime minister, the superlatively named Shalom Gapuj Israel Wargharkhan, from the Bene Israel community of Maharashtra. On Janjira, the Muslim Afro-Indians were culturally nourished by a rich store of unorthodox religious and tribal memory. Shrines were built to the Sufi saint Bava Gor, in whose honour a repertoire of drumming, chanting and singing was developed. The drumming-chanting-dancing ritual of the *goma* was a direct transcription of the Swahili-culture *ngoma*, performed in east Africa.[1]

When the British abolished their slave trade in 1807 and mobilised the Royal Navy to enforce abolition, Arab and European traders, operating from Mombasa, on the east coast of Africa, and Zanzibar, were still permitted to send enslaved people to India. By the time slavery itself in the British empire was ended by act of Parliament in 1833, the East India Company had come to dominate western India and Christian missionaries – often the first footsoldiers in the abolitionist crusade – provided a protected enclave for Sidi runaways at Nasik in northern Maharashtra. But many Sidi either continued to serve as quasi-indentured domestic servants for their Indian, especially Parsi, households, or else found work, often demeaning, within the maddened hive of maritime jobbing that swarmed through harbourside Bombay. By the time that plague struck in 1896, there were Sidi sailors, stokers, dockers, stevedores and porters lodging in chawl tenements close to the docks on the east side of the island city or else drawn deeper into the shadowy criminal underworld quarter of Dongri. Almost all were desperately poor: scraping together a subsistence any way they could, while living in the kind of rudimentary huts in front of which Jacky's mother and father are snapped in the depths of their silent rage.

But really, what was the photographer, Captain Claude Moss of the Glosters – quite possibly a decent man – thinking, when he did this indecent thing: lining up his subjects to pose at the rim of their son's grave? Was this picture meant as a gesture of sentimental ethnography, sahib empathy warming the chilly curiosity of the lens? 'Little Jacky' was just one of 150 photographs shot for a 'Plague Visitation' album commissioned largely as a document of imperial self-congratulation. Most of the images depict the medical Raj at its most efficiently benevolent: nursing sisters (mostly nuns) in starched caps and uniforms, the former Government House at Parel transformed into a plague hospital, men and women patients being (separately) fed breakfast, and so on. In the four surviving copies of the album – three in London, one in Malibu – those images of encouragement and reassurance dominate the opening pages. In the copy at the Wellcome Collection, the death of little Jacky is the concluding image, set there perhaps as tragic epitaph? Nonetheless, there is a shamefulness about the photograph, made all the more painful as soon as one reflects on how it must have come about: the tripod planted on the far side of the grave. The asymmetry of power between the snapper and the snapped is monstrous. Perhaps the parents were persuaded to allow this trespass to take place in return for some paltry recompense, which, visibly, they could not afford to decline. Perhaps they were preparing themselves for an even more upsetting indignity: the inspection of the corpse, required by the most recent plague management regulations. They did what they could to resist the pose, the father standing by his dead son's head, the mother squatting at the edge of the grave. Any cheaply bought expression of sympathy is returned with what wounded force they can muster: their faces become masks of helpless resentment and outraged grief.

The album, with its jarringly ornamental Arts and Crafts frontispiece, was commissioned by the chairman of a four-man, all-British, Plague Committee. After 10,000 had already died from the disease, Brigadier-General William Forbes Gatacre was charged, in March 1897, with imposing a strict regime of inspection, segregation, disinfection and demolition on the infected districts of Bombay. That this assignment was not to be taken lightly was made abundantly clear by the dispatch from Britain of none other than James Lowson, who had made his name from the harsh measures he had inflicted on plague-struck Hong Kong three years earlier. Once more, the medical regime was conceived as an urban military operation. Gatacre was a veteran of campaigns in Afghanistan and Burma, and had earlier commanded military forces stationed around Bombay. During that service, he had been been bitten by a cornered jackal during a hunt of the Bombay Jackal Club. Feverishly sickened by the wound, he had been diagnosed as likely to be only 'temporarily deranged'. While serving in 1895 with an expedition relieving the British fort at Chitral, on the border with Afghanistan, from besieging Pashtun, Jandoli and Chitrali tribesmen, Gatacre had witnessed the extensive use of photography to record the campaign. It was natural, then, for him to commission a comparable album testifying to the efficiency of the next imperial engagement, this time waged against the marauding bacillus. A prefatory page to the 'Plague Album' sets out the bare outlines of mortality receding through the late spring of 1897, with the implication that this was due to the success of the committee's measures, though the toll – some 27,000 – was still grim. The photographs were possibly intended as an illustrative apprendix to the three statistically exhaustive volumes on the 1896–7 epidemic also published under Gatacre's imprimatur. But the terminal date of that history was deceptively optimistic. Bubonic plague would return with a vengeance in the winter of 1897–8 and in subsequent years, before tailing off somewhat after the peak year of 1903 in which more than 2 million died. What unfolded in its first wave was just the overture to one of the most devastating pandemics of the modern era, taking more than 12 million lives before it could be treated by the antibiotic streptomycin in the late 1920s. At the turn of the century, plague would touch down, relatively lightly, in Australia (1900) and San Francisco's Chinatown (1900–4). But since the vast majority of the world's victims were in India, plague seldom made headlines in the European press. Only Adrien Proust, in Venice where the International Sanitary Conference had again convened, insisted plague should be the paramount public health concern of 1897.

Gatacre's preface to the album identifies the photographers. From the wayward focus and erratic exposure of the majority of the pictures, it seems likely that Captain Claude Moss was an amateur, perhaps occasionally employed on a short-term basis by the military, primarily for archival purposes. But the unpolished spontaneity of the small silver gelatin prints is exactly what gives the album its documentary immediacy. In contrast, the compiler and designer of the Bombay plague album, Francis Benjamin Stewart, was a professional, a self-described 'photo artist', based in Poona 90 miles east, whose more usual line of subjects were polo games and portraits. Stewart's medium was the large-format albumen print, five of which are set in the showy middle of the album: carefully composed theatrical spectacles of the Plague Committee going about its business.

'Medical Officers, Justices of the Peace, Search Parties and Disinfecting Staff ready to start on their work' features a pith-helmeted senior officer discussing the route of the day with an Indian magistrate, while a company of Sikh soldiers (probably the Bombay 8th) stand smartly to attention beside the 'hand ambulance' designed by Gatacre himself. This was a steel-framed litter, set atop bicycle wheels furnished with India rubber tyres which were thought to

Medical officers, justices of the peace, search parties and disinfecting staff, ready to start on their work.

minimise painful jolting. The contraption is preparing to carry victims, alive or dead, either to isolation hospitals or to burial grounds and funeral pyres. When hand ambulances arrived, it was, so Indians said, as if a hearse had suddenly appeared. The native people of Bombay are, of course, the silent chorus to all this heavy mustering: the bystanders of their families' and neighbours' fate.

'Flushing Engine cleansing Infected Houses' has at its centre the fearsome Clayton steam engine, pumping and sluicing seawater (thought especially effective as disinfecting agent) through streets and sewage-clogged gullies and over house facades. 'Lime-washers at Work' seems self-consciously artistic in its composition, or perhaps musical: a contrapuntal arrangement of Indian daubers, up and down, to and fro, perched on flimsy bamboo scaffolding like notes upon a bar. Beneath and above them, supervising figures observe the operation, as do ubiquitous onlookers, gathered in front of their shops and houses: people of the streets and tenement chawls. Some may have been dislodged from their dwellings while the buildings are being flushed and limewashed. Occasionally they stare back at the camera; most are captured simply staring at the activity, impassively resigned. If the image has the studied look of a movie still, this is not fortuitous. Stewart would go on to make a career in early movies, hired first by the Warwick Trading Company for industrial shorts and then, more auspiciously, to shoot the Delhi coronation durbar of 1902.

Limewashers at work.

Justices of the peace, who gave their services voluntarily for house-to-house visitation.

All these grand imperial tableaux are contrived to harmonise natives and rulers togther in an unproblematic acceptance of a common need for decisive intervention. Another of the large-format photographs, 'A Justice of the Peace with Search Party', features Gatacre himself with plumed helmet along with a gathering of Indian magistrates, lined up together in an emblematic pose of scrupulously legal cooperation as their 'visitation' work begins. The pose is not altogether self-serving imperial fiction. The JPs were all volunteers: a genuine mix of religious communities, often at odds with each other. Surprisingly, for the time and place, they included at least one woman, who, to judge from those each side of her in another photograph of magistrates, was surely a Parsi.

The Bombay plague album is a vision of British India at its most idealised: orderly, responsible, benevolently custodial, a just and conscientious government on which the sun would hardly dare to set. But what the bubonic plague actually triggered, in the year of the Queen-Empress's Diamond Jubilee, was the beginning of the end of the Raj.

In the Wellcome album, the poignant ensemble of 'The Death of Little Jacky' appears beneath a group portrait of nursing nuns

Bandora Convent Sisters, who nursed at Parel and Mahim.

from the Convent of the Daughters of the Cross at Bandora. While we have no idea of the names of the parents of Little Jacky, the nursing sisters are all identified, left to right, top to bottom: Ceophas, Edith, Francois Xavier, Clara, Ursula, Julia and Hilda. But against the grain of images representing the benevolent care of infected patients, more and more photographs feature the faces and gaunt bodies of the Indian poor, sick, dying, dead or bereaved, emerging from their mass of anonymity. They are no longer a passive chorus of grateful appreciation, but something else entirely.

The 'Widows and Orphans of Hindu Coolies' – given no choice, even in their extremity, but to pose for Captain Moss's camera – fill the frame of one of the most heart-rending of all the photographs. These women and children are the most helpless of the destitute migrant workers, many from the arid, famine-struck Deccan or rural Gujarat to the north. Few of them had the means or the opportunity to exit the 'city of the dead' after half its population of 850,000 had fled by boat, train or foot. The mass exodus made the service of the 'coolie' labourers even more indispensable to the functioning of the city but, crowded together at work and lodged in squalid huts, they were among the most vulnerable to plague, so much so that a temporary isolation hospital had been created

A Group of Widows and Orphans of Hindu Coolies who died of Plague at the Bombay Municipal Slaughter House.

Huts at Bandora. The one on the left had no less than nine deaths in it. All the inmates died.

Karanja. Incantations by Women against the Plague.

Danda Village. Types of Native Fishermen who resisted Segregation.

at the slaughterhouse. Any attempt to separate families altogether would have been self-defeating, so there was none. In return, these poor people were in no position to raise an objection to their being photographed. Equally, however, they have no intention of obliging the lens. The children, some naked, some scarcely less so in their piteous rags, stare back, through scrunched-up eyes, into the blinding sunlight. The littlest of them all, at right, held tight by his mother, is in the grip of tears and howls.

Other pictures of desolation similarly puncture the veneer of official self-satisfaction. Rudimentary village dwellings are pictured with their roofs blown off 'to allow for sunlight and air'. Another shot shows two huts at Bandora. One is half collapsed, the other on the right (the prints are in reverse of the captions) where 'no less than nine' were infected, all of whom had died, is pancaked to the ground, the black death-crosses numbering its victims partially visible on the remains of a wall.

Mixed motives – perhaps horror, laced with a smidgin of humanity – leak into the shots as Claude Moss ventures further into the lower depths of infection, often in the hamlets beyond the metropolitan city itself. At the ancient, ruined fort at Kolaba, on the very southernmost tip of the island, unhoused people, many of them fisherfolk, were camped about the gateway. Moss photographed one of them, a leg bent at the knee, nakedly protruding into the foreground. But lest we assume the worst, the caption reassures us, as in an early silent movie, that the subject was 'left to die, until discovered by the Plague Committee'.

At the fishing village of Karanja, a photograph records 'incantations of women against the plague', their heads bent to the ground, beating the dirt, while their children stand about, and a father, his protective arm around his son, both of them emaciated, scowls at the intrusion. Another image – interestingly, preserved only in the British Library album – has district officers 'warning the headmen' of Karanja, dutifully lined up to receive instructions, 'against resistance to the measures'. That opposition, sometimes tacit, sometimes not, creeps into other photographs. At Khar Danda, three Koli fishermen are described as 'native types', their primitivism presumably explaining the otherwise inexplicable 'resistance to segregation'. Faced with being herded into the holding camp built by the Plague Committee, 400 of the villagers had just taken off on the evening of 28 March, crossing the creek separating Danda from Bombay. Another photo shows the camp they built on their own initative. Lying on the western shore of Salsette island, Khar Danda is in fact one of the easiest places to secure its own isolation, something it accomplished at the start of the present pandemic in 2020 with the result that the village was declared the first COVID-19-free district in the whole of greater Mumbai.

Other photos document the ways in which different communities sought to soften or evade the harshest measures. A caption for a photo taken in a Muslim district of Bombay, and with the same penchant for movie-style drama, reads, 'Mahomedans, on the alert for the Search Party, prepare to move the sick into a mosque.' All

Alibag, Kolaba. Left to die, until discovered by the Plague Committee.

these are but the barest hints of how Indians coped with what the Plague Committee suddenly imposed on their lives. What actually unfolded in Bombay between autumn 1896 and late spring the following year was an altogether more unsparing ordeal.

On 18 September 1896, Dr Accacio Viegas, a Goan-Portuguese physician with a practice and dispensary in the crowded chawl ward of Mandvi, just west of the docks, was called to see a middle-aged woman suffering from a high fever and swellings at armpit and groin. Viegas was not immediately convinced he was looking at a case of bubonic plague, but the swift death of the patient – and the discovery, five days later, of other cases in the same district – persuaded him that this was indeed an outbreak of the terrifying disease. Further house calls in Mandvi revealed that as many as fifty victims had succumbed to the disease during the previous month. But the bodies had been concealed out of well-founded fears of family separations, interference with traditional burial rituals and the destruction of household property or even the dwelling itself. The infallible heralds of epidemic, dead rats, had been seen in some numbers in the alleys of Mandvi. Overnight, entire streets had emptied, residents heading for boats in the harbour or to trains taking them out of Bombay. Rumours spread that plague had been brought by Hindu and Jain pilgrims – *sadhu* mendicants – coming from the Himalayan foothills, where the disease was thought to be endemic, to practise their devotions at the Walkeshwar Temple on Malabar Hill. Many of them, especially the Jains, sought alms in the streets of Mandvi. But since the journey on foot to the Maharashtra shrines and temples would have taken at least two weeks and there had been no discernible evidence of outbreaks along the road, this was unlikely to have been the source.

It was much more probable that, as so often was the case, the routes of commerce carried with them the lethal microbes.[2] Squeezed between the docks and Crawford Market, Mandvi was one of the poorest and most densely packed districts in Bombay – in fact, in all of India. There were 759 people per acre in Mandvi, compared, for example, to 222 in the poorest districts of contemporary London.

The long chawls added storeys as and when needed. On each floor, rooms opened on to a dark central corridor running the whole length of the building, not unlike the structure of a prison. None had chimneys; very few had windows or any other ventilation. A sole latrine could be at the end of a corridor, or there might be none at all. A dozen or even twenty people might occupy a single room since the money paid to Bombay's most indigent workers – sweepers, porters, day labourers – never amounted to what was needed to rent a single room for a single family. In any case, family accommodation was irrelevant for the young, overwhemingly male migrants, who had flocked to Bombay to escape the brutal rural famines of the 1890s. As in Hong Kong, ground floor spaces were often godowns, packed with the rats' choice habitat: sacks of grain, rice and sugar. When the rodents died, the free-loading fleas transferred their blood-feeding opportunities to the close-packed inhabitants of the chawl lodgings above. As a bonus for the bacillus, many of the residents of the buildings were Jains, whose religion forbade them from killing any living thing, including rats, and for that matter, fleas. A frustrated official with the sanitary service complained that 'all attempts to catch rats were opposed and threatened. It is difficult to persuade people with more regard for the lives of animals than for the safety of their own kindred that it was right to kill or capture animals either to stay the pestilence or to obtain knowledge of the disease. So bitter was the opposition to the capture of rats that monstrous stories were invented of throwing live rats on a fire to give pleasure to men.'[3]

For a few weeks, the government in Bombay desperately hoped that the infection might be confined to Mandvi, which they contemplated sealing off from the rest of the city, though proximity to the waterfront presupposed an exhaustive examination of the myriad vessels docking and departing, day and night. The thought of complete closure and quarantine threw the British authorities into consternation. Bombay was the commercial powerhouse of the British empire; not just the greatest entrepot of south Asia but a manufacturing giant, especially its cotton textiles, which competed with and often outsold globally, and especially in China, the best

Lancashire could produce (not least because its labour costs were cheaper). Famine in the Maharashtra and Gujarat countryside had already disrupted food supplies to the metropolis, and the 1896 monsoon had been more violent than usual, turning lower-lying areas of Bombay and the creeks separating its original seven islands into sluggishly choked hollows of mud and raw sewage. Plague was just the latest item in this Biblical catalogue of misfortunes. The port would be shut, shipping quarantined; no one knew for how long. So there was an intial reluctance (as there always is) to admit the gravity of the situation or if indeed there was an epidemic at all. But on 29 September, the governor of Bombay Presidency, Lord Sandhurst, resigning himself to the inescapable truth, wrote to the Earl of Elgin, the viceroy in Calcutta, informing him that there was indeed a state of bubonic plague in the city. Residents were fleeing the contaminated metropolis en masse, taking, it was assumed, the disease with them into the surrounding rural areas, to coastal hamlets and cities like Poona, where plague duly took off a month and a half later.

At the time the epidemic broke out, Bombay had a single hospital, on Arthur Road in the Byculla district, that was capable of accommodating, treating and palliating the victims of infectious diseases. As yet, Arthur Road Hospital had no professional nurses, though nuns of the nearby All Saints' Convent selflessly volunteered their services and very quickly turned into indispensable nursing sisters. The hospital's accommodation was rudimentary: open sheds with earthen floors and surface drains fed by water from Lake Vihar.[4] For the time being, it had just the one doctor, though Nasarvanji Hormusji Choksy more than made up – in diagnostic intellect, untiring application and humane compassion – for what he lacked in colleagues. Choksy was already a home-grown hero of Bombay medicine, appointed to a lectureship in anatomy at Grant Medical College by Henry Vandyke Carter, who was the spectacular illustrator of *Gray's Anatomy.* A champion of smallpox vaccination in a region where that disease made periodic comebacks, Choksy was also a passionate campaigner for Indian medical education and professional recognition. As editor of the *Indian Medico-Chirurgical Review* and founding president of the Bombay Medical Union, he had fought an uphill battle to have Indian physicians recognised in parity with their British counterparts. Throughout the plague crisis, he coped with an impossible and frequently overwhelming situation at Arthur Road. The likelihood – evident already in October 1896 – was that the building would quickly fill with the dying, not least, as Choksy pointed out, because most of the patients were only brought in when they were visibly beyond help. But more temporary sheds were erected to take the sick and dying, roofed with canvas slung over bamboo poles.

At the same time, the municipal government thought it imperative to isolate, if possible, streets and districts which had produced plague victims, expeditiously disposing of the dead while attacking the physical conditions in which it was supposed the disease had been generated. As so often is the case, experience of the last epidemic dominated approaches to the next one. The received wisdom that contagion was bred in poor sanitation was a legacy of the battles with cholera. Bacteriology had suggested something altogether different, even though it would be another two years before the intemediary vector of flea bites was definitively revealed by Simond. In Calcutta, H. H. Risley, who as secretary of the Financial and Municipal Department of the city was not the most obvious choice to head its Plague Committee, complained that 'the whole question turned upon modern bacteriological researches which were extremely puzzling'.[5] Without time or inclination to acquire a grounding in the new discipline, the sanitary establishment of the Indian Medical Service did what it knew how to do best: declare war on 'filth' (a word which recurred with predictable vehemence in James Lowson's reports). To be sure, there was much to clean. The gullies of Mandvi and the rest of the Port Trust estate, clotted with sewage, were gut-heavingly noisome. The first epidemic combat instructions were to unclog them, sweep them and flush out the noxious mess with seawater. Sewers were opened and pumped, the rate of action becoming more frantic as the city government became gripped by the terrifying possibility that the *bhigarris* and the *holkhadars*, the largely immigrant workers who did the most repugnant jobs – draining cesspools; emptying night soil baskets hanging down the sides of tenements, their liquid leaking down the walls – might all depart for the countryside or die from their proximity to the infection. An excrementally apocalyptic vision haunted M. E. Couchman, who wrote one of the most exhaustive reports on plague and who was also one of the compilers of the photographic album. 'On their presence or absence . . . depended the safety or ruin of this vast and important city . . . were they all to remove from the town for a fortnight Bombay would be converted into a vast dunghill of putrescent ordure.'[6]

Although the muncipal commisssioner (effectively the head of Bombay's local government), C. P. H. Snow, reported that his office 'was besieged every day by natives of all classes imploring that nothing drastic should be done', on 6 October their 'extreme terror' was realised when much of the same regime that had been imposed on Hong Kong went into full force. Ostensibly it was applied throughout the city and its coastal and rural satellites, but of course it fell most swiftly and heavily on the poorest districts where plague had already been identified. Aware that the sick and dead had often been concealed by families distressed at being prevented from burying their dead with rites appropriate to their religion or caste, searches were instituted of suspected chawls and whole streets. When a case was discovered, the letters 'UHH' (Unfit for Human Habitation) were painted on a door or wall: a death sentence for a home and a source of mortifying shame for those who had lived there. In some cases, iron rings were attached to exteriors of bigger dwellings indicating the number of cases discovered and removed. Neighbours fled at the sight of them. Bedding, clothing and upholstery were taken into the streets and burned, as was the content of ground-floor godowns and storage spaces. But this meant the loss of the entire stock of merchants and shopkeepers who saw their wares, along with personal property, go up in flames. Official promises of compensation for unavoidable losses were initially made by the municipal government but, in the hellish nightmare of the emergency, were seldom believed. The merchant castes – Banias and Bhatias – dwelling and working near the docks were among the hardest hit though unlike the poorest of Bombay's people, many had the means to depart the city.[7]

A foreign pair of eyes was carefully observing all this misfortune. The eyes belonged to a portly 62-year-old Frenchman: Professor Dr Adrien Proust. Neither age nor parental responsibilities (Marcel was then fifteen) nor sheer physical danger had deterred him from going in person to witness and record the worst of epidemics.[8] There was to be another International Sanitation Conference in Venice in March 1897, and Proust already wanted it to concentrate exclusively on plague. He arrived in Bombay in January and was taken aback by the scale of distress. As a third of the population left, villagers who were in any case starving to death from famine were actually coming in to the stricken city. Those migrants were the poorest of the poor, and Proust was deeply distressed to see the homeless sick, stretched out on the streets with not even the modicum of strength it took to beg. Many of them, he noted, died in full view in those streets and alleys and were left where they lay.

While scrupulously compiling, as was his wont, excess mortality statistics, Proust also noticed that something at the core of Hindu society – caste – was making effective medical responses to the plague very difficult. Anger at the violation of traditional social and religious norms was widespread, fierce and instantaneous. Upper castes, especially Brahmins, were horrified at the thought and the reality of being sent to a hospital where they might have to mix with different, possibly lower, castes, and even share the same diet. Requirements that women be examined, especially in the most private places of the body – armpits and groins, where buboes characteristically formed – outraged purdah, common in both Hindu and Muslim communities, and became all the worse for the absence of Indian women physicians to perform the examinations. The requirement that women remove the *padar* end of their sari so that their armpits might be inspected for swellings offended traditional sensibilities, as did post mortem inspections of deceased women, a gross violation of decencies. The newspaper *Kalpataru* spoke for many when it insisted that 'native ladies will prefer death to the humiliation of having their groins examined by male doctors who are utter strangers to them'.[9] Others thought it intolerable that, even in the name of checking a pulse, a wife should have her hand held by someone other than her husband. Family separations were resisted, sometimes violently. In an episode which threatened to turn into a full-scale street battle, an eighteen-year-old Sidi boy examined and diagnosed as a plague victim was prevented from being taken from his house by his mother, who insisted he was merely suffering from the effects of an all-night Sidi dance gathering. When he was removed anyway, a crowd quickly gathered, armed with sticks and stones to assault the removal party. Only the timely intervention of the local 'king of the Sidi' and his queen stopped the situation getting out of hand. But rumours spread nonetheless that admission to Arthur Road Hospital was tantamount to a death sentence, and the truth was that for all Choksy's and the All Saints' sisters' best endeavours, the mortality rate there in the first season of the plague was 71.4 per cent.

Not just death, but punitive murder, was said to lurk at Arthur Road. Indian lives were thought to have been taken in retribution for offences committed on the statue of Queen Victoria: a necklace of slippers set over her royal person; the stone figure smeared with tar. 'Our Sovereign Lady the Queen has demanded five hundred lives of the people of Bombay', one pamphlet claimed, 'to appease the wrath caused by insults to her statue.' It was even rumoured that the hospital had been ordered to cut out the hearts of patients to send to Victoria for her vengeful satisfaction.[10] Reports of family panic spread around the city: frantic mothers threatened to kill themselves rather than allow their sick children to be taken from them. The Muslim community (about 20 per cent of Bombay's population) was especially anguished that hospital wards would preclude the dying facing Mecca and hearing, in their last hours, a comforting recitation of passages from the Quran, not to mention being in the presence of family as the patients reached the last extremity. 'The *masjid* [mosque] is our hospital' became a common cry of protest in their districts. On 23 March 1897, Kazi Ismail Muhri delivered a petition to Governor Sandhurst signed by 15,000 protesting against the plague measures, especially segregation.[11] Arthur Road became demonised as a place of cruelty and iniquity. A woman from the Khoja Muslim community (converted from Hinduism centuries before) died after jumping out of a window rather than being taken to the plague hospital.[12] At the halal slaughterhouse at Bandora, one distraught worker threatened to cut his wife's throat and then do the same to himself rather than see her taken to hospital.

Resistance moved from individual protests to collective action. On 10 October, rocks were thrown at the hospital, smashing windows and causing terror inside. Rage flared, feeding on itself. On 29 October, a crowd of millhands, estimated to be around a thousand strong, scaled the walls of Arthur Road, overran the temporary sheds that had been erected in the grounds of the hospital and set about assaulting the staff with stones and clubs for what they said was the murder of patients. In the terrible heat of the moment, though, the patients themselves were attacked. What Choksy must have thought as he witnessed this tragic terror can only be imagined.

More ominously for the British authorities, Muslims and Hindus, often at odds with each other, were making common cause in the riots and protests against the measures imposed and enforced by the city government. For the first time, there was talk in both communities of a hartal: a planned withdrawal of labour. Law and order seemed to be disintegrating fast. An extortion racket had arisen in the chawls by which criminals posing as police or municipal officers demanded payment on pain of reporting that chawl to the authorities as a plague-stricken dwelling.[13] Worse (and most decisive in determining a shift in policy), the sanitary apocalypse which so terrified the authorities seemed to be at hand. Bombay's health officer reported that:

> [the] scavengers, the drivers of [night soil] carts and the halakhors [*sic*] were very restless. They spoke of leaving the city saying that if the Sahibs left they also would require a change of air. I had no doubt that a riot would occur and as eager as I had been to remove the sick to hospital I admitted at once that if the policy was continued we should not have men either to segregate the sick or clean the city and that we should be left alone . . . The total interruption of sanitary arrangements would have made the city uninhabitable and the plague would have raged unchecked.[14]

Thus the two founding elments of popular Indian nationalism, the forces which ultimately would break the Raj – social and religious outrage, and mass strikes and demonstrations – were both born in the epidemic. Some, in local government, were alert to the warning signs. Shaken by the scale and ferocity of the reaction, the municipal government went into reverse, easing up on the enforcement of household separations it had imposed earlier in the month. There could yet be another way to close down the epidemic without aggravating religious and political turmoil, a prophylactic which would obviate the need for brutally interventionist measures: a vaccine. After his documented success with cholera, there was an obvious choice for just who might deliver this urgently sought alternative.

In the first week of October, at the height of the crisis, Waldemar Haffkine arrived to find Bombay paralysed by official consternation and the distracted terror of the poor. Though it was the viceroy, Lord Elgin, who had sent him from Calcutta, Haffkine remained uncertain about the extent of his authority. He had returned to India mostly, but not entirely, recovered from malaria, with the blessings of the British medical establishment bestowed on him. The previous December (1895), the *British Medical Journal* had characterised him not just as a pioneering scientist but as a modern saint and saviour.

> Dr Haffkine's work is of the highest scientific value and promises to confer a great boon on our Indian Empire. It has been carried out under circumstances of the most remarkable self-sacrifice and

> devotion to the interests of humanity and science. He has given many of the best years of his life to this research and has with unwearying industry and transparent sincerity worked out in India all the details which can test the value of the new gift of science to life . . . prepared without fee or reward other than his own conscience, his love of humanity and scientific devotion.[15]

His head had been turned; his sense, almost mystical, of personal vocation had never been stronger. Some time in late summer or early autumn of 1895, Haffkine had gone to see Louis Pasteur 'on his death bed', as Haffkine later characterised it, at Pasteur's house in Marnes-la-Coquette, a bosky retreat at the western edge of Paris. But if Haffkine had hoped to get some sort of valedictory benediction from his old *patron*, he must have been disappointed since Pasteur had been profoundly disabled by a series of strokes that would shortly lead to his death on 28 September. There was something a little disingenuous about Haffkine's confession to the Royal Colleges of Physicians and Surgeons that should any credit be forthcoming for the success of vaccinations against lethal contagious diseases, it should ultimately go to Pasteur, rather than himself, since he clearly felt that, in India at least, he was acting as the principal apostle and executor of what had begun at the Institut three years before.

Haffkine's return to India in March 1896 had been in striking contrast to his landing in Calcutta three years earlier. Notwithstanding Dufferin's warm letters of introduction, he had then appeared as a nobody from nowhere, a rank outsider in the eyes of the Indian Medical Service, promoting some sort of French-developed vaccine of dubious and unproven reliability. Now, he came accredited by almost all the great men of late Victorian medicine and biology, given appreciative hearings by the Royal Colleges of Physicians and Surgeons, even listened to at Netley as the exemplar of what the new science of bacteriology could offer to the public health of the empire. Instead of being ignored, left to fend for himself, rescued only by the foresight and comradeship of Ernest Hankin in Agra, Haffkine had barely set foot in Calcutta when he was inundated with urgent requests for his presence. Congratulating him on 'the triumph and success which attended your noble efforts in this country', a local health official asked for his immediate personal presence.[16] There was a reason for this change in attitude. A deadly cholera epidemic had had broken out in north-eastern and central India. Compounding the crisis, the mass infection coincided with a brutal famine. The starving made their way to locations where limited relief supplies were being distributed, but the press of their numbers only accelerated the progress of the epidemic. A letter to a newspaper from 'An Aggrieved' reported bitterly that 'I believe a medical man is to come out some time, probably when the last person has died. In many villages not a drop of water is available within a mile or two.' Mortality rates in the spring of 1896 were indeed terrifying. In the vicinity of the railway station in Narail District, twenty two victims, out of a total of forty seven cases, had died in a single day.[17] In a neighbouring cluster of hamlets, thirty seven had died out of thirty eight people who had gone down with cholera. Two quite different kinds of population were at particular risk: migrant labourers, 'coolies' in coal mines and tea gardens; and confined workers or inmates, such as seamen aboard incoming traffic to Calcutta (where Haffkine was particularly concerned to provide comprehensive vaccination)[18] and the unfortunates in prisons, hospitals and asylums. 'I should be greatly obliged', wrote the health officer responsible for a mental asylum, 'if you could arrange to have all the lunatics here inoculated at your early convenience.'[19]

Haffkine delegated vaccination to whatever assistants could be mobilised at short notice, but most often he was asked to come in person. Elaborate travel instructions involved trips by water, road and train. 'You would . . . wait at the station until day-light,' he was informed by an official bringing him to Jessore, '[but] there is no rest house near . . . there is a small waiting toom at the station.' Sometimes, Haffkine's instinct to use emergencies as measurable opportunities to demonstrate the efficacy of the treatment led to understandable distress among the unvaccinated control group. At Dharbanga jail in Bihar, where, in spite of orders to the contrary, parched prisoners drank from a tank that had been found, years earlier, to be thick with comma vibrio, and from where a large group had been removed to a temporary camp site, Haffkine and his assistant inoculated every other prisoner volunteering for the treatment. No cases were recorded among the eighty six inoculated; six cases among the unioculated, all fatal. Unsurprisingly, then, as the Superintendent-Surgeon Harold Brown reported, the latter group 'were far from pleased at having been passed over and, to our surprise, they rose, almost to a man and begged to be inoculated, nor were they satisfied when told that the medicine had been exhausted.'[20]

Capitalising on this indispensability, Haffkine once more asked the government of India to establish a laboratory where medical officers could be trained up in bacteriology and the practical work of inoculation. A start was made in Purulia at the western edge of Bengal, a location chosen by Haffkine for being at the crossroads of coolie labour migration. Similar labs were begun in Madras and Assam (the last funded by the tea planters).[21] Haffkine even tried to persuade the government that the best way to celebrate the Queen's golden jubilee in 1897 would be the creation of a great Institute of Public Health. Predictably, this fell on deaf ears in Calcutta and London. For all the ad hoc calls to him, and the admiration showered on him in Britain, Haffkine's position in India remained tenuous. There was no official promotion. He remained merely a 'Health Officer of the Calcutta Municipal Corporation'.

And despite convincing data from the cholera campaign, questions were still asked about the cholera vaccine's safety and efficacy. Haffkine was dogged by a prolonged investigation into the single case of a sepoy, Jajajit Mal, in a regiment stationed in the Lushai hills in north Assam who had died in a military hospital four days after inoculation. Edward Christian Hare, the local captain-surgeon was adamant that it was 'inconceivable that the cholera microbe injected subcutaneously could reach the intestines alive' (where they were found in the dead man's stool). Nor could there have been some sort of contamination during the preparation of the vaccine. But the case of Jajajit Mal generated an enormous, lengthy correspondence within the Indian Medical Service, including questions about whether or not the single inoculation recommended by Haffkine had somehow compromised its efficacy. Haffkine confirmed that 6,000 had been inoculated around that time under the same protocol in Assam and 22,000 in Punjab and the North-Western Provinces with no one contracting cholera. But even as Hare robustly defended Haffkine and the vaccine, he allowed that in some official circles the case could be made to seem 'ugly'.[22]

Some of those doubts followed Haffkine to Bombay. But to the higher-ups, faced with the sudden, terrifying outbreak of plague, hesitation yielded to urgent necessity. The governor, Lord Sandhurst, in particular, was all for the speedy creation of a prophylactic which might preclude the need for military enforcement of removals and segregation. His and Elgin's hope was that this different approach could also be a vaccine against that other disturbing plague: nationalist politics. But the lower-downs, especially in the Indian Medical Service, suspicious of the new science and facing an immediate emergency, were unconvinced. Pushed by the viceroy, Governor Sandhurst had suggested a Plague Research Committee in which Haffkine was to play a major role. Its work would proceed with the help of medical representatives from the Indian communities, especially Dr Nariman Bhalachandra Krishna for the Hindus and Dr Ismail Mohammed for the Muslims, who were already advising the municipal government. Best of all was a reunion with Ernest Hankin, who travelled from his laboratory at Agra. The brief of the Research Committee was to report on the aetiology of the disease, its infectiousness, the possibility or impossibility of treatment and the consequences of that analysis for containment and control. The aetiology side of the assignment presupposed that not everyone was convinced that excremental water-and-sludge 'filth' explained the eruption of plague; but from the start, the deliberate and scrupulous work of science was at odds with the default instinct of the entrenched IMS and the civil engineering section of local government, which remained obsessively sanitary. Besides, for many of the moustaches, Waldemar Haffkine, for all the lectures he had given in Britain, was still an unknown quantity: a jumped-up, know-it-all 'Russian', and you know how *they* were making mischief in the Great Game, Cossack hordes camped somewhere over the Khyber. Worse, he seemed to be some sort of Frenchified Russian (you could hear it in his accent, at exactly the time those two powers had made a military alliance); yet worse, a Frenchified Russian *Jew* with, it was said, a shady revolutionary past, spells in prison, that sort of thing. It was not impossible, the English press suggested, that this un-doctor might turn out to be a Russian spy.[23] And even if not, then a decidedly queer fish.

Unsurprisingly, then, Haffkine was given the bare minimum of professional assistance. Ushered into Grant Medical College, its Gothic turrets and lancet windows resembling an Oxford college, he was allotted a single room plus a short stretch of the adjoining corridor. His supporting staff consisted of a single clerk and three servants. 'None of these persons knew anything of bacteriological methods and had to be trained by M. Haffkine.'[24] And trained very carefully since the extraction of pathogenic matter from the buboes of the stricken was deliberately done at the height of their symptoms. As few survived more than four or five days following the initial detection of swellings, this had to be done with maximum humane tact and care. In a detailed report circulated in December, Haffkine outlined, step by step – for colleagues in other endangered cities who might want to make their own vaccines – the procedure, which, since extraction was done either with a grooved needle or by pipette suction, needed to be scrupulously followed to avoid accidents. Pipettes had to be physically destroyed, usually by burning, on completion of lymph extraction and satisfactory transfer of the matter to a slide. Since bubonic plague (unlike pneumonic) was not transmitted through respiratory droplets, very few attending nurses, orderlies or physicians in the hospitals ended up catching the disease. But should they or the bacteriologists get nicked or scraped, lethal disaster was possible. One of Haffkine's colleagues on the Plague Research Committee, Surgeon-Major Dr Robert Manser ('beloved and devoted to duty'), died early in January 1897, probably from a rare case of pneumonic plague, four days after being called to treat the raja of Akalkote, who also perished, as did a Nurse Joyce assisting Manser. Acutely conscious of the frightening mortality of the disease, Haffkine decided that, unlike the cholera vaccine, where he had routinely used both attenuated and 'exalted' cultures, the plague bacillus required 'devitalisation' before it could be safely used as vaccine. Accordingly, he laid out stringent protocols: the matter required heating in a spirit flame to between 60 and 90 degrees Celsius to ensure sterilisation.[25] The IMS's preferred medium of sterilisation was carbolic acid. New admissions to isolation hospitals were immediately bathed in carbolic solution, houses of the infected were doused with it inside and out, and it was indicated as the gold standard sterilisation for experimental vaccines. Haffkine mostly followed those protocols while believing that heat sterilisation was just as effective, if not more so. Even that reticently expressed touch of scepticism and his independent deviation from official norms in response to the exploding demand for vaccines would end up doing Haffkine's career irreparable harm.

By December 1896, infections and deaths in Bombay were rising steeply. And plague was colonising country and towns around the city and well beyond. The epidemic had become virulent in Poona, Surat up the coast, Hyderabad and even Karachi, 550 miles away. Pressure on Haffkine to come up with a vaccine was mounting, although official anxiety did not, somehow, translate into either a bigger laboratory or more assistants. Still, there was much he could accomplish within those confined spaces. Haffkine's experimental procedures followed the lines of his work against cholera, which in turn owed much to the Pasteur protocols. The first step, as always, was to generate a viable and stable supply of the micro-organism. Matter extracted from buboes was cultured in goat broth, with a skin of ghee or agar on the surface. Goat was chosen to avoid offence that would be taken by Muslims and Hindus should there be any suspicion of matter taken from pigs or cows. But because such suspicions lingered, one of Haffkine's junior colleagues, Maitland Gibson, succeeded in making a culture using wheat flour broth.

Usually within forty eight hours, the silky threads appeared that came to be known as 'Haffkine stalactites', leading shortly after to a rich growth of the bacilli. Periodically the flask needed to be shaken to reactivate growth. Once 'devitalised' by heat, the vaccine was delivered subcutaneously into rabbits and pigeons. Survival rates, even of animals developing some of the symptoms of plague, were sufficiently encouraging for Haffkine, on 10 January 1897, to enact his drama of getting himself inoculated with 10 millilitres of the vaccine, nearly triple the dose given to most patients.[26] As had been the case with the cholera vaccine, the occasion was a performance, enacted in the presence of the principal of Grant College, George Maconachie, professor of ophthalmology and veteran of African and Indian campaigns, none of which necessarily qualified him to preside over a teaching and research institution in mid-epidemic. For Haffkine, the purpose of the demonstration on himself was to establish trust among the curious students of the college and the openly sceptical personnel of the IMS. 'A large number of physicians watched the results [of his real-world trials] with eager interest', he wrote in 1897, 'many with great readiness for criticism.'[27] But acting as his own first human experiment was also meant to impress the people of Bombay, and beyond, all of whose imperilled lives he wanted to preserve. Nothing untoward happened following his vaccination: slight pain at the injection site, gone by the end of the second day; intermittent pangs of headache; and a one-day fever of 102 degrees.

Off Waldemar went, once more, into deep India, to the sort of places that had served as trials for the cholera vaccine: prisons, cantonments, villages, slums. He made sure to be accompanied by Indian, as well as British and Portuguese-Goan, assistants, not just as translating intermediaries but to reassure local people that the vaccine was benign and not the poison denounced by street and market orators. Haffkine's tireless ally in the campaign of reassurance was the Parsi doctor Nusserwanji Surveyor, whose services were, after adamant insistence, paid for by the municipal government. Surveyor steered Haffkine through inner and outer Bombay, enabling the cooperation of prison wardens and factory supervisors.

Haffkine administering street vaccinations, Bombay, *c.* 1898. To his left is Captain Milne, with Alice Corthorn under the umbrella.

Drs Krishna and Mohammed were likewise vital for vaccine acceptance. It was a truism for Haffkine that without active help from Indians capable of calming local fears and suspicions, no vaccination campaign would ever fulfil its promise.

As with cholera, Haffkine vaccinated only volunteers. But also, as before, he selected locations where, by virtue of confinement and crowding, residents had good reason, whatever their anxieties, to step forward. When he got to the first site of human trials, Byculla jail, not far from his lab, in the last week of January 1897, there had already been nine cases of plague, five of them fatal, so it could not have been surprising that 337 prisoners of both sexes came forward as volunteers. Half of them received the vaccine, the other non-volunteering half acting as control. Later in the campaign, there were times when Haffkine was torn between the success of a human experiment and the price that might be paid for it by the unvaccinated. On at least one occasion, conscience won over experimental yield when he returned to vaccinate, after urgent persuasion, the whole population of a village. At Byculla, the results, while encouraging, could only be provisional because of any number of independent variables: some of those inoculated had already caught the infection and were showing early symptoms; some had caught it but were as yet asymptomatic until a few days later. The vaccine was billed as prophylactic, but it was not yet clear (as is the case now with COVID-19) whether this meant a shield against infection altogether or against fatal severity.

As the trials went on, it was the latter definition which Haffkine used as vindication of the vaccine. On a six-week tour undertaken with Nusserwanji Surveyor to the north and east, its effects became more dependably measurable. In the village of Undera with a population of just over a thousand, lying on the the northern edge of Baroda, the capital of the princely state of the same name, thirty five families had one or more infected members. Of seventy one vaccinated patients, just three died; of twenty seven unvaccinated cases, only one survived.[28]

Haffkine had been invited to Baroda by its ruler, Gaekwad Sayajirao III. Descended from the most resolute Maratha enemies of the East India Company, the current incarnation was an ardently westernising maharaja: colleges and schools opened; hospitals created; libraries stocked; shiny novelties like bicycles (and in 1898 an early automobile) shown off in princely style.[29] But the initiative to bring Haffkine to Baroda is likely to have come from his chief justice, Abbas Tyabji, a future militant nationalist and close associate of Gandhi, but in the 1890s still a believer in gradual liberalisation and often vocally loyal to the Raj. At the centre of Tyabji's reformism was the emancipation and education of Indian women and his particular bugbear was purdah. So when he decided to have the maharaja and himself inoculated by Haffkine by way of example to the people of Baroda, he made sure not just to have his daughter Sharifa vaccinated as well, but to put her at the centre of the photograph recording the event. In his right hand, Haffkine is holding his syringe, while his left hand is on the girl's shoulder: exactly the kind of physical contact which triggered horrified outrage in Bombay but

in this instance was possible in Baroda. On the other hand, it's also noticeable that none of the men who crowd the picture are smiling.

Haffkine's freedom of action was always circumscribed by politics. Increasingly, and to his dismay, he found himself in the middle of a bad-tempered official debate about how best to contain the plague which, in late winter, was taking hundreds of lives each week. The net of infection was stretching farther and farther through the territory of Bombay Presidency and beyond. But since cases were being found in districts of Bombay itself which had been comprehensively, not to say fanatically, disinfected, it was dawning on the local medical establishment that limewash and carbolic might not be the effective answer. Though James Lowson never abandoned his conviction that Asiatic indifference to the basic norms of sanitation was responsible for spreading, if not causing, the disease, and continued to inveigh against the showers of spit and mounds of shit which he found criminally horrifying, even he was prepared to consider that something other than filth might be generating the germs of infection.

So it was becoming apparent that Haffkine, who was offering the light of science rather than the theology of sanitation, should be given more space and support. In the early spring of 1897, he was moved out of his impossibly tight quarters at Grant College and into the more spacious, government-owned Cliff Bungalow on Nepean Sea Road, at the tip of a peninsula jutting out at the south-western edge of Bombay into the ocean. Malabar Hill was swept, on both sides, by damp, somewhat cooler, salty breezes. There were topiary hedges enclosing other bungalows and houses, a hanging garden (still there close to the gated residences of Bollywood billionaires). He had space, now, for two doctors and three lab assistants who also doubled as clerks.

But Haffkine also found himself spending time at what had been the old residence of the governors at Parel but which had been turned into an emergency plague hospital. Originally a Jesuit monastery built on the site of an older Maratha fort, the priests had been ejected by the East India Company when Portuguese guns were fired at them from the monastery roof. In the eighteenth century, the twin pavilions of Parel House with their toy-fort crenellations, arched windows and shady portico formed the architectural facade behind which the raging greed and military power of 'John Company' grew and grew like some monstrously insatiable hatchling. After the defeat of the Marathas, Parel's twin pavilions were closed together by a continuous bricky facade centring on a Strawberry Hill Gothick Revival tower (more crenellations and dentillations). Applied features and window forms were given the 'Indo-Saracenic' touches that historicising architects, in thrall to fashionable Aryanist ethnology, thought reflected an imagined common stock of Saxon and oriental traditions. The vision of Sir Bartle Frere, governor in the mid-1860s of a Bombay that would become a world-class commercial metropolis, also found room for a heavy dose of ceremonious protocol. Parel became duly imprinted with the fittings and fixtures of pomp on demand, so that by the time that Frere returned to India in November 1875 as the major-domo of the Prince of Wales's tour, Bertie would be greeted by lion's head finials on the balustraded banisters of a great ceremonial staircase and a vast 'Durbar Hall' scooped out of the interior. Between watching rhinoceros fights at Baroda, cheetah-hunting, dining in the Elephanta Caves, visiting the Sassoon Academy for Young Women and ogling the Parsi and Hindu girls dancing and singing for him, the prince received maharajas in sumptuous grandeur. Lest he suffer the inconvenience of travel from Bombay, a railway was laid to Parel and a customised station built well away from the dust, the crush, the smells and the shouting.

Not long after Bertie's departure, death arrived at Parel. In 1882, Lady Olive, the second wife of the then governor, Sir James Fergusson, died of cholera on the premises. The Fergusson family saga was an epic of colonial doom. Sir James's first wife, Edith, 'of a delicate constitution', passed away in Australia after the birth of her fourth child. Fergusson himself would perish in the Jamaican earthquake of 1907. Lady Olive's death may have hastened the decision to move the governor's residence (also the core of the administration) to a bigger and, it must have been thought, a more salubrious site, further south on Malabar Hill, close to the Walkeshwar Temple of the Sihara kings and the Banganga tank said to be filled with the purest water of the holy Ganges.

When the governor moved off to his new residence in 1883, Parel was left as the repository of the vast archive of a century of British administration: tax and military records; commercial contracts and customs; censuses; and, not least, records of public health documenting serially calamitous epidemics and famines. But in February 1897, statistics gave way to stricken humans. Once Arthur Road Hospital ran out of room, even in the improvised sheds in the grounds, Parel House provided another 150 beds, with space to increase the number to 220. In came the plague victims at all hours; out went the dead on their way to Muslim burial grounds or the Hindu funeral pyres that burned day and night. South Bombay was turning into a hellish place of grief and terror, but Claude Moss photographed nurses with parasols standing before the colonnaded portico as if they were about to play a round of croquet at an elegant country house.

There too, at Parel, was Waldemar Haffkine, vaccinating already ill patients and observing their fate. In an adjacent ward was Alexandre Yersin, having arrived in Bombay, as had been the case in Hong Kong, without support from the local government but also with informal blessing to do what he could with the serum vaccine, worked up at the Institut Pasteur. A third version was being offered by the Austrian-Italian pathologist Professor Alessandro Lustig, based in Florence but with roots in another cosmopolitan world – Jewish Trieste.[30] The descent on Bombay of epidemiologists and bacteriologists followed the Tenth International Conference on Sanitation, held in Venice in March 1897.[31] That the gathering of delegates from twenty countries, including Proust, concentrated exclusively on plague was in part thanks to his eye-witness reporting in Karachi as well as Bombay, which was published not long after his return. It was also the first time that the Conference explicitly recognised that the source of the plague was microbial: the specific bacillus identified by Yersin and Kitisato in Hong Kong. While acknowledging that defensive measures would have to be internationally coordinated, and that the lessons of the new science needed absorbing, the delegates were less clear about what that would entail. Disinfection, segregation and quarantine remained orthodoxy.

Nonetheless, the Venice Conference triggered an almost unseemly race among the Great Powers to send their own bacteriological teams to India as fast as possible so that their fatherland or mother-land might claim the credit for saving humanity from the new Black Death. The Russian delegation to Venice announced they would send to Bombay Professor Wyssotkowitsch, Dr Redrov and Danylo Zabolotny, whom Haffkine knew from their student days together in Odessa, where the younger scientist had done time in prison for student activism.[32] The Germans immediately countered by expediting a formidable trio of Kochian scientists: Georg Gaffky, who had identified the salmonella bacillus as the source of typhoid; Richard Pfeiffer, who had developed the vaccine for it; and the military physician Adolf Dieudonné. Once Robert Koch himself had completed exploratory work on rinderpest in South Africa, he was promptly shipped off to Bombay as well, making Imperial Germany, for many international observers, the A-team to beat in the Great Vaccine Race. Despite continuing to work principally on large animals, Koch and Pfeiffer were the only bacteriologists whose portraits were included in the 'Plague Album'. There is a photograph taken of them sitting together with Gaffky in the Portuguese-Indian port city of Daman, where the vaccine had impressive results (36 deaths among the vaccinated to 1,482 of the unvaccinated), and another where Koch and Haffkine are seated together as if, finally, the younger man had been accepted as a peer.

Scientific research has seldom been immune from the habits of power. Both Yersin and Haffkine had had dramatic histories at the Institut Pasteur. Both had worked with Emile Roux, but Haffkine's opportunity only came about because of Yersin's abrupt, and to many at the Institut, mysteriously churlish departure for Indo-China. But in 1895, or 1896, Yersin had been taken back by Roux, precisely to help create a Pasteur plague serum. However devoted both Yersin and Haffkine were to the prevention and alleviation of the immense suffering brought by plague, both were also in competition for the legacy of the *patron* in the immediate aftermath of his death and immortalisation. Now, in the grim spring of 1897, they were rubbing shoulders in the wards of Parel, vaccinating the newly admitted, then watching to observe if what they had done made any difference to their patients' chances of survival.

Two children had been brought into Parel, three days apart, in the last week of April 1897.[33] The boy, Ardesir Jijibhai, treated by Haffkine, was ten; the girl, Kasi Satwaji, treated by Yersin, was four. Their homes and their young lives were separated by just 3 miles at the far southern end of Bombay, and were both steeped in the waters that sluiced through that tip of the town. Ardesir, 'the schoolboy', lived with his father in Pydhonie, its name meaning 'foot-washers': not an occupational description but an allusion to wading through the swift-flowing creek that had originally separated two of Bombay's seven islands. Pydhonie was bright with both architectural and human colour. Koli fishermen came to Pydhonie from their huts at Kolaba and Danda to seek the blessings of the mother goddess Devi at her temple, built in the time of the Maratha princes. Mumbadevi Temple was surrounded by tall houses, each storey decorated with the *jharokha* balconies and verandahs beloved of the Gujarati and Rajasthani migrants who had settled in the neighbourhood.

Little Kasi's infancy was spent in Sewri, an impoverished fishing village just a morning's walk away from Pydhonie. In January, Ernest Hankin had visited the hamlet and claimed to have found the plague bacillus in an insanitary pool used as a latrine by the villagers. Stagnant ponds were not of course where *Yersinia pestis* arose, but it could indeed have cultured there in the fetid heat. There were compensations. Kasi grew up amid a different kind of migrant: flocks of brilliant birds, flamingos especially, tens of thousands of them, nesting and strutting on the mudflats, the oozy surface dull grey as if a foil to the rosy showiness of the stilt-legged birds. Making space as best they could amid the flamingos were black and white pied kingfishers, tawny godwits, and Indian ibises with their downward-curved bills like Indian daggers and a dash of scarlet like a little skullcap at the back of their heads. The girl would have watched them all stabbing the mud for shrimps and crabs, or perhaps, on some day in early October, stared wide-eyed as a cloud of flamingos exploded aloft into graceful flight.

But Sewri was also a sump of infection. Between mid-December 1896 and late January 1897, fifty two cases of plague were recorded: around one twelfth of the entire population. Kasi Satwaji became another victim. By the time she got to Parel, she was very sick. Two days after her symptoms had first appeared, Yersin inoculated her with the Pasteur serum he had first tried on horses. She must have seemed a desperate case because she got a large dose in her flanks, a good deal more than was usually prescribed for young children. The nursing sisters would have tended her with constant kindness; this we know from many Indians of any religious persuasion who praised the nuns' ceaseless devotion. But Kasi's fever rose and fell and rose again, the painful swellings becoming more so.

In the boys' ward, one partition over, Haffkine inoculated Ardesir with his devitalised vaccine on the day of his admission, 24 April. To begin with, there seemed no ill effects; just a slight fever and a little muscle pain, much as Haffkine himself had experienced. But on the 30th, things took a turn for the worse. Buboes appeared in Ardesir's groin and armpit. His temperature rose alarmingly. But forty eight hours later, the fever broke and, on 2 May, the boy was able to sit up, eat some eggs and drink a cup of milky tea, and was moved into the recovery ward. On the 7th, his father came to take Ardesir back to their home in Pydhonie.

Two days earlier, Kasi had died. It is unlikely she would have been cremated as that rite was reserved on the whole for upper-caste Hindus. Whatever became of her remains, they would not have been interred in the picturesque cemetery at Sewri, strictly reserved as it was for the British and Portuguese.

Of the twenty three cases treated by Yersin at Parel, thirteen died and ten recovered, a discouraging result for the Pasteur serum-based vaccine. It would have been beneath their dignity and calling for the two ex-Pasteurians to make a crude contest out of this, but Gatacre was not above tartly remarking in his report that the Pasteur serum had come with 'the guarantee of M. Roux and M. Yersin'. Less justifiably, he added that 'the patient [Kasi] did not benefit in any way . . . on the contrary fresh buboes appeared' (after the inoculation). But so had been the case with Haffkine's patient, who recovered. But there was no doubt that, as the plague finally began to wane in June, it was Haffkine who was winning converts among the IMS doctors[34] and, possibly still more important, the good opinion of the governor of the Bombay Presidency.

Lord Sandhurst and the viceroy, Lord Elgin, were putting their faith in Haffkine's mass vaccination programme as a way of forestalling the return of the much harsher regime of inspection, isolation and quarantine that had been abandoned in late October 1896. But in the immediate emergency, so a report defensively claimed, 'it was found that none of the medical officers employed

Top: Bandora Railway Station, medical inspection on arrival of a train; bottom: Sion Railway Station, medical inspection.

on plague duty had leisure to attend to the inoculation'.[35] In London, the secretary of state for India, Lord George Hamilton, imperialist to the hilt, believed the relaxation of coercive measures had been a foolish error which had led directly to the painfully steep rise in both cases and mortality through the winter. The inability of the Raj to contain and control disease, coming on top of the Indian famines, was not merely an imperial embarrassment in the year of the Diamond Jubilee; it was also an economic and commercial disaster and not just for Bombay. On 21 February 1897, Hamilton wired Calcutta and Bombay that 'the continuance of the epidemic in the present state is a steady menace to the health of the whole continent of India'.[36] The viceroy and the governor, Hamilton thought, were too nice about the dangers of alienating Muslim and Hindu social and religious sensibilities. The Raj ought not to be in the business of humouring 'superstition' and 'prejudice' (the words recur in all the reports of the Indian civil and medical services) if it interfered with getting the urgent job done. Earlier in February, the Epidemic Diseases Act had been passed, with the Bombay plague specifically in mind, which in effect set aside the regular procedures of law and consent, should any local emergency require drastic intervention. Refusal to vacate a house or chawl in which infection had been found would be met with enforced eviction; anyone picking up rags or objects set in the street while such evictions were under way would be arrested. And while an inventory of confiscated and destroyed property would be faithfully kept for purposes of possible compensation, such reimbursement was at the discretion of the authorities since the act added, ominously, 'that no person shall be entitled as of right to claim any compensation whatsoever'. Shortly after, on 2 March, Gatacre, who had been in command of Bombay's military forces, was made 'chairman' of a Plague Committee of just four, none of whom had a medical background. This committee could now reimpose military-backed 'visitations'; family separations; the immediate creation of segregation camps; personal and property disinfections; demolitions; plus stringent inspections at railway stations, in docks and aboard small ships in the harbour. All these measures should be enforced, another government memorandum stated, 'even in the face of ill-will and opposition'. To show he meant business, Hamilton sent James Lowson, who on the strength of his work in Hong Kong had been appointed to a newly created position, 'medical director-general' at the India Office, to India as 'special plague commissioner', a one-man strike force, to manage the campaign. He arrived in Bombay on 1 March, more than ready for action. That was Gatacre's watchword too (unless written for him by the iron-willed Special Commissioner Lowson). Others may observe; a time had come, Gatacre said, to act.

Although justices of the peace – Indian as well as British – who had volunteered for the work were authorised to run the house-to-house 'visitations', their presence was often a rubber stamp for what had already been determined by the Plague Committee's oversight. From March to June 1897, Bombay was effectively under medical martial law. Nearly half of its population fled, but the committee belatedly wanted to prevent the infected from spreading plague further into the countryside and coastal towns. In fact, by late March, people were actually returning to the city, but the inspection barriers Lowson had set up on the main Mahim–Sion causeway road engendered maximum aggravation with minimum prophylactic effect. The same was true at railway stations, where second- and third-class passengers, including women, were summarily inspected on the platform or beside it and, if suspected, bundled off either to isolation hospitals or to the grim tented isolation camps which had been hastily established in fields and empty spaces about the city.

Needless to say, first-class passengers, on the other hand, were examined in the privacy of their curtained carriages. In the harbour (again following the precedent in Hong Kong), a medical barge was put into commission and all vessels, steam and sail, were supposed to tie up close to its anchorage at one of the bunders and submit to inspection. This was easier said than done. Small boats in their hundreds, loaded with fruit and vegetables destined for Bombay's markets, docked between midnight and the earliest hours of the morning, necessitating nocturnal inspections.

But neither the boatmen nor the municipal authorities, anxious about food supplies, had much of an interest in holding up deliveries to the city markets. Cover of darkness offered many ways to evade inspection. Not infrequently, then, the sun would rise over Bombay harbour and the host of boats that had been bobbing around in the small hours had evaporated over the watery horizon.

Once mobilised, the Plague Committee regime seems not to have paused to consider the evidence of experience. Plague persisted and returned to neighbourhoods already exhaustively disinfected. But instead of raising questions about the causal connection between dirt and contagion, the instinct was to order another round of limewash and carbolic. Smoke rose over Bombay from two kinds of fire: the burning of bedding and the burning of bodies on the Hindu funeral ghats. At Mandvi, normally a busy seaport, 350 miles to the north, the disease had taken a terrible hold. When Lowson got there in late March, 70 victims were dying every day in a town of just 25,000. Posted there, Surgeon-Lieutenant-Colonel James Sutherland Wilkins, who had grown up in the Suffolk market town of Eye, found himself in a burning hellscape, cows wandering between bodies – some dead, some dying – left outside houses. Of the Brahmapuri Hindu Hospital, situated outside the walled town and conveniently close to the site of funeral pyres, Wilkins wrote:

> It is impossible to describe in words the hospital and its sights. The two long wards, full of the sick lying side by side in every stage of this dreadful disease; the nurses going about here and there on their merciful work, the ward orderlies and other attendants outside; the constant admission of patients in carts and ambulance carriages; in one angle of the square, the dead lying in numbers prior to removal to the Burning Ghats nearby. It was a mournful sight that met one's eyes every morning and evening when we had to pass near the Burning Ghats and saw the numerous fires which told of the heavy mortality.[37]

Sonapore, interior of burning ground with corpse on pyre.

Faced with such wrenching scenes of human misery, Wilkins the military physician, like Moss the military photographer, abandoned imperial cosmetics. Their reports and images are *de profundis,* from the lower depths of tragedy, and the figures of those who populate the universe of the plague-struck transcend any crass division into rulers and oppressed. This is never more true than when those figures are women, like the nurses at the Brahmapuri Hospital: Sister Elizabeth from the Convent of the Filles de la Croix, or Nurse Horne, both of whom died after catching the disease. The urgent need for women who could perform inspections on female patients meant that British nurses were shipped into Bombay. They are everywhere in Moss's photographs: smiling as they hold recovered Indian children; taking the temperature of patient groups of women; in one deeply affecting shot, standing on one side of the bed of a desperately sick child while his father is seated on the other (for relatives were now accommodated in the grounds of the bigger hospitals like Parel). The nurse stands formally as if just asked to pose, her face grave, the uniform laundered to a standard of blindingly white purity, but her body language tells a different story. One hand is in her pocket; the other is placed on the feverish

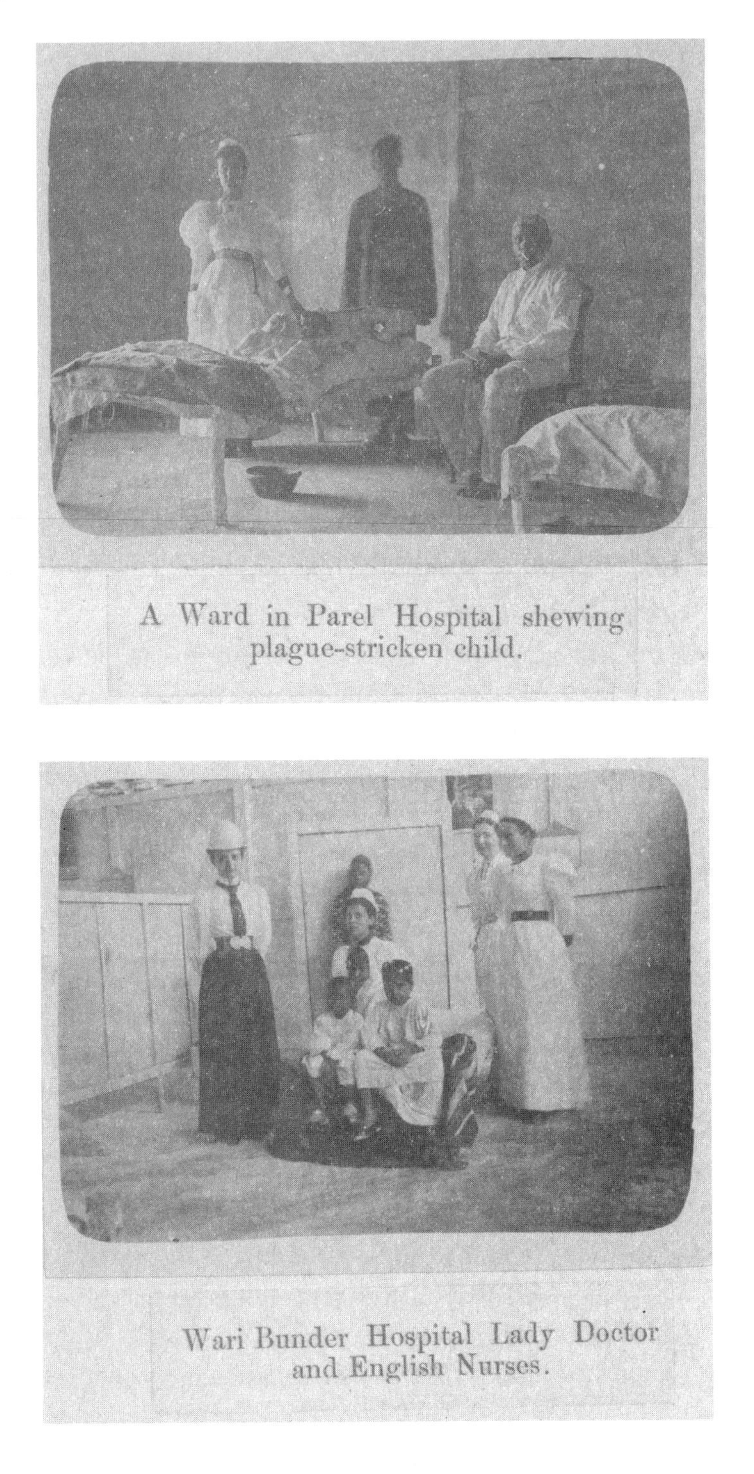

A Ward in Parel Hospital shewing plague-stricken child.

Wari Bunder Hospital Lady Doctor and English Nurses.

brow of the little boy in a gesture that is both medically watchful and tenderly humane.

At least one Album photograph survives of a woman doctor, dressed in the standard signifiers of male authority: pith helmet, woven tie, scrutinising monocle. Was she the 'Miss Cunha' singled out for praise for her work among the sick of the Port Trust estate in the Gatacre report? In that case she could also have been the daughter of the Goan medic and scientist J. G. da Cunha, who wrote the first authoritative treatise on dengue fever. But of other women medics mentioned in the report as working in the same district hospital – Mrs van Ingen, Mrs Walker and the 'lady student' Miss Ferreira (another Goan name) – it seems difficult, as yet, to know more. Together, though, these names and the many more listed in the Gatacre report amount to a breakthrough in what work with the human body was thought proper to and possible for women. The plague imposed its own exigencies, including an urgent need for women physicians who could examine, diagnose and treat female patients. That, in itself, was a radical liberation.

At the centre of this professional leap forward was Edith Pechey-Phipson. In her forties, and despite suffering from severe diabetes, she had come out of invalid retreat to work at the hospital for Kutchi Memon Sunni Muslims, located in an impoverished district of Bombay. But then Pechey's whole life had been an epic of feminist determination. The daughter of a Baptist minister and a mother who had taught herself ancient Greek, Pechey had been one of 'the Seven', a group of young women determined to crash through the barriers placed in the way of their medical education. When Sophia Jex-Blake's application to Edinburgh University was summarily rejected, she called for others to join her in reapplying. Pechey was the first to respond. Predictably, the presence of 'the Seven' at Edinburgh in the 1860s – the first female medical students in any British university – was bitterly resisted. The young women were subjected to humiliations meant to crush their resolve. Separated from male students, they were made to pay punitively higher fees on the grounds that they had forced lecturers to add customised

Frank Meadow Sutcliffe, photograph of Edith Pechey-Phipson, albumen cabinet card, 1870s.

classes to their teaching load. When Pechey won the annual scholarship awarded to the top student in chemistry, it was denied her on the grounds that such an honour was bound to provoke male resentment. On appeal to the university senate, the women students were granted certificates of completion of their chemistry courses, but Pechey still did not get the scholarship she had earned.

Sometimes, the women faced physical, as well as verbal, violence. Arriving at Surgeons' Hall on 18 November 1870 for the final examinations, its iron gates were slammed in the women's faces by a hooting mob of male students, 'passing about bottles of whisky . . . while they abused us in the foulest possible language which I am thankful to say I have never heard equalled before or since'.[38] Dresses stained from pelted mud, Pechey and her comrades managed to slip into the examination hall thanks to a sole sympathetic male student who opened the gates. But the abuse did not end there. Pechey became the target of stalkers, following her home at night, yelling 'medical terms in their abuse to make their purport more intelligible'.[39] The front doors of the women's lodgings were pounded, handles and knockers wrenched off; fireworks were set off in the front yard. Obscenities yelled at them day and night became routine. It was enough to make Pechey and her comrades despair of Edinburgh, especially when it became clear that the university medical faculty would do little or nothing to promote their right to education or protect them from the relentless harassment. Pechey, Jex-Blake and Annie Clark went to Switzerland to complete their studies, graduating from medical school at Bern; and they received their formal licences in Dublin. Then Vienna provided Pechey with an opportunity to take courses in surgery. Without an official English licence, Pechey nonetheless practised as house surgeon in Birmingham Hospital for Women, and opened her own private practice in Leeds. In 1878, at the age of thirty three, she delivered the inaugural lecture at the London School of Medicine for Women, founded by Jex-Blake and Elizabeth Garrett Anderson, the first Englishwoman to receive a medical licence (albeit from the Society of Apothecaries).[40]

Four years later, India came calling on Pechey in the person of George Kittredge, an American businessman living in Bombay who, in 1882, had founded the Medical Fund for Women in India. Like the Countess of Dufferin's fund, created three years later, that initiative was in the first instance meant to serve upper-caste secluded zenana women, forbidden, or discouraged, from seeking medical help as long as it was provided by male physicians.[41] But quite quickly the ambition to create a cadre of women doctors outgrew those social limitations. A Parsi philanthropist, Pestonjee Hormusjee Cama, had provided generous seed money (to the tune of 100,000 rupees) to build a hospital for women and children, run and staffed exclusively by female doctors and nurses. Funds had been raised and a site identified. The hard part was finding women doctors prepared to make their career at the Cama Hospital. Elizabeth Garrett Anderson recommended Pechey to Kittredge, but warned him that she had also been approached for a position in Vienna, where she had had her surgical training. Why would anyone in their right mind opt for the tropical over the Mittel-Europa empire, the double conservatism of zenana seclusion work, coupled with a hostile Indian Medical Service, over experimental Vienna? Never one to shrink from a challenge, Edith Pechey did just that. This in itself was surprising, and not just for Garrett Anderson. It was one thing for Waldemar Haffkine to take his cholera vaccine to India where it could be tested in real-world large-scale trials. Even facing IMS hostility, Haffkine could call on the support of scientific collaborators like Ernest Hankin who were already established in the sub-continent. Pechey, on the other hand, was bound to face every kind of prejudice – against her sex, her work, her shocking presumption. And her determination not to confine herself to matters medical, but take to take a public, vocal stance on the fate of women in India guaranteed that she would acquire a whole new set of enemies.

Cama's funds had been donated on condition that the hospital's operating costs would be met by government. Predictably, the Indian government was unenthusiastic about supporting an institution entirely run by, and intended for, women and children. But the Cama Hospital could draw on additional philanthropy, much of it coming from the Parsi community, including the already prodigiously rich Tata family. While waiting for the hospital to rise in its full Venetian-Gothic splendour, east of Marine Drive, Pechey plunged into her new life, working at the Suleiman Jaffir Dispensary, launching training courses for Indian nurses and taking a crash course in Hindi. Enthusiastic commitment, though, did not pre-empt the usual institutional foot-dragging. Her salary had been set at 500 rupees, well below the rate earned by male physicians. Outraged, Pechey argued that this was, in effect, to degrade women practitioners to the status of second-class physicians, which would in turn undermine their status and their right to respect, not to mention the likelihood of women patients seeking their care. And that, after all, had been the whole point of the Cama Hospital.

If anyone imagined Pechey would now content herself with

Rukhmabai Pandurang

running the first women's hospital in India, or a service for the secluded, they were speedily disabused. She plunged headlong into the contentious debate over child marriage, which she judged inseparable from the overall physical welfare of her sex. The *cause célèbre* of the opponents of child marriage was Rukhmabai Pandurang, married at eleven to a nineteen-year-old relative of her stepfather. Barely a year later, Rukhmabai's husband claimed the right of marital consummation, at which point her stepfather, Dr Sakharum Arjun, and Rukhmabai herself flatly, and by contemporary standards shockingly, refused. A series of judgments handed down by English judges declared that Rukhmabai had been married 'in senseless infancy', posing the issue of whether British or Hindu law should determine the outcome. When, after a series of appeals, and notwithstanding an article written by Rukhmabai herself for the *Times of India* under the signature of 'a Hindu Lady', she was ordered to 'restore conjugal rights' on pain of a six-month prison sentence, the girl declared she would rather face prison than comply. In 1888, the husband abandoned his claims to conjugal consummation. His price for doing so was 1,000 rupees. The result was far-reaching. In 1891, an act of Parliament raised the age of consent from ten to twelve. Pechey had wanted it set at fifteen, but this, at least, was a modest improvement. For their temerity, Pechey and Rukhmabai were demonised in the press, especially by the Poona journalist Bal Gangadhar Tilak, who accused the women, and by extension English law, of violating 'sacred Hindu traditions'. Pechey's response was to break another taboo, raising funds to send Rukhmabai to the London School for Women in Medicine. On her return to India, fully qualified, Rukhmabai was immediately hired as house surgeon at Cama.

In 1889, Pechey had married a like-minded reformer, the wine merchant Herbert Phipson, whom she had met at the Bombay Natural History Society, which he had founded. Together they opened a convalescent village near their own country house north of Bombay, meant for recovering patients from the city's poor. In 1894, the deterioration of her diabetes made it impossible to continue to run the Cama Hospital, but Pechey replaced herself with Annie Walke, who was in her own right a phenomenon, all the more impressive for having started in modest circumstances. Born in 1865, Walke was a dock-girl, living above a chandler's shop with her harbour-pilot father, Jonathan, her mother Laura, and six siblings.[42] While Jonathan Walke worked his way up to senior dock master, Annie became a prodigy. Denied a medical education at Grant College, she nonetheless graduated in mechanics, the kind of learning said by men to be a challenge to the female mind. Like her sister pioneers, Walke had perforce to get a medical education in Europe (in her case at Brussels University) before returning to Bombay and Grant College. Pechey hired her as house surgeon at Cama in 1892 and made her second physician in 1894. Walke treated the poor at Cama while also building a practice for begums at princely courts like Baroda. Though Pechey had stepped out of her country retirement to help with the plague emergency, Walke could see the signs of debilitating illness in her mentor. But in fact Pechey would outlive Walke, who died at just thirty five, not from plague but another unclassified infectious fever. An enormous crowd from all of Bombay's castes, classes and religions attended the funeral. Sixty carriages followed the black-draped hearse to Sewri, her father's birthplace, where Annie Walke was laid to rest near the nests of flamingos.

There were so many reasons for young and newly qualified women medics not to go to India: dangerously infectious disease; the resistance of the native population to the very idea, much less the practice, of female doctors; the condescension or outright hostility of British imperial society. And yet they continued to come. In 1898, Alice Corthorn, forty years old, arrived in the desperately plague-stricken district of Dharwar in the western Carnatic (Karnataka), the only woman (identified by official reports as 'Doctor Miss') in a team of four peripatetic physicians sent to deal with the epidemic.[43] Green though she may have been to western India and to bubonic plague, Corthorn quickly became a local legend, holding, so Haffkine later said, the record for numbers vaccinated in a single day at one thousand.[44]

She was the daughter of a Camberwell wine salesman much given to sampling his wares, both on the job and off. One of six Corthorn children, the early death of her mother forced Alice to join the legions of young women for whom the only escape from menial trade or house service was the grind of a governess. Through that demeaning, captive time (as she thought of it), Alice clung fast to the belief that education would put liberated independence within reach. To her father's fury and horror, she moved out of the family home and took lodgings in Great Russell Street, marking her – as far as father Corthorn was concerned – as a degraded woman. It's possible that some of Bloomsbury's attraction was proximity to the British Museum and its habitues. As it turned out, this was no idle fantasy. Outside the museum she met the feminist Olive Schreiner, who, denied a medical education herself by poor health, went on to fund Corthorn's. Inside the vast, domed Reading Room, she made the acquaintance of Schreiner's friend and fellow radical Edward Havelock Ellis. Though he was the learned and notorious investigator of sexuality and clearly taken with Alice Corthorn, it may be, as her daughter Olive Renier suggests, nothing went beyond the holding of hands. Just before moving to India, though, Corthorn

made a bonfire of letters and diaries which, she said, would be of no interest to anyone, while also admitting there was 'much I would not want my family to see'.[45]

In the plague-riddled towns of Hubli, Gadag-Betageri and Dharwar itself, Corthorn became, as Haffkine's eulogy suggested, a ubiquitous vaccinator, riding around the town to make house visits on a customised horse-drawn carriage driven by herself. But at the same time that she followed her practical routine, Alice Corthorn was alert to anything that could help advance understanding of the origins and transmission of the disease. In Gadag, macaque monkeys and grey-striped palm squirrels were dropping dead from the trees in front of her carriage. Samples taken from the carefully examined animals were cultured in Corthorn's improvised lab and sent to Haffkine in Bombay. Though compromised by inevitable impurities, the results still showed the unmistakeable presence of the plague bacillus. The possibility of a wider range of mammals than hitherto observed acting as reservoirs of the microbe became an ominous, unavoidable conclusion. That 'animals such as monkeys and squirrels', she wrote in a published paper, 'are also susceptible to natural plague infection in addition to rats, cannot be over-estimated especially in relation to the carrying of infection from one district to another.' Since those animals co-existed in enormous numbers with humans in almost every town in India, the potential for wider mammal-to-human transmission was suddenly made more alarming and vaccination all the more crucial.[46]

Corthorn's daily modus operandi was three-fold, which may well account for her impressive numbers. Mornings, she inoculated at clinics set up for that purpose in all three towns of the district. In the afternoon, she drove her medical buggy to make home visits – charging a small fee – and in the evenings she attended a lecture promoting the benefits of vaccination, read in Marathi translation to large gatherings of Indians, including workers on the South Maratha Railway, labourers and – not always deemed safe or acceptable – small children. At the end, she would offer on-the-spot inoculation. In Dharwar, the numbers availing themselves of the offer could be as many as a hundred. As many as 180,000 were inoculated in the three major towns of the district in 1898 and 1899, and the campaign's statistically measurable success in reducing plague mortality helped shift official attitudes towards adopting vaccination in place of disinfection and segregation.[47]

The success or failure of a mass vaccination campaign turned on such acts of persuasion. And that, inevitably, was a challenge. Heroic as this founding generation of women doctors in India was, they were, inevitably, caught in a cultural trap. Annie Walke may well have been a dockland girl but she, like Edith Pechey and Annette Benson, who looked after Pechey in her last sickness, and Alice Corthorn, were all British. To the extent that they were creating a profession committed to any Indian women who needed their help, not just the secluded of the *zenana*, they risked being accused of insensitivity to both Muslim and Hindu traditions. What they imagined to be a moment of liberating change could be – and was – represented as just another act of imperial presumptuousness. It did not help that so much of their work in plague was unavoidably intrusive: the discovery of buboes in the intimate parts of the female body. Angels of mercy the British nurses and doctors may have been, but they were alien angels nonetheless; in their perfectly bleached uniforms and stiffly starched caps, as white as white could be.

Indian vaccinators, allies and champions stepped forward. But the non-medical persuaders acting as exemplars to their community by having themselves (like Haffkine) publicly and ceremoniously inoculated were also the most westernised among their minority communities. The young, recently widowed Farha ('Flora' to the British) Sassoon, and the twenty-year-old Sultan Mahomed Shah, Aga Khan III, were both reform-minded, scientifically knowledgeable and liberally educated. Vocal opponents of purdah and child marriage, they were also passionately in favour of the education of girls and women. This might have immediately put them at odds with their communities. But both the Jewish millionairess and the forty eighth imam of the Ismailis were not antagonists of their respective religions but champions and practitioners of their core traditions. The same was true of the Parsi tycoon Jamsetji Tata, who staged an exemplary inoculation ceremony on the eve of his son Dorabji's wedding to a woman from the almost equally powerful Bhabha dynasty, in which the bridegroom received his own inoculation. Much of the lives of these communal and business leaders was committed to the possibility of reconciling modern knowledge, especially scientific learning, with religious faith. Following her husband's death, Farha Sassoon took over the active management of the family business – and of an exceptionally ramified enterprise at that, extending from the core industry of textiles to shipping, trading and banking. But at the same time as she was a working business executive, Sassoon lived as a strictly observant Jew, employing a ten-man minyan to accompany her whenever she travelled so that the halakhic law governing the minimum number of males needed for daily services would never get in the way of her prayers. And though she was famous throughout Bombay for the spectacular scale and magnificence of her banquets and entertainments, her in-house ritual slaughterer ensured they would always be strictly kosher. Sassoon's devotion to Judaism went beyond legal compliance to scholarly commentaries on the Talmud and Midrash, mastery of Aramaic and Arabic as well as Hebrew and Hindi, and publications on medieval commentators such as Rashi. It would, in fact, have been easy enough for the poorer Bene Israel Indian Jews, who mostly came from Konkan, to see Sassoon as just another Baghdadi plutocrat from the dynasty that was spoken of as the Iraqi-Indian version of the Rothschilds. But instead of keeping her distance from the Bene Israel, Farha Sassoon obeyed the injunction of tzedakah, meaning the duties of both justice and charity. Much of her work among the Bene Israel was pastoral and unifying, supporting the maintenance of the Shaarei Rahamim (Gates of Mercy) synagogue, which had been moved to Mandvi in 1860 and become known among its non-Jewish neighbours as Juni Masjid. More immediately, Sassoon provided the funds for a small hospital where the Jewish poor could get treatment free of charge. Her public devotion to the totality of Bombay's Jews gave Sassoon the credibility she needed when championing Haffkine and urging her co-religionists to follow her example and get vaccinated.

The forty eighth imam of the Nizari Ismailis faced a similar challenge. Only seven and a half when his father, Aga Khan II, died in 1885, his traditional Muslim boyhood instruction arranged by his mother, the Begum, was followed by a very untraditional education at Eton and Cambridge, so that when he returned to Bombay (where his father had been appointed by Governor Fergusson to the Legislative Council), Sultan Mahomed Shah was already committed to a reconciliation of Islam and liberal modernism. A cultural base for Islamic reform already existed in the Anglo-Oriental College founded in 1875 by Sir Sayed Ahmad at Aligarh (though unlike the young Aga Khan, Sayed Ahmad dismissed the importance of educating women). But like the Jews, Bombay Ismailis were divided not just by social fortune, caste and

Farha Sassoon, undated photograph.

Leslie Ward (Spy), cartoon of Aga Khan III, *Vanity Fair*, 10 November 1904.

occupation, but by origin narratives and collective memories. The majority were Khoja – Indo-Iranian descendants of Hindu Rajputs, converted to Shi'a Islam in fourteenth-century Persia.[48] Originally concentrated in Gujarat, the ordeals of the late eighteenth and nineteenth centuries – endemic warfare and brutal famine – had taken them to the relative safety and economic opportunities offered by British Bombay. Some had prospered as rice and opium traders and even as cotton textile industrialists. But many more were modest shopkeepers and market vendors crowded into the alleys of Dongri where mendicant fakirs sat in doorways and Sufi mystics tried to raise their chant above the hubbub.

As with the rest of Bombay's communities, the poorest Khoja were hardest hit by infection. And like Farha Sassoon, the Aga Khan, though barely out of his teens and much interested in fine tailoring, bloodstock horses and English poetry, radically reconceived his duty as simultaneously spiritual and scientific. At this point, Pasteur meant more to him than polo. 'I knew that something had to be done,' he recalled in his 1954 memoir, 'the impact of the plague among my own people was alarming . . . I had to act swiftly and drastically . . . I knew something of Pasteur's work in France [and] I was convinced that the Surgeon-General's Department was working along the wrong lines. I bypassed it and addressed myself directly to Professor Haffkine. He and I formed an immediate alliance and a friendship that was not restricted to the grim business that confronted us . . . It was in my power to set an example. I had myself publicly inoculated and I took care to see that the news of what I had done was spread as far as possible and as quickly as possible . . . My followers could see for themselves that I, their Imam, had, in full view of many witnesses, submitted myself to this mysterious and dreaded process and there was no danger in following my example. The immunity, of which my continued health and activities were obvious evidence, impressed themselves on their consciousness and conquered their fear.'[49]

For the Aga Khan, 'the Russian' represented a third way, neither obstructively traditional nor repressively imperialist. Fascinated by bacteriology, he could see, in ways the Indian Medical Service establishment could not, that inflexible faith in sanitation was beside the point when it came to preventing plague or mitigating its severity. 'Twenty years old I ranged myself (with Haffkine of course) against the orthodox opinion of the time – among Europeans no less than among Asiatics.' So when the Aga Khan went well beyond just exemplary vaccination and offered his 'vast and rambling palace' of Khushru Lodge, located just north of Mandvi and thus close to the Khoja poor of Dongri, as lab and vaccine production space for Haffkine, it was in effect a reproof to the hide-bound, even ignorant, default sanitation obsession of the British. It was only when, to their surprise, the British saw Haffkine move into his generous new quarters, followed by the measurable success of the vaccination of the Khoja Ismailis, that they took him seriously and gave him the respect – and eventually – the space his work needed.

The Aga Khan described his alliance with Haffkine as a deliberate 'test' of his young authority, a demonstration that scientific knowledge and religious solidarity need not be at odds with each other. And, as he correctly recalled, 'it vindicated itself in a new and

Haffkine in his laboratory at Khushru Lodge,
Mazgaon, Bombay, 1897.

perhaps dramatic fashion . . . My followers allowed themselves to be inoculated, not in a few isolated instances but as a group. Within a short time statistics were firmly on my side; the death rate from plague was demonstrably far lower among Ismailis than in any other section of the community; the number of cases caused by contamination was sharply reduced and finally the incidence of recovery was far higher.'

This was not an empty boast. The thousands of Bombay Ismailis – both the vaccinated and the control – who came forward for inoculation at the Aga Khan's urging represented by far the largest vaccination trial yet attempted not just by Haffkine but anyone anywhere. In February 1898, 3,854 Khoja were inoculated, of whom just three subsequently died from plague, while 77 of the 955 declining the vaccine perished over the same period. The results were decisive enough to establish the credibility of the 'Haffkine lymph'. The young Aga Khan felt vindicated in his confidence and in return Haffkine was grateful for the trust – and the working space at Khushru Lodge. But he was not so secure, especially with Yersin in Bombay, as to be above hoping that the Institut Pasteur would take note of his success. When the Aga Khan next travelled to Paris, he carried with him a letter of introduction from Haffkine to Emile Roux.

As many thousands as they were, the Ismailis were a small minority of Muslims in Bombay, let alone elsewhere in the presidency and beyond. Haffkine travelled far and wide in western India, but he could not be everywhere plague had struck, and he did not invariably enjoy the trust of the local British authorities or communal leaders. In many places, the choice of action facing Indians was either to suffer the daily trials of inspections and segregations, or else to mobilise resistance to violations of customs said to be inflicted under the pretext of epidemiological necessity. Farha Sassoon, the Aga Khan and, for that matter, Waldemar Haffkine, marginal as they were to the great waves of mass agitation stirring in late nineteenth-century India, could not alone prevent bacteriology and public health from being seen as the poisoned gift of empire. Thus it was that the war against infection mutated into a culture war in which militant nationalism struck a first, bloody blow.

Poona was the battleground. Ninety miles to the east of Bombay, the city was the seat of the 'Monsoon Residency' of the governor of the presidency from spring to autumn. In keeping with its imperial pretensions, Poona was (as it had been under the Marathas) a military town, its barracks the base for tens of thousands of Indian and British troops: the 17th Poona Cavalry, Lancashire Fusiliers, Maratha Light Infantry and the rest. Though, like many towns in Maharashtra and Gujarat, Poona turned out traditional printed silks and cottons, its big industry was not textiles, but words: paper and print. Word-hungry schools, newspapers and journals flourished there. And just because the military was so visible in the town, those who hated what native and British troops alike represented expressed their contempt in bitterly intense rhetoric. The tide of words flowed freely in many languages: English, Marathi and Gujarati especially, but other vernacular languages too, especially Hindi. Twenty per cent of the city's population of 150,000 was Brahmin, so Poona was passionately, and disproportionately, literate. Since it had been the capital of the Maratha Confederacy, the epic history of its warrior princes, above all Shivaji, the formidable enemy of both the Portuguese and the British, reappeared on the stages of the popular drama which flowered in Maharashtra at the end of the nineteenth century. In addition to the city theatres, travelling companies, the Sangeet Natakas, came through Poona, delivering spectacular acts of speech and music that married up history, myth and fantasy. Extravagantly costumed romances featuring bloody battles and thrilling, sensual dance had mass audiences outdoors and indoors roaring. Though the Bollywood movie cameras had yet to turn, Poona was a city where rapidly modernising media gave new, theatrically vivid life to pre-imperial sources of pride. It was a place where very old things lived cheek by jowl with very new things: ayurvedic medicine alongside microbiology, where the instructions of modernity – including what to do about plague – collided with the defiance of rejuvenated tradition.

No one embodied this divided cultural personality more dramatically than the sometime maths teacher turned prophet of Indian nationalism: Lokmanya Bal Gangadhar Tilak. And at exactly the historical moment when the term 'populism' was being invented thousands of miles away in America, no one understood more shrewdly how the perception of violated traditions could be weaponised for modern politics. It was in this spirit that, through his polemical paper, *Kesari*, in 1892, Tilak drew on Hinduism to invent a festival of the elephant-headed god Ganesha, presented of course as a restoration of an ancient tradition. From the beginning, the Ganesha Festival became as much a political as a religious event. Religious reinvention went hand in hand with history. In the hands of Tilak and Hindu-romantic writers, Shivaji became Hindu India's prototypical nationalist hero – the scourge of both Muslim Mughals and European imperialists.

When plague struck Poona early in 1897, Tilak initially criticised the British authorities for not doing more to contain it. But when the former assistant collector at Satore, Walter Rand, whom Tilak called 'a sullen, suspicious and tyrannical officer', applied the interventionist measures recommended by the Plague Committee, Tilak tacked hard in the opposite direction. On 15 June, he published two articles in *Kesari* denouncing the plague regime as a pretext for the destruction of Indian homes and families and Hindu sanctity. 'The tyranny of the Plague Committee . . . is too brutal to allow respectable people to breathe at ease.' Everything held most dear was now being violated in the spurious name of 'sanitation'.[50] Poor people were being robbed of their only shelter; families stripped of their homes and possessions; floors dug up, allegedly in the name of destroying rat habitats, or just because it was claimed that infection bred in the soil. Household shrines were being violated, domestic puja worship desecrated; sometimes the figurine idols were damaged or even confiscated. Worst of all, the modesty of bodies, especially female bodies, Tilak claimed, was being routinely violated. Columns were devoted to stories of women in purdah dragged from their homes, or the elderly forced to strip for inspection and made to go through humiliating ordeals of repeated sitting and standing while soldiers stood around clapping and laughing and even dancing. 'The plague is more merciful to us', Tilak thundered from the pages of *Kesari*, 'than its human prototypes.'

The most shocking accusations, of rape by soldiers, found an immediate echo in Britain, where they were taken up by Tilak's political rival and critic Gopal Krishna Gokhale, there to testify to a parliamentary commission on financial payments to India. In most respects, Gokhale, also from Poona, was Tilak's polar opposite.[51] In common with much of the leadership of the Indian Congress, he was a gradualist, a believer that gross injustices could be corrected, and the cause of Indian self-government advanced, by appeals to liberal norms then circulating in British politics. But the reported excesses of the plague regime in both Poona and Bombay radicalised Gokhale, who made common cause with Dadabhai Naoroji, the first Indian MP to sit in the House of Commons.[52]

Though Naoroji and Gokhale were both believers in a Liberal-constitutional road to equality, justice and independence, they were coopted by the radical Marxist social democrat H. M. Hyndman, for whom appeals for social reform and representation in local government were just so much begging for crumbs from the table of the Raj. The deluge of congratulations to Queen Victoria on her Diamond Jubilee that came pouring in from Indian princes allowed Hyndman to turn the hyperbole upside down. In speeches and rallies that he organised for republicans, trade unionists and anti-imperialists in London and the industrial towns of the Midlands and North, Hyndman demonised Victoria as the 'Queen of Famine and Empress of the Black Death': a monster of haughty indifference. Discarding their usual deference to the norms of British political decorum, Gokhale and Naoroji joined Hyndman in claiming rape (along with larceny and judicial murder) to be the standard modus operandi of the British empire. The Liberal MP for Banffshire, William Wedderburn, one of the founders of the Indian National Congress and its spokesman in Britain, introduced Gokhale in a conference room of the House of Commons where he spelled out all the violations of Indian privacy, property and person inflicted by the military during plague searches. Gokhale repeated those allegations in an interview with the *Manchester Guardian* on 2 July. Uproar ensued on both sides. In Poona, where the storm raged most furiously, Walter Rand's report insisted that all complaints about misconduct during searches had been scrupulously investigated. Only a handful of cases had turned out to have substance, none of which involved any kind of physical assault, but rather acts of petty theft and bribes taken by soldiers. Gokhale was now asked to provide details and sources corroborating the charges of rape, but on returning to Bombay, he admitted, 'It did not take me long to discover that substantiation was out of the question.' It may be, of course, that the shame of such violations precluded victims from coming forward. But the mortified Gokhale clearly believed the reports of rape had been invented and, in November, he made an embarrassed apology to the press for having given them any credence.

It was too late for Rand to have taken any satisfaction from the retraction. On the evening of 22 June, as he was being driven home on Ganesh Khind Road following a celebration of the Queen's Diamond Jubilee at the governor's residence, a bullet tore a hole through his lungs. Seconds later, another bullet blew away the back of the head of his military escort. Lieutenant Charles Ayerst died on the spot. Rand was still alive, but barely. On 3 July, he succumbed, as it was said, to his terrible wound, in a bed in the David Sassoon Hospital.

The assassins were brothers: Damodar Hari Chapekar and Balkrishna Hari Chapekar, respectively twenty seven and twenty four years old at the time of the murder. They were born and grew up in the village of Chinchwad, on the edge of Poona (now a gleaming, upscale suburb). When Tilak cursed English constitutionalism and its obliging collaborators in the Congress, the brothers cheered. But, especially after observing him playing the grand man at a political meeting, they came to believe that Tilak was not, in fact, purely Hindu enough. How could he be when he mixed freely with 'beef-eating' Muslims and believed, contrary to strict Hindu law, that widows should be permitted to remarry? All he did was talk. They would act. When they were small boys – around six and eight – their grandfather Vinayak Chapekar, Sanskrit-speaking, living off inherited wealth, took the entire extended family, all twenty of them, plus two servants and three carts, on a pilgrimage to the Kashi Vishwanath Temple at Benares where they bathed and drank the holy waters of the Ganges and tried to overcome sorrow for the death of a sister en route in Gwalior. Little by little the family fortune dwindled, the ever more eccentric and wayward grandpa went to live in Indore and the brothers' father, Hari, became of necessity a *kirtankar* – an itinerant performer of religious epics, mostly written by himself. As muttering and worse grew that *kirtan* was not the kind of occupation proper for their caste, the uncles peeled off, and the boys took over the music. Balkrishna played a harmonium, Damodar an Indian variation of the hurdy-gurdy, while their little brother Wasudev beat the *tal* drum and sang in his child's voice.

Kirtan was not enough for a living. But other occupations failed them. They attempted to enlist in a regiment but were rejected. That humiliation, Damodar later wrote, instilled 'an implacable hatred' for the English. Their father taught them ayurvedic medicine, so that one of them might become a traditional *vaidya* physician, but that ambition petered out. There was door-to-door begging; petty theft; drilling with a martial youth gang the Chapekars created but with whom, predictably, they fell out. Obsessed with Christian missionaries, whom they imagined forcing conversion on Hindus, they twice attempted to burn down missions. Not only was there nothing about the Raj to make the Chapekars in any way grateful for the blessings it claimed to have bestowed on India, it increasingly seemed to them a tyranny designed to uproot Hinduism from its native soil. When plague arrived in Poona early in 1897, the brothers saw the measures taken to deal with it as a plot to destroy their religion. Nothing fed their outrage more powerfully than stories of British soldiers committing sacrilegious trespass on domestic puja and idol rooms. When patrols ended in late May, as plague ebbed, the urge to commit an act of vindicating retribution grew. Believing Walter Rand, the principal plague commissioner, to be responsible for all this, the Chapekars stalked him, so that the discovery of his manifestly evil ways would harden their resolve. But as Damodar (rather winningly) admitted in his autobiographical confession, Rand turned out to be something other than Tilak's caricature of a tyrannical petty despot. He had, they discovered during the three months of following him around, 'no vice, no meanness . . . we had even seen him play lawn-tennis at Gymkhana' where he advertised his propriety by 'not playing with the ladies', just the sort of prudishness that appealed to Damodar and Balkrishna. But 'as he had made himself an enemy of our religion we deemed it necessary to take revenge on him. We could not help it.'[53]

The brothers already had an animus towards Queen Victoria, whom they regarded as a mother guilty of dereliction of duty by not coming to the aid of the famine-stricken, instead wallowing in imperial celebration. They duly defaced a statue. What could be a better occasion to make the point even more forcefully than her Diamond Jubilee celebration? Too bad about the nice but plain Mr Rand; he would have to go. How to kill him, though? Surely with their five-chamber revolvers. But just in case the situation should call for it, say with guns jamming, the brothers made sure to bring some other hardware along: a sword each, plus a hatchet carried by Balkrishna. This turned out to be a bad idea. Amid the heaving crowd (no Brahmins though, Damodar noticed), bonfires lighting the hilltops, the bulky blades made it almost impossible to walk freely. They were bound to attract suspicion. The swords and hatchet were ditched in a culvert in case they should be needed. But there was another problem. Although after stalking him for months they knew very well what Walter Rand looked like, the dimming light and the fact that the carriages entering the residency were, from their point of view, facing the wrong way, meant that the brothers were unsure they had the right conveyance and the right man. Better perhaps to do the deed as the traffic was departing. While champagne was poured, toasts were made and a military band struck up 'God Save the Queen', the Chapekars retired to a field to wait.

The sun set over Poona. Carriages bearing merry, politely tipsy celebrants began to exit the grounds of the Raj Bhavan; top hats, frock coats, turbans and saris. Damodar had posted himself close to the main gate. The plan was that, on recognising Rand's carriage and the man in it, he would shout '*Gondya ala re*' ('Gondya has come') and run behind the carriage, meeting up with Balkrishna, who would be sprinting to meet him, and they would discharge their shots together, fraternally. As usual with such events, this one did not go exactly as planned. Rand was recognised; 'Gondya' was yelled; Balkrishna set off towards the gate, but he fell badly behind, so Damodar lifted the rear flap and fired through Walter Rand's back and into his lungs. In the melee, barely catching up, unsure who was who, Balkrishna took a shot at another man, another carriage, and exploded most of Charles Ayerst's head.

In the general panic – perhaps surprisingly – the Chapekars managed to disappear into the enormous, now wild, crowd and make an escape. It took three months before the police finally caught up with Damodar in Bombay, informed on by those he had trusted to keep him safe. But in his confessional autobiography – assuming, of course, that it is reliable – Damodar gives the impression of a man waiting to be caught so that he could reveal his unrepentant glory. After writing his lengthy life story, Damodar was tried, convicted and hanged in April 1898. Balkrishna managed somehow to elude capture a while longer, but he too was arrested and executed in January 1899. If they wanted to remain at large, it was probably not a good idea for the youngest brother Wasudev and two friends to try to kill the informers who had betrayed his brothers.

Though Damodar had come to despise Tilak as a compromiser (not a charge often levelled against him), much in his statements about the British contempt for Hindu tradition and observance could have been lifted straight from the fulminating pages of *Kesari*. When news came in of another successful Jubilee Day assassination at Peshawar, the British laid the attacks squarely at the door of Tilak's inflammatory polemics. He was arrested, tried for sedition – by, as he immediately pointed out, a European-majority jury – and sentenced to eighteen months' jail time. The imprisonment of course only reinforced Tilak's reputation as a hero of Indian nationalism, in dramatic contrast to Gokhale, who had grovelled before the British in making his agonised 'apology' for the unsubstantiated rape accusations. Henceforth there would be two ways for plague-born Indian nationalism to evolve: the way of Gokhale's acceptance of the norms of liberal gradualism, or the way of Tilak's uncompromisingly militant Hinduism. Nathuram Vinayak Godse, who murdered Mahatma Gandhi on 30 January 1948 in the name of defending the purity of Hinduism, came from, and grew into his fanaticism in, Poona. The two ways still divide the politics of contemporary India. In the time of Narendra Modi, the Chapekar brothers have been good Bollywood box office, the latest movie version of their lethal adventure appearing as recently as 2016. A museum, the Chapekar Wada, honouring their memory has been established at Chinchwad, and every year, on the anniversary of Damodar's execution, there is a local commemorative march. In 2021, Yerwada jail, where the brothers received their sentence and wrote their confessional life-and-death stories, was selected for a star role in a new programme of 'jail tourism' designed to familiarise Indians with the history of the heroic incarcerated.[54]

Violent resistance to the plague regime was not, however, confined to Hindu militants. The removal of plague patients from their homes was still a source of anguished fury for Bombay's Muslims, the vast majority of whom were not Ismaili. On 23 March 1897, fired up by preaching at Friday prayers, 15,000 Muslims demonstrated against separations and isolation on the grounds that those measures violated a fundamental Islamic principle of family caring for the sick. More ominously for the authorities, purely religious objections were increasingly coloured by economic and social grievances. On 9 March 1898, a crowd of Muslim mill weavers in the northern industrial district of Madanpura prevented plague-searchers from removing a twelve-year-old girl to hospital. The protest swiftly turned into a full violent riot, involving Hindus as well as Muslims, despite the fact that just five years earlier in 1893 the two communities had been at each other's throats. Calling in troops to quell the violence resulted in four deaths in Madanpura and another two in a neighbouring district. When random attacks on Europeans began to take place, the situation took on a new and, for the British, ominous note. Cavalry was summoned from Poona, but for a week much of Bombay was shut down by strikes.[55] Shops specialising in cloth and grain, the two commodities most likely to be burned by the plague inspectors, closed their doors; bullock-cart drivers, butchers and municipal clerks as well as cotton weavers and spinners all removed themselves from the labour force, paralysing the city. Even more unusually, dock workers joined the hartal. Thus the blunt instrument of the militarised plague regime managed to bring together both Hindu and Muslim, religion and economic conflict, in a major, history-shaping act of resistance.

The Madanpura riot followed a series of strikes the previous years, especially among mill workers crowded in chawls and huts in the north of the city. Aditya Sarkar has shown how those labour actions became progressively more militant since, as she persuasively argues, the abandonment of Bombay by tens of thousands of workers fearing plague had created a drastic labour shortage in which textile industrialists, already beset with a shrinking Chinese market for woven fabric, seldom had the upper hand. Every day, workers gathered on street corners to see how much their wages might be bid up by rival factory agents desperate to hire spinners.[56]

The beleaguered governor of Bombay, Lord Sandhurst, was thus facing a double crisis: outraged religious sensibilities and an acute labour shortage in which, for once, factory workers, along with small shopkeepers, dockers and carters, were capable of bringing the enormous economic machine that was Bombay to a grinding halt. The whole city seized up. Both the elements of mass action that would, eventually, make the British Raj untenable – the hartal and demonstrations, and religious and cultural protest – thus first saw the light of day courtesy of bubonic plague. Sandhurst reported optimistically but defensively to the government in London: 'I am confident that the disturbance is not the result of any excessive severity in the regulations or in their administration, and I feel sure that when order is restored . . . the work of combating the plague will proceed, every endeavour being made to give no avoidable offence to the customs or beliefs of the people and to induce native gentlemen of position and influence to cooperate with the Government in the performance of this humanitarian duty.'

What Sandhurst needed, above all, was for plague to be over. But it was not over. Each declaration that the epidemic had been contained or beaten back, and was on the point of disappearing altogether, was followed by another outbreak in the autumn. So it was in 1897, and in 1898. The relentless War on Filth; the carbolic hosing; the burnings of objects, inventories and bodies; the separation of the infected from the uninfected; the herding of the former into what one army medic called, without irony, 'health camps';

the launching of a rat extermination campaign with per-rat monetary incentives: none of the above had stopped plague. Back it came; and along with it, more anger, more hatred, more death. It was time, finally, while not altogether abandoning the plague patrols, to try something that might obviate, or at least reduce, the fury and the chaos; time to call for Waldemar Haffkine, not just as a short-term auxiliary, but as an officially government-supported producer and deliverer of vaccine. At least some of the imperial medical establishment were beginning to see it his way. In January 1898, Haffkine went to Poona, to the heart of resistance to imperial disinfection, and, under the auspices of the local surgeon-general, delivered a lecture on the proven benefits of plague inoculation. At Grant Medical College, speaking to the graduating class, which included women, Haffkine had commented that 'the present calamity in Bombay' was an 'illustration of how fatally energy can be paralysed by want of knowledge'. The knowledge he had in mind was the bacteriology that ought to have made the campaigns of disinfection and segregation redundant.[57]

In the second week of August 1899, Viscount Sandhurst travelled from the Raj Bhavan at Poona to the old residence of his predecessors at Parel House to preside, on the 10th of the month, over 'an assembly of notables, both European and native'. The occasion was the inauguration of a new Institute of Research, directed by Haffkine, to be lodged at the old Government House where there could be ample space for laboratory work, vaccine production and storage. Haffkine, of course, knew the building well from his regular hospital visits and inspections. And he had been happily productive at the Aga Khan's Khushru Lodge. But that extra space had been a gift of patronage. He was still denied official imprimatur and Haffkine suspected, correctly, that even though he was about to become an official member of the Indian Medical Service, its establishment would never really accept him as one of their own. But the government was becoming desperate. So, finally, he was being given the keys to his own much bigger working space, as well as official recognition. His staff would number more than fifty. He had become an institution.

At the inauguration, praise was heaped on his head. Sandhurst was pleased to say that the plague appeared to have abated a little. (They all said that in August.) But should there be an unhappy recurrence, he believed, Bombay and everywhere else in India visited by *Yersinia pestis* would now be spared horrifying rates of mortality. The reason for this upbeat prognosis was the vaccine created by Mr Haffkine. Describing what was done in the lab in graphic detail, Captain-Surgeon Charles Milne taxed the composure of some in the gathering as he graphically described the procedure by which the vaccine was produced. There was the lethal microbe materialising in its bath of goat's meat broth, developing drooping chains of 'long, silky stalactites', followed by their clotting into the 'jelly seed'. Fans fluttered a little faster. Queasy ladies and gentlemen reached urgently for seconds of champagne. But the governor made it plain that even if it were the case that people of their sort seemed to get ill at a far less deadly rate than the native population, they should not imagine they were immune. Remember poor Surgeon-Major Manser, and his attending nurse, Miss Joyce, who had perished while treating patients. Were he in the position of doctor or nurse, he affirmed, he would most certainly hurry to be inoculated. Clumsily aware of the many Indians in his audience, Sandhurst proceeded, in his effortlessly though unintentionally insulting way, to exhort *native* ladies and gentlemen not to delay. 'If I were to advise certain individuals of a class liable to contract the plague I should certainly say, "Be inoculated!"' A 'hear, hear' rose from the assembly, followed by a wavelet of applause from those beneath the broad-brimmed hats, the whiskered men in their perfectly pressed linens, who would have been joined by the guards and bandsmen in their dress whites were not their hands otherwise dutifully engaged, and perhaps too from the Indian ladies, were the protocol not unclear. But clapping rose from the ranks of those who (notwithstanding the strikes) imagined they made Bombay Bombay: the textile and tea merchants, the shippers and chandlers, the doctors and bankers, the Parsis, Muslims, Jains and Hindus and certainly from the Sassoons there on the lawn.

'Well now,' Sandhurst went on, 'Major Bannerman has made a fitting reference to the genius of Professor Haffkine . . . Major Bannerman tells me he is now making full use of the genius of that distinguished scientist . . . When the time comes for the history of this plague to be written, Professor Haffkine will occupy the foremost place . . . The discovery of Mr Haffkine has unquestionably, as all the world knows, been instrumental in saving an immense number of our fellow-subjects in India from death.'

At this point, Haffkine should have been rosy with diffident blushes. But Haffkine was not there. Instead, he was on leave, as far away as you could imagine from the plague cities of India; up on windy Boars Hill, being photographically beautified by Angie Acland when he was not being publicly eulogised by the great and good. In fact, as he later suspected, the condition of his praises being sung, especially by his second-in-command, Surgeon-Major Bannerman, was that he was not, in fact, present in person to hear them. Was this opening *in absentia* planned as such? How else to explain its odd timing? Was some sort of temporising going on, a 'reconsideration'? Before he had left India, invited to lecture to the Royal Society by Lord Lister, there had been a suggestion that once Haffkine was satisfied with the running of the Bombay laboratory, he would be appointed as director of an all-India institute of research. But while he was in England, the new Viceroy, Lord Curzon, whom Haffkine thought approved the plan, was having serious misgivings, obsessed above all with the Russian threat and possibly aware that, during Haffkine's time in Calcutta, some papers had put it about that the Russian vaccinator could in fact be a Russian spy.[58] Was it wise, then, to place him a position of such authority? On 25 May 1899, Curzon wrote to the permanent under-secretary of state for India, Arthur Godley, that, since Haffkine 'is Russian . . . it seems a little bizarre to place someone of his nationality at the head of an Indian institute'.[59] To set an example, George Curzon – by some distance the most imperially self-important of all the rulers of the Raj – had himself and his senior staff inoculated. But Curzon let it be known he had not enjoyed the experience. 'They pump into your arm the best part of a wine glass of disgusting fluid which inflames the whole limb, gives you fever . . . you are in acute agony for twenty four hours and in some cases [the vaccination] leaves you seedy for at least five days.'[60]

And then there was William Burney Bannerman, the epitome of IMS military medicine: straight as an arrow; Edinburgh born and bred; posted as army surgeon to Baluchistan and Burma; malaria victim; student of Henry Harvey Littlejohn, the champion of sanitary reform in Edinburgh. In the spring of 1899, Bannerman was deputy sanitary commissioner in Madras. His brief was also to watch for plague outbreaks in the city, which never materialised in any serious number. Later, in the light of what he regarded as a betrayal by his deputy, Haffkine described Bannerman as 'pulled out by the ears by me' from Madras, where he was 'engaged in supervising drains and sewers',[61] and brought to the Plague Research Laboratory in Bombay along with his wife Helen, on the verge of becoming a best-selling author with the publication of *Little Black Sambo*. Haffkine's damning judgement of Bannerman was coloured by the appalling circumstances in which the latter would take over as head of the Laboratory. If it was true that Bannerman's career to date had been that of a conventional hygienist, for whom the martial approach to inspection, segregation and disinfection was public health orthodoxy, it is also true that he underwent a genuine conversion to Haffkine's vaccination alternative. In 1899, he published in the *Indian Medical Gazette* an account of 'The Inoculation of an entire community with Haffkine's Plague Vaccine'. The 'community' was a cantonment of sepoys based near the town of Belgaum (Belagavi) in the western Karnataka. In December 1897, Haffkine inoculated 1,665 men out of a total of 1,761 (and subsequently their families), resulting in a completely plague-free population while the disease was raging through the town a few miles off. 'The above results,' Bannerman wrote, 'seem to point to the truth of Haffkine's expressed opinion that in his prophylactic we have a means of controlling an epidemic of plague and converting it to a manifestation of sporadic cases only.'[62] The caution in

Bannerman's choice of words is telling. He was not about to follow Haffkine's assertions that recurrent outbreaks of plague in towns and villages which had been exhaustively disinfected, sometimes more than once, rendered the orthodoxy futile. Haffkine was beginning to make these challenges explicit. 'Evacuation', he wrote, 'though it may check an outbreak in a particular locality is not a measure which will be of assistance in checking the extension of plague from one locality to another . . . Disinfection [unlike inoculation] is also a measure that cannot be efficacious.'

To the hygienists of the IMS, even those who countenanced or encouraged vaccination as an auxiliary approach, such sentiments were heresy. In their view – set out in the volumes of the Plague Commission under the chairmanship of yet another Edinburgh medic and toxicologist, T. R. Fraser – the two approaches were complementary rather than alternatives. It seems likely that it was the commission that was influential in settling on Bannerman as acting head of the Plague Research Laboratory during Haffkine's leave in Britain, but it could also have raised the possibility that the Scot might be an altogether more suitable chief than the 'Russian'. As far as Haffkine knew, Bannerman was just holding the fort until his own return from leave. It was a devastating shock, then, on arriving back in Bombay, to be informed by a telegram from the government of India on 27 November, that Major Bannerman 'will remain in charge of the Laboratory', setting Haffkine 'free for scientific work'.[63]

Waldemar was no innocent; he understood straight away that 'freeing him' for research was a disingenuous way of displacing him as overall head of the lab. Dismayed and indignant, he sought immediate clarification. To begin with, Surgeon-General Harvey in Calcutta apologised for the brusquely misleading message of the telegram. It had been 'issued in ignorance of the arrangement between the Viceroy and Lord Sandhurst by which your position as head of the Bombay Laboratory was assured pending further employment in a more important branch of scientific research'.[64] Harvey, who addressed Waldemar as 'my dear Haffkine', then came up with a solution by way of job titles. Haffkine was to be director-in-chief; Bannerman, superintendent, in charge of day-to-day administration. The implication was that Bannerman was to report to Haffkine. But Haffkine remained wary that his authority had been compromised. Increasingly, he believed that when those he called 'the gentlemen' wanted to have someone less 'quirky', less 'touchy' keep an eye on you, they did it deviously, inventing new positions and job descriptions. He might have been made a Companion of the Order of the Indian Empire in the Golden Jubilee Honours, but that didn't signify much, since that same decoration had also been bestowed on someone 'whose sole work was to check the groins and armpits of passengers in the Bombay dock'. In 1900, he had become a naturalised subject of the Queen. But that, he well knew, was also no guarantee of social and professional acceptance. Lord Lister might well write in *The Lancet* that 'even opponents have been compelled by the logic and facts to admit that Haffkine's system has triumphed'.[65] In Berlin, Robert Koch sang his praises. But back in India, in the empire of gentlemen, it was evident that when all was said and done, and despite everything he had accomplished, Waldemar Mordekhai Wolff Haffkine was still thought of as a foreign body. His fight, he wrote, was 'not with the rules but also with men who do not like success'.[66]

At some point during the summer of 1899, when he was the cynosure of England, Waldemar Haffkine scissored out newspaper reports from a French court case in Rennes that attracted his attention and, as was his wont, carefully archived them along with material about inoculation. The case was a retrial of a French army officer accused and previously convicted of treasonable espionage, name of Alfred Dreyfus.

viii

Carbolic

In the first years of the twentieth century, Punjab was, in some ways, what it had always been, but in other respects – the pressing concerns of life and death, for instance – not at all the same. It was still the roadstead to the northwest frontier, troops slogging along the Great Trunk Road. Railway platforms were seldom without the clatter of army boots, carriages jammed with Sikh infantry, their *dastars* brightening up the bobbing sea of khaki pith helmets atop the boiled faces of British soldiers. Tribal rebellions on the north-west frontier in Waziristan and Baluchistan were becoming more frequent. And what the Russian empire might or might not do in the next round of the Great Game was never far from the official business of the day in Calcutta or Simla. Beyond the throbbing cities, Lahore and Amritsar, beyond busy towns like Hoshiarpur and Ludhiana, fields of wheat, barley, millet and sugar cane stretched away, watered by the tributaries of the Indus basin and providing farmers with their subsistence, except, that is, when the monsoon failed and drought cracked the earth. This happened twice within five years at the close of the nineteenth century – in 1896 and again in 1899 – not quite as catastrophically in Punjab as further south and west in the Bombay Presidency and the Deccan plateau, but seriously enough, nonetheless, to turn men, women and children into traipsing, fleshless skeletons. The Raj at the height of its self-ordained pomp was, then, also an epic of human misery. The viceroys – especially Lord Curzon – wrung their hands and

Dispatching department, Plague Research Laboratory, Parel, 1902–3.

Media preparation service, PRL, 1902–3.

offered expressions of regret, while making it clear that public assistance must never unbalance the finances of the Raj or where would that lead? As accounting niceties were scrupulously maintained, feral dogs and jackals gnawed on the ribcages of dead cattle, goats and humans, lying this way and that by the roadside.

Like much of western and north-western India, Punjab was on the rack. 'After famine comes the fever' ran a local saying and when the monsoon returned in 1900, buzzing swarms of anopheles mosquitoes bred, bit, and produced an epidemic of malaria. Then, inevitably, came bubonic plague, returning from what had already been a serious outbreak in 1897–8. The first wave of the infection travelled north from Bombay up the coast through Gujarat to Ahmedabad and Karachi. But then, as the rat-and-flea vector moved along with sacks of grain, it penetrated inland country villages, ruralising the epidemic. Bodies weakened by years of dearth were especially vulnerable to plague. Excess mortality, which had ebbed somewhat in 1900, spiked with a vengeance the following year.

At the Parel Plague Research Laboratory, vaccine production was swiftly and spectacularly ramped up to meet the urgent demand. Nominally director-in-chief, Haffkine was still troubled about the extent of his true authority, but it was he who had created the world's first large-scale vaccine production line. By the autumn of 1899, half a million doses of the vaccine had already been delivered; by June 1902, the figure was 2,877,038.[1] More than 2 million of those doses had gone to Indians, while – often overlooked – 200,000 had been exported to Africa and 110,000 to Mauritius, where the plague had struck the island in 1899. Between April 1902 and June 1903, a further 3 million doses had been produced, 1,373,880 in the month of February 1903 alone.[2] This was an astonishing and unprecedented achievement. The Plague Research Laboratory had begun with a staff of just 53, expanding in two years to around 200, though at least 20 of those were boys employed in the not unimportant job of bottle labellers.[3] Nothing on this scale had been achieved anywhere else in the world, let alone in such a short period. But for Haffkine, this was just the beginning. Though there

Senior staff at the PRL, 1902–3: William Bannerman, third from left, standing; M. K. Pansare at far left, standing; R. J. Kapadia at far right, standing; J. P. Pocha at far left, seated; Nusserwanji Surveyor front row right; Haffkine, middle row second right.

was no hope of the government agreeing to the staff of 7,000 that he calculated was needed for the expansion, a plan was drawn up in 1902 to supply 6 million more doses over the next few years: a breathtaking ambition for the early twentieth century.[4] Albert Calmette, director of the Lille branch of the Pasteur Institute, produced the serum Yersin had created in Bombay, which turned out to be more effective therapeutically than as a prophylactic and was used in a plague outbreak in Porto in 1899.[5] Haffkine himself recommended the Yersin-Calmette serum as a 'curative' should the vaccine not succeed in acting as prophylactic in some patients. But the Lille output was modest compared to Haffkine's at Parel. Other bacteriological laboratories and stations – the Pasteur in Paris, the Russian version in St Petersburg, Koch's laboratory in Berlin – were all making vaccines against rabies, anthrax and, more recently, typhoid, while working on a possible vaccine against tuberculosis.

A small laboratory in Kasauli in the Himalayan foothills was producing rabies antidotes, entitling itself to be known as a Pasteur Institute. But none of those establishments compared in scale and purpose with what Haffkine had created in Bombay. By late 1902, 10,000 doses were being produced at Parel every day. A photograph from the period shows a large room, emptied of all the stacks of bureaucratic files which once climbed the walls – reports on customs revenues and land taxes, harbour-dredging, the extension of roads and railways, troop movements and famine relief – all now given over to long tables on which stood thousands of incubation flasks growing the cultures from which the plague vaccine was prepared. Other pictures – evidently commissioned by Haffkine – document the non-stop production.

And though the men at the top were, predictably, white IMS officers – Bannerman, Liston, Gibson – Haffkine hired four Indians too. These were M. K. Pansare and three Parsis: J. P. Pocha, Nusserwanji Surveyor and, as senior surgeon, in effect his scientific deputy, R. J. Kapadia. If he had had his way, Haffkine would have trained up more Indian colleagues, not least because they would be valuable in helping persuade the reticent and because, as in the case of Surveyor, he instinctively thought of them as allies amid the overbearing presence of the IMS men. Psychologically, he never lost the uneasiness of the outsider. Nonetheless, under his directorship, Parel was going international. Shipments of the 'Haffkine lymph' were going to ports around the Indian ocean, south Asian world where plague had broken out: Ceylon (Sri Lanka), Zanzibar, South Africa, the Caribbean, Port Arthur, Japan, Hong Kong, Australia, New Zealand and Hawaii. Faced with news of a plague outbreak in the Siberian east and Manchuria, and anxieties that it might travel west with fur traders, the doctor-playwright Anton Chekhov wrote to his friend Aleksei Suvorin in August 1899 that 'the plague is not so very terrible here. We already have inoculations that have proved effective, and for which we are obliged, to a Russian, Dr Haffkine. In Russia this is the most unfamous man but in England he is known as a great benefactor.'[6]

The geographical spread of the disease in its peak years of 1901–2 and its entrenchment in rural India put a premium on production at Parel. Affected communities were not only harder to reach but also could be more resistant to persuasion, especially in Punjab. Temporary bans on pilgrimages, festivals and fairs where traditional healers, hakims, were usually accessible engendered more fear and anger. Folk remedies in these years abounded, especially in the countryside where doctors practising western medicine were often thought to be bringing, rather than healing, plague, exploiting local anxieties to make money.[7] The most tradition-bound communities set a ring of stakes around the perimeter of their villages, topping the wooden spikes with the heads of demons, or they painted animals on walls and doors while camphor and neem leaves were burned as exorcising fumigation. Folk medicine went from *vaidyas* offering ayurvedic remedies, including cow dung or urine as disinfectants, all the way down to the application of transversely slit frogs directly on to the darkly swelling buboes. The most notorious case of charismatic therapy, successful enough to develop into a minor cult in 1903, was that of a woman known as Bhagirathi who claimed to be an embodiment of the goddess Kali and who treated plague by biting buboes and then pressing her toes hard on the wound. Her following was serious enough for her to be arrested as a menace to public health, despite being defended by at least one Indian newspaper as a practitioner of the traditional art of healing as she understood it.[8]

Appreciating that the all-important work of persuasion could be sabotaged if folk remedies were all grandly dismissed as the product of Indian backwardness, Haffkine made a point of resisting the usual stereotypes and caricatures. 'The natives here are, of course, not the prodigies of imbecility, superstition and ingratitude which some good people amuse themselves in representing them . . . stories of superstitions and fabrications attributed to the people are sometimes manufactured, sometimes favoured and exaggerated.'[9] As much as the regimen of the IMS, for which he now formally worked, allowed, his staff at Parel were mostly Indian. The same necessarily held true in inoculation missions to regions and districts remote from Bombay, not least because the diversity of Indian languages necessitated local assistance. But it was also crucial to have mediators on the ground who could reassure locals of the safety of the vaccines. Persuasiveness was the critical factor, with vaccine acceptance presented as the opposite of draconian, unscientific and ultimately futile regimes of evacuation and segregation. At Cawnpore in April 1900, the segregation camp had been violently attacked, with five dead before the riot was suppressed. Riots had also broken out in Sialkot and Gurdaspur against segregation and evacuation. At Shahrada, there had been a pitched battle between sepoy soldiers, reinforced by the police, and a crowd of 300 Jat Sikhs wielding swords and knives. At Sankhatra, a medical officer and two hospital assistants had been killed and the entire plague camp of segregated people burned down.[10]

In the light of what was turning into a medical civil war, Haffkine did his best to argue that only a mass vaccination campaign could pre-empt those incendiary confrontations. 'Inoculation is the *only* measure that can hope to induce the people to adopt [measures] on a scale at all adequate to the situation. Inoculation can be brought within reach of those whom we have to protect from the plague.'[11] He did his best to instil that spirit into the vaccination teams working in the field. At one point in 1901, beating Alice Corthorn's standing record in Dharwar, Surgeon-Captain Edward Wilkinson vaccinated 3,200 – the entire population of a village – in a single day.[12] So there was good reason to feel confident about the results of the campaign in the hard-hit Punjab. Comparative evidence from the 1897-8 epidemic was irrefutable. In Jullundur district, for example, of 134 inoculated, there had been only one case, while among 1,357 non-inoculated, 42 had contracted the disease. (There was still at least a 70 per cent mortality rate.) In another village, Maral, of the 865 inoculated, there were just six cases, none of them fatal.[13] The surgeon-general who reported these statistics, C. H. James, was also becoming a sceptic of the usual approach to plague containment, not least because he, like others in the medical service, were noticing that, if anything, intensive disinfection triggered an exodus of plague-bearing rats to neighbouring villages. 'What is the good', he wrote, 'of segregating and keeping under rigid police supervision a few infected people when rats are carrying the disease to almost every part of a village?'[14] It was even believed (not altogether fancifully) that the rats could communicate with each other on ways to evade danger. Challenging another of the reigning orthodoxies of disinfection, Haffkine had his researchers at Parel examine hangings, clothing and furniture in plague-struck streets and buildings for the presence of the bacillus, and, needless to say, found none on any occasion. In 1907, in a speech to the Liverpool School of Tropical Medicine, Haffkine would witheringly dismiss the obstinate assumption that disinfection and segregation could be an effective way to contain the plague as no better than a 'placebo'.[15]

What temerity! Negative evidence was not going to move the IMS establishment to revolutionise the public health practices established as hygiene dogma throughout the last century, replacing military disinfection with missions of mass inoculation. In other respects too, colonial hierarchy held firm. There would always be a British chief medical officer on site and often another local IMS attending physician as well. And by 1901, William Glen Liston and Maitland Gibson been joined by three more Netley-trained British medics. 'The gentlemen now in power will keep out any man not of their set,' Haffkine later wrote.[16] He also knew that approaches by the rulers of two princely states, Maharaja Rana Ram Singh of Dholpur in Rajasthan and Maharaja Bhupinder Singh of Patiala, to create their own independently well-funded institute with Haffkine as its head had been received by British officials as more evidence that the 'foreign' scientist was impatient to do things his own way.

The dramatic spike in plague in the winter of 1901–2 meant that Haffkine had more urgent things to think about than the authority of his position. He may have succeeded in making 10,000 doses a day, but the demand was now for 80,000. 'I have daily calls from the population begging to be inoculated,' he wrote. There were some elements in the procedure that he evidently thought could be changed for the better without compromising safety or efficacy. Following Institut Pasteur protocols, the bacillus was now cultured in a medium of agar and water rather than goat broth. But more significantly, devitalising the vaccine with the addition of a 0.5 per cent carbolic solution was replaced by heat sterilisation. Pasteur procedure – which for all his differences with Alexandre Yersin, Haffkine still regarded as the gold standard – had demonstrated conclusively that heating the culture to around 60 degrees Celsius for fifteen minutes was sufficient for reliable sterilisation. The exact degree of heat was a fine judgement. Too hot and the capacity of the devitalised vaccine to engender an immune reaction might be compromised; not hot enough and the vaccine could be dangerous. The disadvantage of depending on carbolic phenyl was the additional time: at least fifteen days before its anti-toxic effect was safe enough to complete vaccine preparation. After careful trials on lab animals, Haffkine took the decision to follow Pasteur protocols established in 1900 and dispense with phenyl. Later, this would be held by the IMS to be a sign of slack procedures; worse, the Bombay health authorities would claim that he had failed to inform them of the change. That charge was untrue. Haffkine had indeed notified them, but hearing nothing back on the matter, took silence as assent.

To the devotees of the church of carbolic, such alterations were heresy, a symptom of wayward conduct. Since it had first been applied in 1864 by Joseph Lister to the site of open fractures to prevent sepsis (hitherto thought of as an intrinsic stage of the healing process), carbolic had been revered, especially in Britain, as a medical miracle. There were other antiseptics – perchloride of mercury for example – but none of them occupied such a central place as carbolic acid in what Lister, an early convert to Pasteur's germ theory of infection, called his 'antiseptic system'. Before Lister, carbolic had been used to weaken the stench from animal and human cadavers, and lessen the reek of raw sewage that had made Londoners gag in the Great Stink of 1857. But once the antiseptic system was seen to save lives, its indispensable ingredient, produced from coal tar, was thought of as industry's inadvertent reparation for the effluent-soaked world it had brought about. Whoever in Lever Brothers dreamed up the brand 'Lifebuoy' for the soap they first put on the market in 1894 knew what they were doing. This was not just the stuff of vigilant hygiene; Lifebuoy could make the difference, its brand promotion claimed, between life and death. Every time that hand-washing becomes a matter of social and medical urgency, the 'soap with a mission' (as it was called on its recent reintroduction in the UK) gets a new lease of life.

In India, the brilliant coral-coloured carbolic soap has never gone away, surviving the demise of empire, notwithstanding the fact that its vermillion shade was the tint of sahib swagger. It is still all-Indian, Bollywood film stars booming and squealing its salutary virtues, inhaling the bracing scent that says CLEAN. In the early years of its arrival in Asia, it was held up as the typical gift *par excellence* that the imperial home made to its far-off subjects, subsisting in unwholesome grime. Before the aetiology of infection was properly understood, Asiatic filth was thought to be the matrix in which plague, not just faecally contaminated, cholera-generating water, was bred. Away with it, then, in a thorough sluice of carbolic. The first thing that happened to infected patients, or even those suspected of harbouring infection, when they arrived in segregation camps was a bath of red carbolic. The same dousing took place for any survivors on their departure. Meanwhile houses, warehouses, inside and out, were all saturated with carbolic before being allowed to dry and the disinfecting completed with limewash or whitewash. Even the disinfection gangs that had worked a street or district were themselves carbolicised at the end of the day before being dismissed.

Dispensing with carbolic, then, *especially* in favour of some French alternative promoted in Paris, was, for the guardians of the antiseptic system, not just irresponsible protocol; it was hygiene heresy, a flirtation with dubious foreign experiment. There were bound to be casualties.

In the autumn of 1902, around a thousand souls lived in the Punjabi village of Malkowal. The cultivators raised the usual north Indian cereals – wheat, barley and millet – watered by the river Beas and its tributary creeks and ditches. Malkowal's one-storey mud-brick houses were built close together, forming small compounds in which goats and chickens roamed along with the village children. The houses were made for shade rather than light or air, their windows no more than simple openings, unglazed and mostly unshuttered. Wooden double doors were often painted pale blue and fitted with curling iron handles. Earthenware pots of different sizes, the largest very big indeed, stood inside and out, holding grain: a kitchen call for the local population of rats. This virtually guaranteed that plague would visit this corner of north-eastern Punjab, as indeed it had in 1897–8, when the Hoshiarpur district, in which Malkowal was located, suffered one of the severest outbreaks in the province.

In October 1902, with the epidemic raging once more in central and northern Punjab, an IMS team, under the supervision of Surgeon-Captain A. M. Elliot, formerly civil surgeon at Bijapur, but who had two years' experience performing inoculations, arrived in Malkowal to deliver doses to as many in the village as would accept the vaccine. As was nearly always the case in such small communities, there was some initial reluctance to take up the offer, but moved by the willingness of the village elders to set an example, 107 received the vaccine. On 30 October, in a field on the edge of the village, Elliot began the inoculation.

A week later, beginning on 6 November, some of the vaccinated began to show the unmistakeable symptoms of tetanus poisoning: violent muscle spasms locking the back into an iron arch; arms rigid; faces frozen in *risus sardonicus,* the rictus of lockjaw; and, before a day or two had passed, the cardiac paralysis which ended their lives. By the time the last such case appeared on 9 November, nineteen of the people of Malkowal were dead. It was immediately noted by Elliot that all the fatally affected victims had received vaccine from a single bottle marked 53N. None of the remainder of the inoculated had suffered any ill effects whatsoever. But the entire village was gripped by terror and grief. Elliot knew right away that the disaster might have killed not just nineteen country people but the entire Indian vaccination campaign.

It took a further week before news of the calamity reached Parel. It fell on Haffkine and his staff like a thunderbolt. Notwithstanding the fact that 80 per cent of those inoculated at Malkowal were healthy, an immediate halt was put on vaccination. The grand plan to inoculate a further 6 million was abruptly abandoned. A committee of inquiry was set up consisting of just three members, all drawn from the heart of the British-Indian establishment. Two of them, the chairman Sir Lawrence Jenkins, the chief justice of Bombay, and the surgeon Gerald Bomford, principal of Grant Medical College, had no bacteriological expertise, either theoretical or practical. The third member, Lieutenant-Colonel David Semple, was the head of the rabies laboratory at Kasauli. Semple had been a student at Netley and had worked on a typhoid vaccine. But his own rabies vaccine, developed from churned sheep's brains, would have such alarming side effects, including paralysis, that it would end up being discredited as dangerous.

What had happened at Malkowal inevitably turned political. The 'Haffkine lymph', which had saved hundreds of thousands if not millions of lives, was now thought by those who had all along been sceptical to be dubious if not lethal. Government and the IMS concluded that since it had always been difficult to persuade Indians of the safety and efficacy of the vaccine, the task was now well-nigh impossible. It might even foment political discontent from all the sections of the Indian community who had had doubts about British plague management from the beginning. Lord Curzon raged as if Haffkine were personally responsible for undermining not just his own championship by example of the vaccine, but the whole claim of the British Raj to care for the welfare of its Indian subjects. 'Haffkine', he wrote to Arthur Godley at the India Office, 'ought to be hung for his folly. If this appalling catastrophe comes out there could be an end to all inoculation and India and the cause of science set back . . . In my opinion he deserves to be tried for his life.'[17]

But such was the wildfire spread of plague in Punjab that in 1903 vaccination was resumed, and a 'Health Exhibition' staged in Bombay to reassure the anxious. But a scapegoat for the Malkowal disaster was required and, of course, presumption of guilt immediately fell on Waldemar Haffkine. Even before the committee had begun to take evidence, an official wrote that 'it would very probably be necessary to dismiss him'. After hearing, in great detail, from Elliot, the Jenkins Commission reported to the Indian government in April 1903 concluding that 'the contamination was introduced before the bottle was opened'. Their reasoning was simply that if the vaccines had infected nineteen victims, it was necessary for there to have been an abundance of the toxin 'for some considerable time'. The flasks including bottle 53N had been prepared and sealed forty one days before the Malkowal vaccinations, and twenty six days after the vaccine had begun to be shipped from Parel. The clear implication was that the contamination had taken place at source, and although the brief was not to pass judgement on who should be held responsible for this fatal lapse, it was noted that the adoption of agar water culture and the omission of carbolic acid from sterilisation had been the decision of the director-in-chief. No one seemed to wonder, out loud at least, why other bottles which had an identical form and preparatory timing proved harmless at Malkowal.

In his testimony to the Commission, Haffkine had challenged these idle and, as he evidently thought, ignorant assumptions.[18] Given the daunting challenge of raising vaccine production from 10,000 bottles a day to 85,000, the most expeditious means of manufacture had been adopted, but without, he stressed again and again, any compromise to safety. Agar-based culture medium was 'the simplest, easiest and quickest'. Carbolising the vaccine after heat sterilisation at 70 degrees Celsius, he patiently explained, was not only gratuitous, but 'would have involved opening bottles', not only holding up production but potentially exposing the vaccine to greater, not lesser, risk of contamination. The anti-plague serum developed by Albert Calmette reliably used heat sterilisation alone, an observation which, predictably, failed to impress the Jenkins commissioners.

What of other testimony from colleagues at Parel? As Haffkine would discover, only Maitland Gibson had argued decisively in favour of the likelihood of contamination having occurred in the village, rather than at production source. Liston and Bannerman had equivocated and then, shockingly for Haffkine, had gone along with the premature judgement of the Commission. But Bannerman, not so long ago his trained protégé and travelling assistant, at least in Haffkine's understandably agitated mind, had turned from a colleague into a usurper. During May and early June 1903, relations between the two men descended swiftly into unedifying, bitter intra-lab warfare. Suspecting Bannerman of wanting to take advantage of his predicament to elevate his own position as superintendent, Haffkine invited him to spell out exactly what he thought his job specification was. Bannerman fired back wondering what kind of detail he was supposed to specify: hours, inspections, whereabouts? He then became testy when Haffkine failed to reply to three additional notes. 'Am I to understand that you do not mean to answer at all or are you taking time to consider matters? . . . you have not even acknowledged receipt . . . I shall be compelled to lay my case before the Surgeon-General pointing out the unworkable nature of the present arrangement.'[19] When he did so, Bannerman made it clear that he was going to assume 'entire control' over everything that was not pure research: 'discipline' of staff, finances, orders for equipment, oversight of procedures and protocols, provision of lab animals and, most important of all, final say over appointments. At this point, the last remnant of mutual trust disappeared. Haffkine accused the superintendent of side-stepping him by communicating directly with the surgeon-general in Calcutta. Bannerman counter-accused Haffkine of undermining his position by telling staff to ignore anything coming out of the superintendent's office and report to him instead. To Bannerman's insistence on 'entire control', Haffkine wrote to Surgeon-General Harvey that, aside from his research and vaccine production, he thought of himself as a teacher, committed to training a whole cohort of Indians who would follow. But as long as his every move was circumscribed by managers like Bannerman, his teaching could not be properly sustained. As for Bannerman's insistence of 'entire control', 'not only can none of the above [requirements] be for one moment entertained but the presence in the lab of a man with Major Bannerman's demands and aspirations has unavoidably as will be evident to anyone produced a disorganising and paralysing effect . . . it will be evident I hope to Government that this arrangement has to be put to an end.'[20]

It did indeed end, but not the way Haffkine wished. Malkowal had fatally damaged his authority and he was in no position to make managerial demands. At the end of 1903, desperate to lay his side of the story before the Bombay government, Haffkine wrote to the governor's military secretary asking for 'Private Entry' to the formal 'Levée' to be held on 22 December, perhaps to have a chance to bend the ear of Lord Lamington.[21] Supposing this happened (and it would have been unlikely), it was, in any case, of no avail. Haffkine was told to take a leave of absence for one year on half pay, pending the conclusion of further investigation. But official Anglo-India had made its mind up. Lord George Hamilton, the Secretary of State for India, began a letter to Curzon claiming that he would be 'sorry if you parted company with Haffkine. He is a genius with quite exceptional powers of research', but then added, as if the rogue scientist had learned his lesson the hard way and might become more respectful of IMS conventions, 'I think he will be so humiliated and depressed as the result of his own folly that you will find him for the future a valuable servant as he has to wipe out the discredit which will for years be attached to his murderous serum.'[22]

Hamilton was right about one thing: Haffkine did indeed feel his authority and reputation now lay in ruins. Just two years before, he had been a figure commanding admiration, almost to the point of reverence: invited to take on the founding and directorship of an all-India Research University, funded by the Tata endowment. The government of India had gone along with the scheme to the point of asking him to draw up detailed plans for its operation.[23] Now the Viceroy wanted him tried, convicted and executed. But this personal blow was as nothing (so he told himself) compared to the destruction of his life's mission: the saving of untold numbers with vaccines. His battle all along had been to replace one kind of authority by another. What he had encountered in India was the empire of drastic disinfection, applied, for the most part, invasively, indiscriminately and coercively. He had hoped to replace it with the authority of science, specifically bacteriology, delivered through inoculation to those who consented to receive it. Persuasion rather than coercion, together with the measurable demonstration of protection and mitigation, would transform the lives of those most vulnerable to terrifying waves of infection: the poor of Asia.[24] Now, despite the resumption of the vaccine campaign in Punjab, the long-term fate of mass vaccination was in dire jeopardy. And not because of a single disaster, horribly fatal though that had been, but through the faulty reasoning and unjustified conclusions of the inquiry that had laid blame for the deaths at Malkowal at the door of the laboratory producing the vaccine.

This was all the more galling because the evidence presented to the committee of inquiry seemed overwhelmingly to point to contamination not at source, but during the village inoculation itself. To begin with, none of the other four bottles from which the vaccine was drawn produced any ill effects, yet all had come from the same batch. Second, tetanus toxin which had been allowed to develop for any extended time – certainly the forty one days since batch 53N had been completed at Parel – invariably gave off an unmistakeably strong, tell-tale odour. But Elliot, who routinely examined newly opened bottles for just such signs of contamination, had stated that there had been no such smell coming from the toxic bottle. Someone on the committee had then suggested that since odours of different kinds came off from vaccine bottles, it was no wonder Elliot did not smell anything untoward. Startled by the obtuseness, which must have seemed wilful to him, Haffkine had to point out (three years later in his long letters to Godley) that *differences* in odours were not the point. Bottle 53N had had no odour *at all:* thus it could not have had a toxin introduced at Parel, growing steadily for the period of more than a month before the inoculations were administered. Third, the one-week delay before the victims showed signs of tetanus strongly suggested that its growth had begun on site. Had there been a lapse at Parel, the correspondingly abundant growth of bacteria would have produced an immediate onset of symptoms and certainly not longer than a day after the inoculation.

But the evidence which ought to have clinched the case that the fatal mishap occurred in the field at Malkowal, and not in Haffkine's lab, was revealed during Elliot's detailed account of the preparation of the vaccines prior to inoculation. Between the opening of the first bottle and the filling of syringes, there had been what might have seemed a minor mishap. The compounding assistant, Narinder Singh (assigned to bottle-shaking, opening and syringe-filling), had dropped the dissecting forceps used to open the bottle before completely removing the stopper. He then retrieved them from where they had fallen on the field, giving them a perfunctory rinse in carbolic before continuing to extract the rubber stopper. Though few had formal medical or scientific training, Indian compounders like Narinder Singh were an indispensable part of the peripatetic vaccination team. They acted not only as assistants to the travelling, usually (though not invariably) British lieutenant or captain-surgeon, but also as interpreters. The willingness of local village elders, the *tahsildars* and the *qadis*, to accept inoculation for their village or small town, and feel reassured that frightening rumours were misplaced, depended critically on these men and their familiarity with the local community. And there was no reason, in fact, for Narinder Singh to have supposed he was doing anything wrong since he was following the directions of the locally supplied *Plague*

Manual for Punjab, which dealt with such occurrences. Whoever wrote those instructions could not have been aware that it took at least fifteen hours of immersion in carbolic acid before a contaminated object was safely disinfected. Haffkine's little blue book of instructions meant to be supplied to all inoculating personnel could not have been clearer (or more exhaustively detailed) on the utmost importance of on-site sterilisation. It specified that the forceps 'are heated in a flame of the spirit lamp immediately before use and guarded from contact with any unsterilised object'. Likewise, 'the stopper and neck of the bottle are passed through the flame of the lamp so as to cause a slight singe on the stopper which is then withdrawn with the heated forceps . . . After the bottle is opened any contact between its mouth and any other unsterilised object is to be carefully avoided and if contact inadvertently occurs to the mouth is to be heated again in the flame.'[25] But it would emerge later that the Haffkine blue book was not invariably distributed to travelling inoculation teams, and even when it was available, was often disregarded in favour of the older government-issued manual with its absolute trust in carbolic. After they had been picked off the dirt of the Malkowal field, the forceps had not been sterilised in the spirit lamp flame. But it took three more years before Haffkine would learn of this critical detail.

To put to rest any doubts about the conclusions of the Jenkins committee, a second investigation was commissioned from the Lister Institute of Preventive Medicine in England. Experiments were conducted in England using lab animals and the remains of the original contaminated bottle, though by the time that the Institute got hold of it, further impurities vitiated any of their conclusions. No matter. In November 1904, they reported to the government that they concurred in most salient respects with the report of the Jenkins commission and that 'in all probability the tetanus was at time of the inoculation [already] in the bottle'. Just how or when it got there, however, they could not say. Oddly, but tellingly, given the IMS conviction that the omission of carbolic had something, perhaps everything, to do with the accident, the Lister Institute was also asked to make a judgement as to whether Haffkine's adoption of agar medium and heat sterilisation was as safe and effective as the 'standard' (so they called it) procedure. The Institute wrote – unsatisfyingly to Haffkine's critics – that their animal experiments suggested that indeed it was.

The concession was too guarded and too late to affect Haffkine's fate. On 24 November 1904, he was removed from his post as director-in-chief of the Plague Research Laboratory. William Bannerman, recalled from Madras, resumed his post as superintendent, but this now, doubtless to Haffkine's chagrin, clearly meant unchallenged head of the lab. Unable to defend himself in India, Haffkine shipped back to England, moving into rooms at the St Ermin's Hotel, a recently converted mansion block close to St James's Park. He was not giving up or giving in. Over the next year, he returned to the two cradles of his science. In Paris, he gave a course of lectures at the Institut Pasteur on his plague vaccine as though nothing had compromised its value or his own standing. In late 1905, he travelled to Odessa, visiting the university where his life as a scientist had begun. But the most compelling reason for his presence there was the horrifying pogrom that had taken place in October in which 600 Jews had been murdered, thousands brutally assaulted, homes and property burned or demolished. Haffkine would have remembered, tragically, the part he had played in communal self-defence in 1881.

Back in England, in February and March 1906 he wrote two long letters to the governors of the Lister Institute pointing out the glaring discrepancies between the evidence of their own investigations and their conclusion. In fact, the Institute had already started backtracking from its original endorsement of the Jenkins committee, albeit in the most equivocally mumbling fashion. While still maintaining that the vaccine fluid 'probably became contaminated with tetanus *before* opening', they could 'not exclude the possibility' that it had happened at the village. In May, in response to Haffkine's detailed exposure of these inconsistencies, the Lister governing body shuffled backwards even further, though not altogether repudiating their original verdict. 'While of the opinion that tetanus impurity was primarily in the fluid [that is, before opening the bottle] the governing body did not feel justified in asserting this as proven fact.' Quite so. Furthermore, 'the governing body regrets that in their report of the 24th November [1904] they inadvertently referred to their conclusion as the same as that of the Commission.' They were prepared to consider the possibility that tetanus spores might have settled between stopper and the rim of the bottle (without alluding to the dropped forceps) even though such considerations, they still felt, were 'cancelled out' by others (not listed). In the circumstances, they thought that 'Mr Haffkine was right to claim the benefit of the doubt'.

Three successive sets of conclusions, each one retracting something from the last, was embarrassing for the Lister Institute investigators, but their last concession was still a long way short of exoneration and it did little good for Haffkine's cause, other than to make him even more determined to battle for vindication. He now realised that official instructions ordering inoculators to abandon the Parel-Pasteur heat sterilising protocols in favour of carbolic had not only been unhelpful but had actually *caused* the disaster.[26] In the light of the Lister quasi-retraction, Haffkine fired off letters to the India Office; one of them from the Pasteur Institute in Paris, a lengthy *apologia pro sua vita* including a moving statement of vocation about 'the saving of life which has been given to me to see achieved', but still 'insignificant' 'compared to the enormous amount of mortality that might be obviated' if he could but be given the resources, and trust, and the power to appoint junior scientists and physicians he judged up to the mighty task.[27] Predictably, these kind of high-minded, impassioned appeals ran into the brick wall of official imperturbability. When Arthur Godley at the India Office declared that it should suffice that the government would publish the findings of the Jenkins Commission, Haffkine, astounded, replied that that report had endorsed the findings *against* him, and indeed the facts of the matter, so how could he be expected to be content with its publication? Godley was unmoved. It got worse when Haffkine, accepting that he would be unlikely to be allowed to return to Parel, made it clear that he expected to be in charge of whatever institution to which he might be transferred. Moreover, 'in the event of my work necessitating assistants and Government allowing these to me it is of importance . . . that their recommendation for the appointment come from me.'[28] It was swiftly made clear to him that nothing of the sort would be forthcoming. Godley wrote that Haffkine's salary would be restored; he was free to do research and publish, provided he report findings to the Indian government and seek their consent. But, 'they are unable to agree that you should be a scientific adviser in direct communication with them or to improve your salary, status and prospective pension or to empower you to appoint and exchange your assistants.'[29]

All this – the banishment from the Bombay laboratory and vaccine production facility he had created, the cancellation of any prospect of his leading an institute, his demotion to the status of a researcher attached to Calcutta General Hospital – was the best Haffkine could expect from the government, even as the policy of segregation and coercive disinfection to which it had reverted was doing nothing to arrest the spread of plague. Seen from the Viceregal House, it was useful, after all, to have a foreigner, even a naturalised one (since some people only naturalise on the surface), be held accountable for calamity.

But officialdom then made a serious mistake. On 1 December 1906, the same day as Godley's dismissive letter, the collection of documents relating to Malkowal, including the Jenkins report, was, as promised, published as a special supplement to the *Gazette of India*. Sitting in his panelled room in the India Office, Arthur Godley must have imagined that this would settle the matter once and for all. But this was the first time that Haffkine learned of the dropped forceps and the dramatic, instantly vindicating detail jolted him out of the exhausted demoralisation that increasingly overcame him. Now the injustice seemed more unforgivably outrageous than ever, not least because he had been so painstaking about protocols of sterilisation and had replaced cork stoppers on the phials with India rubber precisely because the former could not be reliably sterilised.[30] At the beginning of February 1907, those same papers were published in *The Journal of Tropical Medicine and Hygiene,* prefaced by a fiercely critical comment by the editor, James Cantlie, on the way in which the whole episode had been dealt with. The *British Medical Journal*, however, took a different tone, echoing the exasperation of the IMS that the matter was still being contested. 'It is difficult', the *BMJ* commented, 'to conceive how the matter could have been investigated more thoroughly and more impartially and we think Mr Haffkine would be well advised to allow the incident to be forgotten.'[31]

The new revelations had the opposite effect. William Ritchie Simpson, who as chief medical officer in Calcutta had been Haffkine's champion, friend and occasional travelling colleague in the cholera vaccine campaign of 1894–5, and who was now professor of tropical medicine at King's College, London, wrote to the *BMJ* that the evidence produced by the Lister report made it clear that the assumption that a long growth of the toxin was necessary for fatal contamination had been proved wrong. He added that since this error was the sole basis for asserting it must have begun at Parel, a 'grave injustice' had been done to Haffkine.

It was probably Simpson who suggested Haffkine write to his friend Sir Ronald Ross, who, in 1902, had become Britain's first Nobel Prize winner for his work on the life cycle of the malarial parasite and its transmission through the bite of mosquitoes. Haffkine was not unknown to Ross. In 1895, when as Surgeon-Major Ross he was struggling to contain a cholera outbreak at Bangalore, he knew, as he wrote in his memoir, that 'W. M. W. Haffkine had arrived in India in 1893 from the Pasteur Institute bringing with him his famous anti-cholera vaccine and was now trying it in India on a large scale'. But much as he would have liked to use the vaccine at Bangalore, 'I was not able to . . . because my laboratory and staff were not nearly developed enough to give it to the large number of people while the people themselves would have objected in those days'.[32] Relations between the two of them were not uncomplicated. In 1897, when Ross was struggling to get Indian government support for his work on malaria and believed (without much justification) that Haffkine had the ear of officialdom, he suspected that reservations on the part of the bacteriologist meant he could not look to him for an influential endorsement. But if there had been some coolness between the two scientists, it had disappeared by the late spring of 1899 when Haffkine went to Liverpool for the inauguration of a tropical diseases department at the University College, with Ross installed as its first lecturer.

In late 1904, Haffkine had written to Ross, studiously avoiding any reference to Malkowal and his own subsequent fate. When he did raise the issue early in 1907, he was careful to do so in the most tentatively respectful tone. Might Ross 'care to contribute' to the debate by 'making known the view you may form on the matter?' He might indeed. And with that 'contribution', everything changed.

Much of what followed happened because Ross had come to see in Haffkine a fellow fighter for the cause of science against what he would call 'administrative barbarism'. Theirs, he thought,

Sir Ronald Ross

was a natural comradeship of outsiders, those whose work had been beset, over and again, by the indifference, ignorance, ingratitude and downright hostility of the imperial establishment: a tyranny of nonentities. 'Governments', he wrote in his memoir, 'love to appoint Tom, Dick and Harry to their most important scientific posts and then look upon them as true prophets . . . [whereas] very often they do not possess the brains either to make sound discoveries themselves or even understand them when made by others.'[33]

On the face of it, though, Ross's career seemed to have been an exemplary imperial progress. He had been born in Almora, in the Kumaeon hills bordering Nepal, in 1857, three days after the outbreak of the great sepoy insurrection which for a while looked as though it might bring down the whole power of British India. His pedigree was Clan Ross of Ayrshire; an ancestral earl had died on the field of battle in 1333. More to the point, like so many of the Scottish gentry, the Ross family fortunes had steeply declined in the seventeenth and early eighteenth centuries, so that when an opportunity to repair them in India had arisen, the Rosses turned into East India Company condottieri. A grandfather had served in John Company's armies and, thanks to the usual bloodily accumulated loot, had bought back some of their land in Ayrshire. Ronald's father, General Campbell Claye Grant Ross, became a commander of a Gurkha regiment seeing action against Pashtun tribesmen. The family was notoriously hot tempered. A great-uncle had been shot dead in a duel on Blackheath by his own cousin.

The Indian Rosses may have been sanguinary imperialists but they were not philistines. When he was not on horse, shooting and being shot at by the Indians, Campbell Claye was busy sketching their native landscape. (Figures, his son added, were not father's forte.) In the evening there was family music: flute trios. Ronald was the eldest of ten children, precociously gifted in every way. When he looked back on his upcountry childhood, he remembered, romantically, the taste of dal and mahseer fish; the gentle whoosh of punkahs; the sight of blooming rhododendrons; the dawn sun burning off the mist in the valley below the hills while a drowsy leopard snoozed on the verandah.

Inevitably, the Indo-Alpine idyll came to an end, in his case at the age of eight when he was sent to the Isle of Wight to live with an uncle and aunt, and attend a nearby dame school, followed by a succession of schools where he made friends with boys called Dashwood and Binns. Ronald's reading, once he had worked through the Bible, was Shakespeare and Marlowe followed by *Don Quixote*, Milton and Homer. He also began to paint. A watercolour of a sailing boat on the Solent, painted when he was about twelve, is luminous with pearly marine light. At his second school, Springhill near Southampton, Ronald was given a cucumber frame in which the already zoologically fascinated boy observed the habits of lizards, frogs and snakes. Though he was judged 'moony' by his teachers, they could see that there was almost nothing Ronald could not do well, except, perhaps, algebra. His geometry, on the other hand, was so excellent that it brought him a book prize called *The Orbs of Heaven* which, rediscovered many years later, made him believe that his true vocation must be to solve the riddles of higher mathematics. At fourteen, he won the Oxford and Cambridge Local Examinations prize for drawing, submitting a hastily dashed-off copy of a 'Torchbearer', then thought to be by Raphael. An artist, then? A writer of romances? A poet?

Inevitably, the general had other ideas. Music and painting were all very well; he dabbled in them himself, but a life must be made of sterner stuff. Very well, Ronald replied, how about the army or navy? But the general was firm in his view that Ronald ought to be a physician in the Indian Medical Service. Ross senior was periodically visited by bouts of malarial fever and his son had witnessed the distressing sight of him shaking with chills before burning with sweat. Ronald had no wish to be a medic of any sort. The muses were calling. But such directions were not his to decide. Off to Bart's Hospital School went the seventeen-year-old in 1874, where by his own account, he spent most of his time writing poetry and prose romances, mastering

Beethoven's 'Moonlight' and 'Pathétique' sonatas, and sketching whatever came his way, including a chameleon given to him by handsome, blond Uncle Charles, whose seductive curls of cigar smoke introduced Ronald to a lifetime's tobacco addiction. Learning was rote, swotting for exams something to be done at the last minute; an arrogance which saw him through one qualifying test for the IMS but failing another. Escape from paternal thunder was a spell at sea as a ship's surgeon on the SS *Asata*, plying the north Atlantic route between England and New York. Aboard were well-heeled passengers in first class and impoverished immigrants, many of them Jews, packed in steerage. No matter that he had no medical degree and no experience; actual surgery was seldom called for as the waves rolled on, just the occasional dose of creosote to settle a heaving stomach. No one could quite resist bright, plausible, young Ronald, not the lounge lizards in the saloon, not 'the American millionaires with drug habits', not the 'reverend gentlemen of various persuasions' and not 'the doubtful ladies'. He was twenty four, handsome, smart as a whip but not really a rover. Another try for the IMS exams had him scraping through, barely, seventeenth in a cohort of twenty two. It was enough.

Stars of the service were generally posted to Calcutta or Bombay. Ross was assigned Madras. Seven years of tedious apprenticeship followed, including a spell in Burma, once General Harry Prendergast's expedition had destroyed the ancient kingdom, an achievement which even by British imperial standards was a spectacular exercise in murderous cynicism, and which set off a decade-long insurrection. Ross was no great critic of empire, just of the bloated arrogance that presided over it, whether from the ceremonious offices of Whitehall or the cantonments and clubs of India. By turns dutiful and scornful, he dragged himself through the usual routines, mostly attending to the well-being of the sahibs: peering at coated tongues and jaundiced corneas, prescribing doses of quinine for the malarial shakes and other remedies for the clap. Whenever possible, he took trips into the country, fishing and sketching and finding placid scenery in which to write his not very good but not very bad poems and 'dramettas'. But he swayed between ennui and anomie and in moments of self-examination asked himself 'What was I doing to perfect the engine of human life?'

Weary of it all, he issued the first of many threats to resign unless a petition of leave was granted. In 1888, he found himself back in London where he demanded a two-month extension for reasons the IMS could hardly deny. Ross signed up for the very first British degree course in public health and returned to Bart's Hospital School to study bacteriology with Emanuel Klein, the Croatian Jewish scientist who had published the first English-language book on bacteriology in 1885. The year before, Klein had confirmed Koch's identification of the 'comma' vibrio, albeit registering scepticism about its causal agency, but he had recently had to fend off furious attacks from anti-vivisectionists on his use of laboratory animals. The student and the teacher became close enough for Klein to give Ross boxes of bacterial cultures to take back with him to India.

In 1892, General Campbell Claye Ross died. The childhood apparition of his malaria-struck father had never left the son. Back in Bangalore a year later, what he called 'the Great Problem' began to preoccupy him. In 1893, he versified his new-found commitment.

'Indian Fevers'

In this, O Nature, yield I pray to me.
I pace and pace and think and think and take
The fever'd hands and note down all I see,
That some dim distant light may haply break.

The painful faces ask, can we not cure?
We answer. No, not yet, we seek the laws.
O God, reveal thro' all this thing obscure
The unseen, small but million-murdering cause.

Ross knew of Alphonse Laveran's discovery, in 1889, of a malarial parasite, a single-celled protozoan organism of the plasmodian family, but was unpersuaded that it could be the agent of the disease. Much later, he wrote that it was the indistinctness of Laveran's drawings that made him excessively sceptical. On the other hand, he also knew how absurd it was that IMS orthodoxy still supposed that malaria – as had long been thought for cholera – was carried in miasmatic vapours arising from corrupted organic matter, or that somehow it arose as a form of gastric poisoning. In the spring of 1894 in London, he became friendly with Alfredo Kanthack, a young Brazilian professor of pathology and microbiology who had returned to Bart's from Berlin, where he had studied with Koch. Surprisingly, with the presence of Klein and Kanthack, the ancient teaching hospital was becoming the centre of microbiological research. Learning of Ross's interest in the aetiology of malaria, Kanthack brought him to see Patrick Manson, who had become a legend in parasitology for identifying the filaria worm causing elephantitis.

What Manson showed Ross was a revelation to the younger man: the 'crescent' gamete parasites he had identified in the intestinal tract of malarially infected mosquitoes. Even more significantly, he pointed out the thread-like 'flagella' which during development detached themselves from the crescent and which, Manson was certain, were somehow the agents of further cell growth and capacity for infection. He took Ross to the Seamen's Hospital and Charing Cross Hospital where he had collected blood specimens from malarial patients and, while walking down Oxford Street, close to his house on Cavendish Square, Manson wondered out loud whether, just as mosquitoes had been shown to infect humans with filaria, they could likewise carry the developing malarial parasite. (In fact, Laveran had already made this suggestion.) Less correct was Manson's conviction that since, as he wrongly supposed, females died shortly after depositing their eggs in water, malaria was transmitted to humans when they drank that water. Ross would spend much time and effort on this wrong track, drinking water in which infected insects had perished – and getting others to do the same – before concluding that Manson was mistaken.

Another possibility dawned: that the parasites developing within the gut of a mosquito might migrate to the salivary gland, from where, during a second feed on the body of a human, they would deliver the disease. He could see, under a small microscope he had adapted to be portable wherever he went, that not long after an infected meal had taken place, the parasitical organisms swelled into spheroid forms with the capacity then of migrating and implanting elsewhere within the mosquito host, potentially the thorax or indeed the salivary gland. But he did not observe this invariably taking place in many of the mosquitoes collected for examination. So Ross began to think that perhaps that development unfolded only in a particular species: perhaps the brown variety he described in high poetical manner as 'dappled-winged': anopheles.

There was, however, a shortage of humans on which to test these hypotheses. On the presumption that what went for the infection of birds would also prove to be the case for humans, he decided to make do with what was more easily obtainable: birds, in particular larks and sparrows exposed in cages to the bite of the insects. The concentrated period of experiment in 1897 turned into a personal ordeal, conducted in a 'dark, hot little office' in the Begumpett district of Secunderabad during the broiling heat of summer, without the benefit of punkah fans lest they blow away the precious reservoir of mosquitoes. One result of that self-denial was that Ross was tortured by swarms of other minute flying insects determined to burrow into his ears and eyes, all making steady microscope observation painfully difficult. As if that were not enough, 'the screws of my microscope [were] rusted from the sweat of my forehead and hands' and his last intact eye-piece had developed an alarming, possibly terminal, crack. Frustrated and exhausted, Ross was on the point of giving up. 'I was tired and what was the use? I must have examined a thousand mosquitoes.' But on 20 August 1897, on what he would call 'Mosquito Day', 'the Angel of Fate fortunately laid his hand on my head'. That same day, in a state of elation he wrote to Manson. 'The dissection [of the malariated mosquito] was excellent . . . I went carefully through the tissues . . . searching every micron with the same passion and care as one would search in some vast ruined palace for a little hidden treasure.' He found what he was looking for in pigmented cells lurking in the stomach of the insect, unmistakeably swollen as they developed. 'Best place to start looking,' he noted on the drawing of that day's dissection. That insect was, then, the intermediate vector, transferring through the life cycle of the parasite the organisms that in turn would indeed lodge in its salivary gland, ready to infect another human when it took its next blood meal.

Surely, a poem was called for.

> This day designing God
> Hath put it into my hand
> A wondrous thing and God be praised
> At his command
> I have found the secret deeds
> Of million murdering death
> I know this little thing a million men will save
> O death where is thy sting
> Thy victory O grave?

There remained, however, an unscientific problem: the baffling and enraging indifference to his work by the IMS; the ha-ha chuckles over tiffin at funny old 'Mosquito Ross'; the infuriating habit of sabotaging his work time and again, either by peremptory orders to non-malarial regions of India or else, as in this case, to work on a completely different disease: kala azar. It was, he later wrote, as if 'Columbus, having sighted America, was ordered off to discover the North Pole'. This systematic sabotage (as Ross saw it) continued even after his breakthrough work had received professional recognition from his peers. In the spring of 1898, Manson read Ross's reported results to the first meeting of the tropical diseases section of the British Medical Association in Edinburgh,

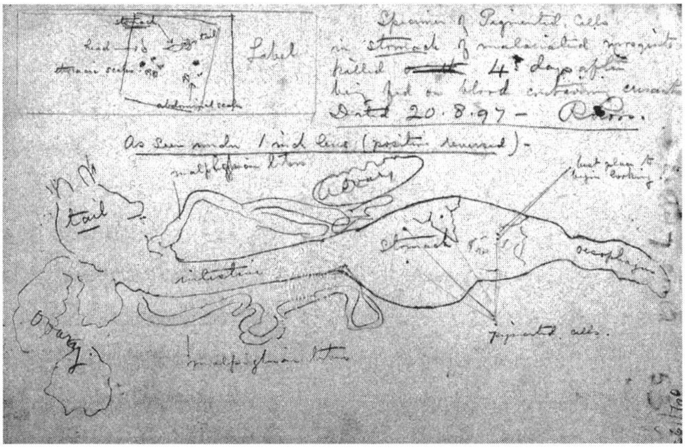

Ross's drawing on 20 August 1897 of his dissection of a malariated mosquito.

where they were greeted with a standing ovation. The assembled members then passed a resolution, Manson reported, 'congratulating you on your great, epoch-making discovery'. This cut no ice whatsoever with the IMS, nor did threats to resign should his malaria work not be better and more consistently supported. So he made good on the threat.

In 1899, the year he published his 'Romance', *The Child of Ocean* (shipwrecks, pirates), Ross was offered, and took, a post at the newly created Liverpool School of Tropical Medicine. The position was something short of what he might reasonably have hoped for: in the first instance, a lectureship rather than a chair (that would come) and at the less than princely salary of £250 per annum. The job was for three years, renewable if all satisfactory. Ross himself thought his treatment was an example of how 'medical men are exploited by committees of businessmen' of the kind funding the post and the new department. Nonetheless, his was the first teaching job at the new school and it called for a celebration. It was there, on 22 April 1899, that Ronald Ross met Waldemar Haffkine, then at the acme of his own fame and honour. Different as their lives, careers and fates had been, they also had this much in common: they saw themselves as outsiders, waging a battle for bacteriology against the myopic conservatism of the IMS and its default search for 'sanitary' answers to infectious diseases, rather than open-minded immersion in the scientific research. Both their missions were at the same time educationally persuasive and socially practical, the results of their hard-earned knowledge the saving of untold numbers of lives.

So when, eight years later, Haffkine wrote to Ross, asking for his help in bringing to professional and public attention the injustice that had been done to him, and the wrecking of his work, Ross immediately understood the common enemy: 'institutional barbarism'; the disrespect for science, the unconscionable *inattentiveness* to its illuminating knowledge, by those entrusted with public health, and those superior gentlemen who set themselves up as the custodians of imperial duty. Haffkine was the innocent victim of a syndrome Ross knew all too well: the sovereignty of the ignorant and the lazy over the persevering and the learned. 'Generals and civilians were made dictators of matters in which they had no knowledge and when their tactics failed they laid the blame on subordinates, the doctors whose advice they had frequently ignored and whose science they had habitually despised.'[34]

Since their last meeting, their stars had followed dramatically different courses. Haffkine was unemployed and, in the eyes of much of the world, disgraced. The *Poverty Bay Herald* in New Zealand was typical in February 1903 in characterising its report of what had happened at Malkowal as 'the story of a great blunder'. Ross had been elected a fellow of the Royal Society in 1901 and, with varying results and levels of support, had been sent to the malarial hot zones of Sierra Leone and Ismailiya on the Suez Canal to further verify his Indian work and make recommendations about reducing the disease. The essential thing, he had long been saying, was to get rid of pools of standing water, in which the anopheles bred. This had been met in India mostly with bemused disbelief. What? The tanks, the reservoirs? Really? A year later, in 1902, Ross won the Nobel Prize for Physiology or Medicine, which came with £8,000. He had long complained about those who did the hard work of revelatory knowledge being shut out of any remuneration from its application. Now, at last, he had something of a reward. As the first British Nobelist (though strictly speaking, Anglo-Indian), his reputation was sky-high. He meant to use it on Haffkine's behalf, but also use the miscarriage of justice to attack those in government for their uninformed disrespect for science.

Beginning in February 1907, Ross and William Simpson went into concerted action, surprising the overjoyed Haffkine with their eloquent determination. They started with professional opinion, medical and scientific – essay-length letters to the *British Medical Journal*, the *Journal of Tropical Diseases*, *The Lancet* and *Nature*. Then Ross brought out the heavy artillery, firing off three 'hot' letters, as he called them, to *The Times*, on 15 March, 13 April and 1 June. Throughout the vindication campaign – for that is what it was – Simpson's was always the more measured voice, scrupulously detailing the Malkowal evidence, giving compelling reasons why the accident must have taken place there and not in the Bombay laboratory, and expressing disingenuous astonishment that any dispassionate investigation could possibly have come to a different conclusion. For both of the campaigners, what was at stake was the integrity and authority of science itself. 'It is a bad day for science', Simpson had written in 1906, 'when truth is dominated by passion and expediency.' Ross echoed that view but, as befitted the romantic writer, was more militant. Thus, his indictments were against not just those who had wronged his friend but the institutionalised ignorance of the rulers of the Raj. Thinking as much of his own battles as of Haffkine's travails, Ross wrote that 'the man who can do is not allowed to do because the man who cannot do is put in authority over him'.

The most excoriating of all of Ross's interventions was published in *Nature* on 21 March. It began by blaming the rush to judgement of the Jenkins committee for a catastrophic interruption of vaccinations against plague. The result of its premature and faulty judgement was the widely shared perception that 'the poisoning was due not to local accident but to carelessness at the laboratory', and had resulted in 'a sudden and wholesale rejection of the invaluable vaccine'. As a result, 'thousands of lives may have been lost to the plague'. Then there was the gross miscarriage of justice inflicted on Haffkine. Since the incident of the dropped forceps was known to the committee, 'on what grounds . . . were the laboratory and its Director indicted?' But perhaps the most serious part of the whole affair, Ross went on, was not the suffering of the few, grievously tragic though that was, but 'the much greater loss which probably followed the suspicion thrown on the prophylactic by the erroneous judgement of the commission, and more even than this a certain ingratitude shown in India to a man who is one of the very greatest benefactors it has ever had'.

Ross then went from accusation to eulogy. Lest people forget,

> Haffkine not only elaborated the method of immunisation by dead culture but where many a man of science would have contented himself with merely writing an article on the subject, he addressed himself on the contrary to the more difficult practical verification. I remember well when he arrived in India with his anti-cholera vaccine and by his energy and perseverance gradually forced his ideas upon the people and Government . . . When the frightful calamity of the plague overtook the country in 1896 . . . and measure after measure failed and people were dying by hundreds of thousands, then Haffkine was the only one who made any successful stand at all against the storm, quickly inventing his anti-plague prophylactic and forcing the authorities along with him. Though he could not control the disaster he at least checked it by saving thousands if not hundreds of thousands of human beings who now owe their lives solely to him . . . The fact that more than six million doses of prophylactic have been issued in India alone attests the success and magnitude of his work. Yet he has received for it less than nothing for services which, compared with his, are really of a trifling nature; all kinds of officials receive . . . pensions, promotions and decorations. As for him not only has he received no adequate recognition for his immense service but he has been blamed for an accident which

> could not have been due to his fault . . . It is doubtful if he will ever return to a country which has treated him – I can only say – so ungratefully. Contemplating this history one cannot help being filled with indignation at it. India seems to be becoming quite notorious for its treatment of scientific works, suggesting an ignorance both of science and the importance of science . . . While all sorts of people climb easily into their seats of honour it seems that men of real merit are fortunate if only they can escape without censure.[35]

Ross's *coup de grâce*, though, in this indictment was to turn upside down the ugly stereotypes and ingrained suspicion of someone who somehow had missed out on the great good fortune of being born British. 'It appears to me a foolish thing for a nation to treat great men as we have sometimes treated ours and the case of Mr Haffkine – to whom, as he is a foreigner, we are doubly bound to show national gratitude – seems to be a glaring example of such treatment.' In his third letter to *The Times* on 1 June, Ross pointed to Haffkine's having been deprived of the pay and promotion which he could well have expected due to him and said that he ought, in fact, to receive royalties on the millions of doses of the vaccine he had, after all, created. Characteristically concerned not to overdo anything, Haffkine wrote to Ross that, nice as that might be, he had no patent in the vaccine.

He was, though, deeply moved by the storm of Ross's vehemence. After the first letter of his champion's appeared in *The Times,* he wrote to him that it had 'made a very great sensation amongst all those I have met today. I have no doubt that it will have very important consequences. Personally I find it impossible to express my own indebtedness to you.'[36] Haffkine was right to believe that bombarding every possible medium of opinion – professional and public – would make it all but impossible for the government, both in Calcutta and in London, to shrug off the exonerating evidence, although he predicted that if the whole matter was raised in Parliament, it would just generate 'inane' statements. But it was nonetheless a subject of comment in the House of Commons and this made an uncomfortable impression on the Liberal government. The secretary of state for India was now John Morley, whose monumental biography of William Gladstone had made him the keeper of the flame of moralising modern liberalism. Together with the new viceroy, the Earl of Minto, Morley cast himself against Curzon's proconsular arrogance. He would, instead, be the bringer of reform and (up to a point) political representation to India. With his acute nose for political sensitivity, Ross knew that a *cause célèbre* was not what Morley or the government wanted. Therefore he pressed the case hard, with Morley on the receiving end becoming increasingly defensive. Even Arthur Godley, still permanent secretary and the epitome of institutional circumlocution, confessed to Ross that 'speaking for myself I feel great sympathy for Mr Haffkine', though of course without suggesting that an unequivocal public exoneration was in the offing. In response to the questions raised in the Commons, a governmental 'Return of Documents' – a 108-page compendium – was placed before Parliament and the public in June. It anthologised all kinds of evidence relating to the incident, including the admission of the compounder, Narinder Singh, that he had indeed dropped the forceps on the ground of the field at Malkowal, though adding that this happened quite often and, as far as he knew, no one else had died as a result. Among the broad public, the 'Haffkine affair' had become akin to a medical Dreyfus case, not least because, faced with conclusive evidence of a gross miscarriage of justice, the authorities resisted any outright admission of wrongdoing. John Maynard Keynes understood this very well. In a letter to Lytton Strachey, he wrote that it was a classic example of government putting 'a tolerable face' on matters, allowing it never to withdraw anything, a habit that Keynes thought 'quite wrong and very dangerous'. Haffkine 'is apparently censured for negligence in his laboratory of which he is demonstrably innocent. But government maintains that position because an entirely different reason renders it undesirable to employ him again in his old position. It is quite clear to me that, whatever they do subsequently, censure for the thing of which he is innocent should be freely withdrawn first.' That even Keynes felt muffled by the very syndrome he had diagnosed was exemplified by his adding to Strachey that he should not say anything in public about his expressed view.[37]

Without that dramatic admission of wrongdoing, official discussions had started on what could be offered to Haffkine to make the embarrassing controversy go away. This was becoming urgent since Ross and Simpson were tirelessly mobilising the good and the great of bacteriology on behalf of the wronged scientist. Ross's teaching institution, the Liverpool School of Tropical Medicine, awarded Haffkine the Mary Kingsley Medal for pioneering achievement in the battle against tropical disease and gave him a banquet to celebrate the award. In his acceptance speech, an emotional Haffkine called the occasion 'the red letter day of my life'. Ross then prepared a nomination for the Nobel Prize and for a Royal Society medal, amazed that Haffkine was not already a fellow (he never would be). The Society of Tropical Medicine and Hygiene elected Haffkine to its council. In July, a fourth letter to *The Times* was organised to rehearse the case yet again but with the additional signatories of ten of the most eminent bacteriologists of the day, including the first president of Rockefeller University in New York, Simon Flexner; Albert S. Grunbaum, professor of pathology at Leeds; G. Sims Woodhead at Cambridge; and, not least, Ross's still controversial teacher at Bart's, Emanuel Klein. Drafted by Ross, it was the most damning yet, insisting that 'the whole case against Mr Haffkine's laboratory has collapsed; the charge not only not proven but distinctly disproven'.[38] Still more seriously, all the co-signatories, when asked by Simpson and Ross whether they would constitute a 'Haffkine committee' for as long as it took to set matters right, responded affirmatively. This ensured that the business was not going to be brushed off.

The government was getting nervous. Godley wrote again to Ross, effectively conceding the case but declining to reopen the matter with another independent inquiry as the heavyweight letter to *The Times* had demanded. It took Morley a month and a half after Haffkine had written to him to respond in a letter that he made public. Even by the standards of British officialdom, it was a masterpiece of mealy-mouthed tergiversation, beginning 'Though views on this matter are not unanimous' before going on, grudgingly, to acknowledge that 'an important body of scientific opinion is favourable' to Haffkine on the 'Mulkowal disaster'. Morley added that his own view of the matter might be construed from his offering Haffkine a favourable position in India. Perish the thought that a government minister should actually admit that an unforgiveable miscarriage of justice had been perpetrated.

If this was not exactly the full-throated exoneration he had been looking for, it was enough for Haffkine to take seriously the offer of a position presented by Morley as equivalent to the one he had left and at the salary he had been paid in 1903. This, too, turned out to be misleading. There was no question of Haffkine returning as director-in-chief of the Plague Research Laboratory at Parel, now being run by Liston. He was, instead, being offered the post of director-in-chief of the Biological Laboratory of Calcutta, attached to the General Hospital. Earlier that year in a letter to Ross, Haffkine had been adamant that he was 'unable to submit to any unfavourable change in the condition of my work in India nor to the slightest personal degradation or loss imposed on me'.[39] But it may well be that, in the summer of 1907, he had moved from the fight and the fury of the campaign for vindication, which he had already all but won, to a more sober reflection on how the rest of his professional life might now be spent. He had been heavily medalled and awarded by the scientific community in Britain but somehow no one was queueing up to hire him; there was little or no chance of a teaching chair at one of the British universities. And he was still drawn to India as he had been fifteen years before when he had listened to Dufferin and Hankin about how the new science might save millions of lives. Putting the best face on it, he accepted Morley's offer, adding 'that it is with pride that my work has been recognised of public benefit' and that his new work 'will lead to results that may be of further benefit to the people in India'.[40]

Aware that this was well short of the full reinstatement they both had asked for, Ross wrote to Haffkine that while Morley's letter 'might have been more generous', it did provide 'assurance in writing that you are not to be held at fault and therefore liberated you from further necessity of vindicating yourself. Now go back and prosper. You will again lead India in the path of science!'[41]

This did not come to pass. There was a moment, just before returning to India, when he felt buoyed again, reading a lecture to the Royal Society of Medicine stating that all the evidence pointed to a reduction of 85 per cent in incidence of infection among the plague-vaccinated population. But almost as soon as he got to Calcutta and was installed as director-in-chief of the Biological Laboratory, he discovered that his grand title meant little. Funds were meagre, staff minimal; it was unclear, not least to himself, what research he could now pursue. What was painfully apparent was that he was being kept at more than arm's length from the business of vaccine production and delivery. When he sought permission to make a devitalised version of his cholera vaccine to meet the resurgence of infection, he was denied, more than once. Dispiritingly, that 'vindication' Ross had spoken of had been left behind in England. In January 1908, he wrote to Ross that he was 'sorry to say that Morley's letter did not alter the position in the slightest degree. The whole of the unjust punishment for Malkowal . . . remains on me quite as before; on every occasion in print and speech it is repeated and kept alive that I was responsible for the case.' He was still a condemned man. The gentlemen had won. Saddened, Ross replied that he was sorry that all their fight had 'failed to exonerate you', adding, feelingly (and from the standpoint of 2022, prophetically), that 'it is a curious thing that the public always hates its benefactors'.

Waldemar Haffkine was just forty eight; in the prime of life. But he was a spent force; the catastrophe of his disgrace and the battle for vindication had used up his hitherto inexhaustible reserves of mental energy. As the young tyro of Odessa, Mechnikov's prodigy, in and out of jail, he had published five big papers in two years. In the six years between returning to India and taking early retirement in 1914, he published just one. His mind had not shut down completely. There are notebooks from this period filled with assorted observations on the effectiveness of serum-based prophylactics, and the development of an anti-typhus vaccine, about which the government would approach him for help when the Great War began in 1914. But his mind was also on other things: the flight of soaring birds among other almost random inquiries. Rather pathetically, the sometime director-in-chief of the Plague Research Laboratory now craved acceptance from the gung-ho society of the Raj. In 1910, after passing a 'Proficiency Examination' testing his knowledge of when and how to clean his carbine and 'when you see an enemy patrol working towards yours . . . what would you do?', he was admitted as an officer of the Calcutta Light Horse, issued with sword, gun, bandolier and, presumably, a mount.[42]

But the tiffins of Trooper Haffkine, as he signed himself to brother officers, were now kosher. As his active life as an experimental scientist ebbed, so his obsession with reconciling devotion and science grew. Increasingly his mind dwelled on Moses Maimonides who, in the twelfth century, had united in one person medical learning and the philosophy of Judaism. But Haffkine's practical vocation as evangelical vaccinator was ebbing. In his speech to the Maccabaeans in 1899, he had invoked an old question, posed in some liturgies during Yom Kippur, the Day of Atonement, as Jews face the judgement of the Almighty: 'The journey we take here on the earth is so short and before we know who we are, we are at the end and called on to answer an inner voice, "Have you finished the work you had to do?" Happy are those who can think they have finished the work.'[43]

Though he deserved otherwise, Haffkine was not destined to be among the happy.

ix

Departures

No. 70527

PASSPORT.

SIX PENCE

We, Sir Edward Grey
a Baronet of the United Kingdom of Great Britain and Ireland, Knight of the Most Noble Order of the Garter, a Member of His Britannic Majesty's Most Honourable Privy Council, a Member of Parliament, &c. &c. &c. His Majesty's Principal Secretary of State for Foreign Affairs.
Request and require in the Name of His Majesty all those whom it may concern to allow
Dr. Waldemar Mordecai Wolff Haffkine
to pass freely without let or hindrance, and to afford him every assistance and protection of which he may stand in need.
Given at the Foreign Office, London the 10th day of January 1916

E. Grey.

This passport is valid for two years only from the date of its issue. It may be renewed for four further periods of two years each after which a new passport will be required.

RENEWALS.

This Passport is hereby renewed.
Valid until 10 [illegible]
[illegible]
British Vice-Consul, Paris

This Passport is hereby renewed
Valid until 10 January 1920. 1916
[illegible]
British Vice-Consul, Paris

This Passport is hereby renewed
Valid until 10 Jan. 1922
H. J. Dorey
for British Pro-Consul
British Vice-Consul Paris

DESCRIPTION OF BEARER.

Age 55 years Profession Retired Bacteriologist with Indian Government
Place & date of birth Russia
Maiden name if wife or married woman travelling singly } 15 March 1860
Height 5 feet 10 inches
Forehead medium Eyes blue
Nose medium Mouth medium
Chin medium Colour of Hair light grey
Complexion fair Face normal
Any special peculiarities
National Status Naturalized British Subject 1899 – of Russian origin

This Passport is granted "with the qualification that the Bearer shall not, when within "the limits of the Foreign State of which he "was a Subject previously to obtaining his "Certificate of Naturalization be deemed to be "a British Subject, unless he has ceased to be a "Subject of that State in pursuance of the laws "thereof, or in pursuance of a treaty to that "effect." 33 Vict., c. 14.

FOREIGN OFFICE 10 JAN. 1916

PHOTOGRAPH OF BEARER.

SIGNATURE OF BEARER.

W. M. Haffkine.

Haffkine's passport

High summer of 1926 found Haffkine in Soviet Odessa, the city where he had twice seen the light: the first time as he arrived in the world sixty six years earlier; the second time when Ilya Mechnikov's microscope had revealed a hitherto hidden universe, the teeming horde of micro-organisms. But it was also the place where, thirty five years before, he had come out from the lab and into the streets, armed with a pistol, prepared to do whatever needed to be done to fend off a pogrom. He understood zeal; he had kept company with women and men the state called terrorists. But really his own fervour had been for his people. Just what the Bolsheviks now wanted for the Jews was an open question.

At first, in 1917, there had been only good news. The Provisional Government in Petrograd, formed after the February revolution, had abolished all restrictive laws confining the Jews geographically to the Pale of Settlement and excluding them from many occupations and professions. But to the Bolsheviks, any idea that the autonomous life of Jews should be encouraged was counter-revolutionary bourgeois nationalism. They were not to imagine that they could be separated from the fate of the proletarian masses; instead they must dissolve their identity into that common lot. The condition of Jewish emancipation, then, was cultural and legal disappearance. Hebrew, the vehicle of that reactionary separateness, and which had been reborn in Odessa (Haffkine had been among its earliest promoters), was therefore banned. The only acceptable language of Jewish workers, Yiddish, would be encouraged, not to perpetuate distinctiveness, but as a tool to educate backward Jews in socialism. Newspapers and magazines like *Emes* ('The Truth') were created to spread that word. Zionism, another nationalist blind alley, was decreed an enemy of communism, Judaism almost as reprehensible. Its retrograde and arcane practices were not, for the moment, banned outright, but neither would they be protected, much less subsidised by the state. The Jewish section of the Communist Party, the Evsektsiya, was the most fervent prosecutor of this programme of cultural liquidation. Under its encouragement, Hebrew schools and seminaries were closed, many synagogues shuttered or turned into sports halls, gymnasiums or workers' clubs, cemeteries converted into parks. Teaching or even studying the Torah and Talmud could earn students a spell in prison or banishment beyond the Urals.

Haffkine was dismayed by this latest oppression. Finding a way to help the Hebrew language and Jewish education and worship to survive in Soviet Union would have to be his new vocation. Though he would be introduced in Warsaw and Moscow as one of the world's famous scientists, and though he had won the prestigious Prix Bréant in 1908, the flame of research, lit all those years ago in Odessa, had burned out. The gentlemen had seen to that. Such work as there had been in Calcutta had been inconsequential, his efforts to create a new, more easily produced cholera vaccine frustrated by the government. When it came to developing an anti-typhus vaccine that might protect troops trapped in trenches populated by both rats and lice, he had been asked to judge other people's work. It was time to concern himself with the fate of the Jews and in particular with a question that – in Israel in particular – has become existential: the vexed relationship between scientific knowledge and religious learning.

In 1915, with reports of Jews in Poland and Ukraine caught between the sledgehammers of the Russian and Austro-Hungarian armies fighting on the eastern front, and suffering, predictably, from a fresh wave of pogroms, Haffkine helped raise relief funds for the shelter and medical treatment of victims. But, increasingly, he brooded on what the future might hold for Jews, or whether indeed there would be one. Collective self-determination had been born in Odessa in 1882 with the foundation of Hovevei Tzion, the Lovers of Zion, committed to bringing Jews back to the land where Judaism, Hebrew and Jewish identity had been formed. The writer of its manifesto, *Auto-Emancipation*, Leon Pinsker, had been a physician. Haffkine shared Pinsker's vision of self-determination, while having reservations about political Zionism when Theodor Herzl brought it into being in the late 1890s. In 1909, Hillel Yaffe, the smuggler of guns during the hot days of pogrom in Odessa, suggested that Haffkine should follow his example, emigrate to Palestine and found a bacteriological research institute there. Yaffe, originally a specialist in diseases of the eye, had become a physician in the Jaffa hospital but had begun the work which would shape his own life and Palestine's: the eradication of malaria. He had spoken at an international conference in Paris, and owed his subsequent mission to drain swamps in the Palestinian lowlands to the lessons learned from the work of Ronald Ross. In 1902, the Turkish government had put him in charge of containing a serious outbreak of cholera. His life was shadowing Haffkine's. But only so far. Haffkine considered joining him in Palestine, but somehow never got on the boat. In 1915, Yaffe's brother-in-law, another Odessa comrade, Haim Mordechai Rabinowitz, also settled in Palestine, raised the subject once more with Haffkine, who seemed warmer towards it. Such a research institute could be part of a Hebrew university, planned for a site on Mount Scopus in Jerusalem. But the war which had come to Turkey and the Middle East scotched the idea for the foreseeable future, and it never arose in Haffkine's mind again.

Instead of going east, Haffkine went in the opposite direction, travelling around the United States reporting on farming cooperatives that had been established by the Jewish Colonisation Association, funded by Baron Maurice de Hirsch. Odessa's Lovers of Zion had been the originators of the idea that Jews might make new lives for themselves individually and collectively, away from the poverty of the shtetls, by reconnecting with the land from which exile had uprooted them. Hirsch, along with Edmond de Rothschild, both of whom Haffkine knew personally, had funded settlements in Ottoman Palestine, but also in Brazil and Argentina. Land was bought in Canada and the United States – in Saskatchewan, Alberta and Ontario; New Jersey, New York, Delaware, North Dakota, Utah and California. Farming – the air, the soil, the livestock, the harvests – was thought to be the social antidote to the unwholesome crowding of tenement sweatshops, the tubercular hovels of the Pale. Sometimes the little colonies put down roots, often they didn't;

sometimes the cultivators stayed, often they moved back to the urban antheap. Haffkine travelled around, to settlements in Cumberland, Salem and Gloucester counties in New Jersey, to poultry farms in Utah and Portersville in California's San Joaquin valley where the Jews tended orange groves and pulled rutabaga from the loamy dirt. On some days, the whole thing seemed absurdly utopian; on other days, a tangible inspiration. But what unsettled Haffkine, far more than the improbability of Jewish American farmers, was the mutation, as he thought of it, that traditional Judaism was undergoing in the great metropolitan cities of New York and Chicago, Cincinnati and San Francisco. Something he thought oxymoronic – Reform Judaism – was teaching that strict adherence to the Torah's 613 commandments needed to be rationalised for the needs of modern life, its more arcane demands discarded. Let there be synagogue car parks. But if a Jewish future would be shaped by some sort of reconciliation between secular and religious knowledge, Haffkine wanted this to come not from convenience but from principle.[1]

He had become an observant Jew. Just when is hard to say, though we know that when he departed from Bombay after his removal from the directorship of the laboratory, he packed his *tefillin,* the phylacteries worn by observant Jews every morning prayer, for the voyage to England. His parents had been Enlightenment Jews in hectically modernising, commercial Odessa. Waldemar had learned the rudiments of traditional Judaism from his maternal Landsberg grandfather and knew Hebrew. But at some point he had undergone a *tshuva* – a Hebrew word that signifies both an act of repentance and a return. Had that happened in India? It is possible that the substantial Jewish community in Bombay, both native Indian (Bene Israel) and Baghdadi like the Sassoons, encouraged his own observance. There was certainly no shortage of synagogues there for him to attend. Farha Sassoon, with her personal kosher slaughterer and travelling minyan of ten prayer-ready men, was a striking instance of a Jewish businessperson who had no difficulty reconciling the demands of piety with the worldly timetables of commerce. It is less likely that, on his countless vaccine campaigns through the Indian countryside, Haffkine would have been able to keep kosher, though Brahminical vegetarianism would have served as well. He would often invoke the similarity between Muslim, Hindu and Jewish dietary rules as an instance of common regard for health alongside piety. Though he by and large had kept clear of Jewish communal organisations in Bombay, at the height of the plague epidemic of 1896–7, he had helped found a small hospital specifically for stricken Jews.

All the same, nothing in that sketchy record suggests the fervour of the ten-page pamphlet that he began to write in New York as a counter-blast to Reform Judaism and which he completed in his Marylebone flat in the spring of 1916. Published in the weekly *Jewish Chronicle* as 'The Endurance of the Jews' and in the American *Menorah Journal* as 'A Plea for Orthodoxy', the essay was an impassioned statement of the belief that Jewish survival depended not on secularist adaptation, much less in believing the hollow promises of equal treatment by Gentile societies, once they had judged Jews to be adequately assimilated, but rather in a cleaving to the ancient texts and rituals set out in the Torah and elaborated over centuries by codifiers and commentators. None of this was in the least bit original, but coming from a scientist, it was unusual. It was not that Haffkine was now repudiating science, whether from a sense of personal wound or not; rather he claimed that science and Torah were not necessarily mutually exclusive. On the contrary, as the life and career of the medieval physician-rabbi and philosopher Moses Maimonides demonstrated, they were mutually strengthening. Often, he argued, what the Torah had laid down, modern science had confirmed. Wisdom of the natural world in this view presupposed some sort of proto-scientific, infinitely intricate Creator. A section of his essay, titled 'Old fashioned kashruth and the up to date microscope', argued that the command to purge meat of any trace of blood anticipated microbiology's revelation that, 'immediately after death', the bodies of slaughtered animals are 'invaded by microbial germs' spreading 'infection throughout the rest of the tissues'. Still more important, the observation of dietary laws and the recitation of prayers in Hebrew secured the communal connection of Jews across space, time and the generations. The arguments are not crudely instrumental. Haffkine builds to a defiant declaration that

> of all religious and philosophic conceptions of man, the faith which binds together the Jews has not been harmed by the advance of research, but on the contrary has been vindicated in its profoundest tenets. Slowly and by degrees, passing through innumerable stages in an analysis of the life of animals and plants and of the elemental phenomena of heat, light, magnetism, electricity, chemistry, mechanics, geology, spectroscopy, astronomy, Science is being brought to recognize in the universe the existence of one power which is of no beginning and no end . . . This sum total of the scientific discoveries of all lands and times is an approach of the world's thought to our Adon Olam, the sublime chant by means of which the Jew has wrought and will further work the most momentous changes in the world.[2]

It is safe to say that not many scientists, not all that many Jewish scientists for that matter, would have agreed. Famously, Albert Einstein's Jewish lodestar was Benedict Spinoza, who had refuted almost everything Haffkine asserted in his pamphlet, other than, perhaps, the original creative act, filling the otherwise inexplicable origination of the universe. But nothing that had happened in the turbulent years following the World War had shaken Haffkine's convictions that Jewish survival depended on preserving, not discarding, traditional practices. His insistence that Judaism and modernism, science and spirituality, were not mutually exclusive but mutually nourishing was, of course, an exercise in autobiography; the two halves of Waldemar Haffkine made whole. But this personal stake only made his advocacy more intense. To liberals like the social philosopher Professor Horace Kallen, whose father had been an Orthodox rabbi in Germany but who had coined and preached 'cultural pluralism' and argued that Torah Judaism could be safely modified without losing its essence, Haffkine replied that, on the contrary, it *was* the irreducible essence. He said much the same thing to others of the Great Men of Jewish America who, in the persons of Louis Brandeis and Felix Frankfurter, had gone all the way to becoming Supreme Court justices and confidants of presidents. What dismayed Haffkine was the apparent ease with which this success was imagined as a good conduct prize for the loss of burdensome ritual or socially eccentric habits. So Jews drove to their grandly decorated cathedral-like synagogues in glossy automobiles; their prayers were chanted in English or German to the swelling chords of grandiose organs. They ate whatever they fancied. And this, as he saw it, hybrid Judaism was now the way in which the most powerful community in the world expressed their Jewish identity.

Back in France, Haffkine became still more adamant. Living in Boulogne-sur-Seine at the western edge of Paris, he had been one of the signatories appealing to the Peace Conference at Versailles for the safeguarding of minority rights throughout the world. The liberal order would surely deliver that much to liberal Jews. But he was not naive. When formal guarantees of religious and social protection granted to Jews in Marshal Piłsudski's Poland not only failed to prevent, but actually triggered, another round of antisemitic violence, Haffkine was distressed but not surprised. Jews in Russia and Ukraine had suffered unspeakably both during the Great War itself and even more brutally in the long and bloody civil wars that followed, attacked mostly by the Whites but, during the last decisive battles in Crimea, by some Red units and especially the violently antisemitic irregulars of Nestor Makhno. Marauding Whites would plunder and burn down a shtetl, and physically attack its fleeing inhabitants. Incoming Reds might offer protection, but on condition that the people un-Jewed themselves. Were they peasants or workers or even artisans? No? Well then, especially if they had businesses, big, small or minuscule, or if their only way of life was teaching or studying religious texts, they were henceforth classified by the

Soviet Constitution of 1918 as *lishentsy* – disenfranchised persons, beyond the pale of legal, civil or political rights: unpersons.

None of which, other than to measure distress and mobilise charitable assistance, was much reason for Haffkine to go to the Soviet Union. But although he had turned down an invitation to attend a bacteriological conference in Moscow in 1924, there was something about a return which tugged at him. In 1925, that vague but insistent feeling would crystallise in an invitation to repeat his tour of American Jewish farming cooperatives, but this time in Ukraine and Crimea. The opportunity arose because, during the New Economic Policy, an alternative to the total cultural erasure demanded by the Evsektsiya had been mooted by circles in the Party and the Soviet government. In 'The Fate of the Jewish Masses in the Soviet Union', an essay written with the essayist Mikhail Kol'tsov and published in 1924, the Jewish agronomist Anatole Bragin proposed to the Politburo an ambitious resettlement of 100,000 Jews from towns and shtetls where they currently subsisted in desperate penury. The Bragin–Kol'tsov proposal was to transform this social waste into productive workers by bringing them to the land, in particular to Crimea and regions of southern Ukraine where, it was thought, there was enough uncultivated acreage to absorb a large population. Even more intriguingly, should this experiment in social engineering succeed, an autonomous Jewish republic might be created around the littoral of the Black Sea linking the Ukrainian and Crimean Jewish pastorals. For a while it seemed that this possibility even had the blessing of one of the few Bolsheviks who actually came from peasant stock: Mikhail Kalinin, the president of the Russian Soviet Federative Socialist Republic – the titular head of the USSR.

Bragin had originally proposed Byelorussia as a suitable region for Jewish agricultural colonisation. But the obvious objection that the local rural population would not take kindly to this intrusion killed the plan. Crimea, on the other hand, was thought to be much less densely populated and with large areas of uncultivated soil. But if both those assessments were right (and they were not completely wrong), it was for the good reason that the soil was mostly a waterless steppe. And this did not mean the region was uninhabited. Crimean Tatars could live the same migratory pastoral life there that sustained their people in central Asia.[3]

On the other hand, sustainable agriculture might be possible with investment in the drills and wells needed to supply irrigation. And where would the funds for that come from? From the heart of capitalism: the United States, indeed from the twin pillars of Jewish banking, Felix Warburg and Jacob Schiff, who in 1914 had founded the mighty Joint Distribution Committee as a relief organisation for the Jewish poor and persecuted. In 1923, JDC money had already helped establish two farms in Crimea intended to train emigrants to Palestine for cooperatives and kibbutzim there.

It was astonishing how quickly the project came into being. In 1925, a special division of the JDC, Agro-Joint, was created to manage, together with their opposite numbers in the USSR, the settlement of 100,000 Jews in ten years on land around the Black Sea. KomZET, a Soviet government agency, would supply the land; the cost of everything else – housing, well-drilling, tractors, machinery, conveyance – would all be provided by Agro-Joint. Why would the Soviets turn the fantastic idea down? Despite furious opposition from two parties – Zionists who saw the Crimea scheme as a damaging diversion from settlement in Palestine, and the Evsektsiya, which denounced it as another kind of dangerous Jewish separatism – the scheme got under way. Another agency, the unenticingly titled Society for Jewish Toilers on the Land (OZET), was created to promote the plan in centres of Jewish population. In 1927, a top-tier creative team – the director Abram Romm, and writers Vladimir Mayakovsky and Victor Shklovsky (better known as a literary theorist) – produced a documentary called *Evrei na zemle* ('Jews on the Land') to sell the dream of the Soviet-Jewish bucolic, in explicit competition with the Zionist original. It helped that Zionists were being arrested in their hundreds at this time. Posters for the film and for the whole scheme featured the usual hook-nosed, heavy-hatted, stooped-back caricatures of Jews transformed by georgic toil into heroically muscled, tractor-driving comrades of the fruitful earth. A start to new settlement was made in 1925 with 346 families, 1,116 individuals, moved to their allotted, as yet largely non-existent, villages given Yiddish names like Naybrod (New Bread), Gartenshtadt (Garden Town) and Beys Lechem (House of Bread), even though bread was in short supply. Predictably, the reality fell short of communist pastoral. Water had to be fetched by hand from afar; lodging was in tents or dugouts roofed with tarpaulins. Seeds were provided but the initial work of breaking the dry earth of Crimea, before the Agro-Joint money and expertise came to the rescue, was unforgivingly back-breaking.

Why, aside from the fact that Odessa and southern Ukraine was his native home, would Waldemar Haffkine, now sixty six, want to be involved in such a scheme? He had, it is true, relevant experience under his belt from the American tour of 1915–16. But he was not going as some sort of agronomist. Drawn by the possibility that the Jewish farming districts would enjoy a considerable degree of autonomy – even if a Jewish republic was not really on the cards – notwithstanding the Evsektsiya war on Hebrew and religious education, it might be possible to implant the ingredients of a truly Jewish life, as Haffkine thought of it, in the agricultural colonies. The people's commissar for education, Anatoly Lunacharsky, had let it be known that some sort of religious freedom, including Hebrew instruction, might not, after all, be outlawed. Haffkine would, then, act as a representative of the Alliance Israélite Universelle, founded in mid-nineteenth-century France to establish schools for Jews throughout the Levant and eastern Europe. The mission of the Alliance was modernising, secular and technical, and Haffkine for some time had been a member of its council. But, increasingly, he hoped to use his status as a scientist – the personification of Alliance excellence – to argue that modern instruction need not preclude religious teaching; that somehow the two kinds of education could nourish each other. The issue remains alive in the worlds of contemporary Jews today.

On 12 June 1926, Waldemar Haffkine stepped out of the Paris–Moscow train and on to the platform at Kursky station. Amid the hiss of the slowing trains, a young woman whom he had never seen in person was hailing him: his niece Moussia. At her side was Rabbi Samuel Kalmanovich, whom Haffkine had met in London. On learning of Haffkine's plan to come to Russia as the representative of the Alliance, Kalmanovich had been beside himself with happy anticipation and had written effusively to the prospective visitor. A window would be opened; through it would come help from the Jews of Europe and America. Not just for tractors, but for the Torah. And help was needed. Four years before, three Talmud students had been arrested, interrogated, physically intimidated and thrown into prison simply for conducting private studies: Evsektsiya tactics. Kalmanovich feared for the future. Haffkine and the Alliance would help find ways to allow Hebrew learning to survive, as it had survived countless ordeals through the long history of the Jews.

As he walked around summertime Moscow, a perennial Jewish truism came to Haffkine, namely 'it could always be worse'. 'In Moscow,' he wrote in his journal, 'life is intense', but the streets were cleaner, the people more sociably open than he had been led to believe or had expected. Haffkine, who always valued politeness, was surprised to find it had not entirely left Moscow. The many rough sleepers, bunking down on benches or hard pavement, he put down to an understandable shortage of housing. Though he knew that many synagogues had been closed or converted, the Great Synagogue itself was full of worshippers. On a later visit, Haffkine complained happily that he had had a hard time finding a seat for the Shabbat morning service. It was all reassuringly familiar, *chazanut* prayer melodies fighting with the constant drone of gossip; beards smelling of tobacco and/or old soup; *schnorrers* hanging around the doors buttonholing incomers and outgoers, begging for attention and handouts, claiming to have relatives in wherever it was you came from.

Meetings with the hard-faced men of KomZET and OZET brought Haffkine down to earth. Disconcertingly, he was getting mixed messages. The newly appointed chief of OZET, Yury Larin (born Mikhail Lurie), was an economist, but his father, Schneur Zalman Lurie, had been a writer of elegant Hebrew essays, as well as an unapologetic Zionist. Surely such a son of such a father would understand? Larin seemed to be encouraging the notion that once settled on the land, the Jewish kolkhozes might enjoy some limited degree of autonomy. But when Haffkine assumed, out loud, that that would include the practice of religion, stony silence told him he was pushing his luck.

On 26 June 1926, Haffkine, the perpetual voyager, was on his way south-west to Ukraine. He was not travelling alone. Along with him was a small team, each of its members commissioned to look over the prospects of farming as social salvation for the multitudes of destitute, legally defenceless, city Jews, at that moment kept alive only by charity. The journey to Odessa, which would take, he noted, thirty-eight and a half hours, started by train. Soldiers filled the carriages and corridors. Haffkine noticed what had been abandoned: locomotives in the sidings and even at disused platforms at Konotop; children – the *besprizornye* urchins – at Kiev. Also, synagogues. In Zhytomir, Haffkine went to look at the handsome building and found it in a woeful state of depressing neglect: 'doors hanging off a beam; not a window left unbroken'. In Odessa, whether he wished it or not, he was accompanied by two young 'farouche' communists who nonetheless turned out to be good companions as Haffkine made his way down the path of memory. The experience was emotional but not altogether happy. The parks and arcades of the beautiful city wore a mournful aspect. There was, he noted, 'no animation' at the dockside where he would have remembered forests of cranes, the endless loading and unloading. Now 'there were hardly any ships at all'. Some of the grand mansions on Deribasovskaya Street with their operatically Italianate roof lines stood intact, but in Moldavanka, the warren of the poor, there was damage and decay alongside the usual squalor. The *besprizornye* were there too in gangs, their faces grimy and their hair matted, rags falling off their little bodies; yet somehow,

Haffkine thought, they looked as though they had managed to find something to eat. Haffkine forbore from asking himself or them how and why, but contented himself with enjoying the urchin circus, the kids carving spaces through forests of grown-up legs, yelling and jumping; the same as any other place any other time, but in Soviet Odessa somehow a raw force for life. It could get too raw. Haffkine was less amused when, later, an arm poked through a half-open window on his side of the car he was in and snatched his hanging overcoat. Wherever he was, we know from his photographic history, Haffkine liked to be dressed just so.

When the cavalcade got under way and began its drive into the countryside, there were more surprises, not least pigs. The two training farms established in Crimea in 1923 were socialist-Zionist, and not much bothered with the niceties of Jewish dietary laws. If there was a market for pork, take advantage of it. Whether the worker-farmers ate it as well as sold it, Haffkine wasn't sure, but he recoiled from the usual Russian greeting embraces lest his kosher face be defiled by a porky kiss. Less disconcerting was the socialist ethos governing the running of the farm colonies: a common store of clothing from which the farmers and their children dressed themselves; shared proceeds and meals, very much along the lines of the Palestine kibbutzim; unforced collectivisation which in the Stalinist years would be monstrously perverted into tyrannical ruin and starvation.

The 'pioneer' farms had been given Hebrew names, echoing counterparts in Palestine: Tel Hai and Mishmar (where the hogs and sows rutted gruntily in the slops). The new settlements, intended as anti-shtetls, in some cases bore Yiddish names which, like Beys Lechem (House of Bread), also had Biblical-Palestinian connotations; others had Yiddishified tributes to Soviet benefactors – hence Kalinindorf or, still innocently, Stalindorf. Some of their fields looked auspicious. But it was a conscious policy that the new Jewish farms should innovate, avoiding the standard cereal grains and instead planting experimental sorghum and Sudanese hay grass. This was all very well, but as yet for many there was still no water, either for irrigation or drinking; housing was under canvas or in the roughest cabins; and, it was noted, a number of the settlers, without any experience of farming, were hedging their bets by not selling off the homes they had left behind in the cities. And Haffkine was not so remote from his earlier life not to see that the settlers had more than their share of sickness; he listed, among others, anaemia, tuberculosis, trachoma, malaria, even occasional cases of typhus. If the settlements were to survive, he wrote to the head of the Alliance in Paris, they would need resident doctors, dispensaries, bath-houses and a regional hospital quite as much as teachers and schoolrooms.

Ironies abounded. While the Black Sea region had been chosen for its relative proximity to areas of the Ukrainian countryside where Jews had lived for centuries, Haffkine discovered that Crimea had been home to a much older Jewish migrant community, known now as Krymchaks, which went all the way back at least to the first century BCE. There was also in Simferopol a community of Karaites, Biblical fundamentalists whose Judaism stopped with the Torah and who rejected the 'oral law' of the rabbis and the entire corpus of Talmudic commentary. As if this wasn't enough of a revelation, it was proposed to bring the Mountain Jews of Azerbaijan, who spoke not Yiddish but Judeo-Tatar, into the mix of rural colonists. Just how this Noah's Ark of Black Sea and Caucasian Jews would get along, not just with each other but with the Tatars, Greeks and Romanies already in Crimea together with ethnic Russians and Ukrainians, was not clear.

Back in Moscow in July, Haffkine drafted the reports for the Alliance which would appear the following spring in its journal. He went to the lively Great Synagogue and sought out the company of Rabbi Yosef Yitzhak Schneerson, the head of the Orthodox Lubavicher community, somehow managing a dangerous balancing act between appearing to abide by Soviet restrictions while maintaining a secret network of hidden yeshivas – seminaries for the study of Rabbinical Judaism.

His work in the Soviet Union was done; but Haffkine was getting signals from his heart – fibrillations, the occasional stab of pain – to suggest that his own allotted time might be running out as well. As the British were always reminding him, he was no doctor, but neither was he a medical idiot. Where was he now in the run of his days? What was there left to do? Where to go? Who could he turn to?

An answer, of a sort, had come to him on a raft-ferry crossing from the northern to the southern bank of the Dnieper between Beryslav and Kakhovka as it broadened in its run towards the Black Sea. The scenery resolved itself into a family memory. His step-brothers, Alexander and Salomon, had lived in those towns at the end of the previous century. Waldemar's family feeling, it seems to me, ran warm and cool. He had only been able to study with Ilya Mechnikov in Odessa, after all, thanks to subsidies from Alexander, yet their lives had pulled far apart. He writes in his journal that he had wanted to see Alexander's daughter Yanina in Kiev, but was so pressed for time with all his appointed working visits to schools and hospitals that the meeting turned into a mere telephone conversation. The same was true for Moosia's brother, Dr Paul Haffkine, the other microbiologist in the extended family. He had made a phone call to Waldemar while the latter was in Moscow trying to set up a meeting, but somehow it eluded both of them. There is something oddly uneasy about all this.

There was still one of the three sisters, Henrietta – Ietta – living about as far away as could possibly be imagined, in the city of Barnaul in western Siberia. It was August, the dog days of summer hanging thickly over Russia. No need to go home just yet, and, in any case, Haffkine was no longer quite sure where home was. He had never really managed to make one for himself. He had once said that 'solitude is my condition'. Even in England at the height of his acclaim, Lord Moynihan had noticed how 'lonesome' he seemed in his handsome solitude. He had not only not married, he seemed to have never come close to it.

There had been a girl – Justine – in Odessa who seems to have said no, and who in any case he left behind when he went to

Geneva. There is another dimly suggested story of being turned down again at the time of his great trouble in 1902. When asked about his bachelor life, Haffkine came up with the unconvincing explanation that his work being so peripatetic and putting him in such daily peril, that it would have been unfair to any prospective wife to put her through such trials of uncertainty.

Was he asexual, then; was he gay? We will never know. I have no hunch to offer, and, unless I have missed something, the archive is mute. But such family as survived he resolved to see, because his heart was letting him know, with the odd hop, stumble and jab, that there was something wrong.

Nothing too exerting on the Trans-Siberian railway, though. Eight days to reach the river Ob. Samovars rolling down the corridors; nights rocked by the rhythm of the carriage racketing over the sleepers; sentinels of silver birch facing the tracks; then, through the boreal forests, hanging curtains of fir and larch. Eventually the train emerged on to the dusty plains of the steppes beyond the Urals and then after a week, as if in a dream, vast, dazzling carpets of wild flower meadows – poppies, daisies and cornflowers – as the train chugged towards the silvery Ob. Barnaul, where Haffkine alighted, was silver too, the mined ore smelted there enabling the communist empire of the Soviets to throw its weight around central and north-eastern Asia much as its tsarist predecessors had done. The air was metallic. Inhaled, it coated the tongue like bad medicine. There had been terrible fighting in western Siberia during the civil wars; shattered and burned-out buildings disfigured the streets. Signs of poverty – beggars and the unhealed wounded shuffled about. When they were reunited, Ietta told her brother she had suffered badly: there had been White Army pogroms; she and her children had been summarily thrown out of their house. Barnaul's once prospering Jewish community had been decimated, its four synagogues all but destroyed. In the midst of all this devastation and his sister's hardship, Haffkine's journal is, for the historian, frustratingly economical about what might have been said. He did not care, he wrote on another occasion, for those who wore their heart on a sleeve, always ready to display their emotions. So we are left with little idea of what brother and sister, separated by many thousands of miles and thirty eight years, said to each other in the time spent together, only that the talk was full of memories and that it went on for some days.

Ietta had a son, Leonid, living at Tomsk. The town had a university at its centre and Leonid was studying for his admission exams. He was evidently part of the future. But Uncle Waldemar didn't want whatever that Soviet future held to come at the expense of his people's tradition. They spent three days together and during that time in Tomsk, without brow-beating his nephew, Waldemar tried to speak to him about Judaism, showed him his travelling tallit, the prayer shawl. There was (and is) a poignant emblem of the fate of Russian Jews in Tomsk: the Soldiers' Synagogue, meant for 'Cantonists' who had been taken for twenty five years of service in the tsar's army and who, on completing their term, had returned to Judaism. Cold-shouldered by the resident Jewish community, the old boys cut loose by everyone had built with their own hands the graceful wooden structure, carving the Magen David, the shield of David, at its entrance. Inevitably, it was one of the many synagogues seized, shut and converted to other purposes by the Soviets.

Autumn fell on western Siberia; daylight was in retreat and Haffkine was back on the week-long train to Moscow. A pity he had not gone further in the opposite direction, east, all the way to Harbin in Manchuria, where there was a substantial Jewish community and his nephew Paul was director of the Plague Institute. Uncle and nephew would have had much to talk about. In 1911, a brutal outbreak of pneumonic plague in Manchuria had killed 60,000 before ebbing at the end of the year. It returned in 1921 and again in 1928. That its fatalities were less in the later epidemics was probably due to Paul Haffkine's work. In 1911, at the height of the first epidemic, the Malay scientist Wu Lien-teh, the first Chinese graduate of Cambridge and subsequent student of Ronald Ross at Liverpool and the Institut Pasteur, had organised an International

Plague Conference at Mukden, in Manchuria, the first to concentrate on the disease since Adrien Proust's Sanitation Conference in Venice in 1897. It was Wu who had discovered that the Manchurian plague was pneumonic, transmitted between humans by respiratory droplets and all the more deadly for it. It was why he modified the standard surgical mask by adding cloth and gauze, making it available for daily use in pandemics. The N95 is not only now mostly made in China; it was the invention of a Chinese scientist a century ago.

A photograph of the delegates was taken. Paul Haffkine is in the second row, pale and moon faced; in the front, in the seats of honour, Kitasato Shibasaburo, now Baron Kitasato, sits next to Wu Lien-teh. In 1931, Kitasato's empire (about which he had courageously mixed feelings) would invade Manchuria and its army take Wu prisoner. To their left in the front row is Danylo Zabolotny, who argued that the reservoir vector for the Manchurian plague bacillus had been the Siberian marmot. Zabolotny had been a prize student of Mechnikov's during the brief period between 1886 and 1888 when he had been the chief of the Pasteur rabies institute in Odessa. Here they all were: Italians, Americans, French, Chinese and Japanese (but no British), an improvised epidemiological League of Nations, just a few years before the world was engulfed in war and another pandemic – influenza – killed millions.

In 1911, Metchnikoff himself was living in Sèvres on the western suburban edge of Paris, not far from Boulogne-sur-Seine, where Waldemar Haffkine would settle down after his return from the United States. But 1916, the year Haffkine came back to Paris, was also the year Metchnikoff moved into what had been Louis Pasteur's residential rooms in the Institut. Metchnikoff was seventy one when a heart attack took him, a decent age but possibly not what he had hoped since he had spent much of his time since winning the Nobel Prize in 1908 researching and thinking about the causes of senile decay. Those were to be found, he was sure, in the microbial wars battling it out deep inside the human organism. Specifically, the agents of senescence were putrefying bacteria lodged within the gastric system which, if unchecked, in time would affect the function of all the vital organs. In his gladiatorial mind-set, Metchnikoff believed that this metabolic attrition could be reversed by the white knights of the gastro-microbial world he called 'pro-biotics'. Why was it, he wondered out loud (knowing the answer in advance), that Bulgarian peasants lived well beyond the life expectancy of other Europeans and Americans? Yoghurt! In particular the sour and slightly fermented yoghurt they consumed daily. He followed their example and pressed it on others, even though the Bulgarian yoghurt habit may have come too late to extend his own span. 'Pro-biotics' disappeared when Metchnikoff did, only to return in the 1990s – though, alas for it is beautifully memorable, his face has yet to appear on any yoghurt packaging.

Metchnikoff had become almost as famous as Pasteur. In his *Doctor's Dilemma,* first performed at the Royal Court Theatre in London in 1906, George Bernard Shaw has one of his characters retort, when grandly informed about the action of 'white corpuscles', 'That's not new . . . that's what a man called Metchnikoff calls the phagocytes.' The award of the Nobel Prize in 1908, shared, not altogether to his pleasure, with Paul Ehrlich, who had quite different views on the operation of the immune system, only broadcast Metchnikoff's fame further. Instead of showing up in Stockholm for the ceremony, the old man went on a grand tour of acclaim, which, for once, included a Russian stage, despite the country stirring mixed memories. A major reason for his trip was to visit Leo Tolstoy, a year before the writer's death. There they are in the grounds of Yasnaya Polyana, two cranky geniuses, each, in their own way, refusing to smile for the camera.

Tolstoy had been one of Metchnikoff's sharpest critics, writing him off as a fetishist of science who dismissed the entire spiritual side of life, which was the saving of humanity from mere mechanical existence. A version of Metchnikoff appears in Tolstoy's last novel, *Resurrection*, to exemplify this higher form of callow narrow-mindedness. But when they actually encountered each other, Tolstoy – like so many others – was disarmed by Elie's still child-like passion, realising that for Metchnikoff science was a kind of sublimity. 'He believes in science', Tolstoy wrote, 'as in Holy Scripture – he is a sweet and simple man but as some weak men get drunk on alcohol he gets drunk on science.'[4] But Metchnikoff's permanent intoxication was justified. His understanding of the function of inflammation, the revelation of cells within the body capable of engulfing invasive bodies, decisively changed pathology and the medical science that depended on its wisdom. Almost as important, towards the end of his time at Pasteur, Metchnikoff showed that syphilis, previously thought to attack humans exclusively, could be present in monkeys. This meant that animal experiments could be used to affect human outcomes. The discovery, anticipated by Alice Corthorn, also spoke to what in our own time has become a truism of epidemiology: that most of the devastating infectious diseases have a zoonotic origin, and startlingly, as we now know to be true of COVID-19, this can be a two-way transmission, with humans infecting the animal kingdom.

Leo Tolstoy and Elie Metchnikoff, 1909.

For all his daily intake of Bulgarian yoghurt, and his *Essais Optimistes* that celebrated the prolongation of life, something died inside Elie Metchnikoff on the outbreak of war in 1914. The year before, at what was considered an advanced age, he had been offered the directorship of the Institute of Experimental Medicine in St Petersburg. He was, he replied, 'deeply moved' by the invitation, but had to decline. He had flourished in the French Third Republic, however imperfect its democracy. And 'though I am an opponent of any politics, I would not be able to witness indifferently that ruining of science which is preferred now with such cynicism in Russia'. There may have been some especially egregious instance of this in Metchnikoff's mind, but he must also have thought back to the time thirty years earlier when he had Haffkine in his laboratory; what he had had to do to spring his protégé from prison, and the relentless political pressure put on him in the years that followed. If anything, the autocratic surveillance of knowledge had got worse, not better. After his heart condition deteriorated, he moved to Pasteur's own apartment to expire. To close the biological arc, he told Emile Roux that when the moment came (as it did on 15 July 1916), he wanted 'to be incinerated in the great oven where our dead animals are burned'. Roux assumed this was some sort of darkly facetious piece of Russian humour. But Olga confirmed the instruction to be in Elie's will. His ashes were gathered and, as also specified, set in an urn that was then stored in a cupboard at the Institut.[5] Olga, who herself had become an experimental biologist, outlived him by many years and went on to publish Elie's biography in 1921. To date, it has gone through fifty editions in multiple languages.

The war threw a profound shadow over another of Haffkine's comrades. On 26 August 1914, barely a month into the hostilities, and following disastrous defeats at Mons and Charleroi, regiments of the Second Army Corps of the British Expeditionary Force were ordered to turn at Le Cateau and engage with the headlong German advance. It was a sacrifice to slow the enemy down and it succeeded, but at a bloody cost, especially for the Royal Scots, in which Ronald

Ross's nineteen-year-old son, Ronald Campbell Ross, served as second lieutenant. Reported as missing presumed dead, it took two agonising years before the parents had their son's death confirmed. Ross went on teaching at Liverpool, but his instinctive sense of fight turned sour. In 1898, an Italian parasitologist, Giovanni Battista Grassi, had published his demonstration of the life cycle of the malarial parasite just a year after Ross. It was originally intended to award the Nobel Prize to both men, triggering a burst of outrage from Ross, who was the sole winner, and the antagonists continued to feud well into the 1920s. Ross became bitter at being denied a proper financial reward for his malaria discovery; he was constantly picking fights with colleagues and overseers of the Liverpool School, from which he periodically threatened to resign; and despite the knighthood and honours, he never felt he had truly been given his due. He continued to churn out bad poetry and fanciful novels including *The Spirit of the Storm*, *Fables and Satires* and *Lyra Modulatu*. His 500-page memoir on the 'great problem' of malaria casts himself, not without reason, as a combatant against the ignorant, complacent and illegitimately powerful people who ran the institutions of the British empire and whose failure to support his campaign of eliminating the breeding pools of anopheles mosquitoes cost untold numbers of lives.

Ross and Haffkine – the one intemperate, the other painfully reserved – became allies because they both cast their lives and their work as a story of struggle against the odds: battlers for the integrity and social usefulness of science against the self-serving habits of power. One a knight, the other a companion of the Order of Empire, they were still, in their own heads, The Opposition. But could one work and live for obvious exploiters of imperial power and still reckon oneself a champion of public good? William Simpson spent much of the rest of his life after the Malkowal disaster working in Africa, first in Sierra Leone and the Gold Coast (now Ghana), not least in trying to protect Africans from malaria. But in 1929 he went to the Chester Beatty Roan Antelope copper mines in Northern Rhodesia (now Zambia) where a brutally exploited workforce was perishing from any number of terrible infectious diseases including cholera. Simpson did what he could, separating the sick and contagious from the healthy, and instituting sanitary clean-ups which were right for cholera and wrong for plague. He did succeed in dramatically lowering mortality among the African miners. Was the result, then, a reprehensible enabling of imperial exploitation or was it the disinterested mitigation of the extraction industry's gravest evils? At any rate, Simpson never swerved from his allegiance to Ross and died in 1931 of pneumonia at, of course, the Ross Institute in Putney that he had helped to found. How the great men of microbiology liked to die among their own!

In fact, not all of them did; and not all thought of themselves as crusaders, even when they had been. In the late 1920s and 1930s, an elderly gent could be seen on Brighton beach, flying model gliders in the breeze, along with a small group of neck-craning helpers. Aerodynamics, whether in birds or in machines, had long been an obsession of Ernest Hanbury Hankin's. But not the only one; of all his peers, friends and colleagues, Hankin was the least single-minded, the least narrow in his fascinations. It was one thing in India to become interested in the toxins delivered both by cobra and opium, and to detect (a breakthrough) that the Ganges contained viral phages that consumed harmful bacteria; another to become a serious scholar of Indian, particularly Mughal, architecture, and the mathematical and geometric web governing ornamental decoration. On all this (and more), both in India and back in England, first in Norfolk, then at Torquay and finally in Sussex, Hankin completed deep research, wrote and voluminously published. There was nothing which seriously caught his attention on which he wouldn't try to systematically organise his thoughts, hence his paper on 'dust-raising winds and descending currents' for the Indian Meteorological Department. Because he had engaged deeply with Indian religions, Hankin had come to value modes of thought and epistemologies usually stigmatised in the imperial west as 'primitive' or 'inferior'.

In 1921, he published *The Mental Limitations of the Expert*, which made the case for intuitive thinking as equally valid, creative and productive as the knowledge-gathering processes conventionally taught in the western tradition. This, half a century before the fast and slow thinking models described by Kahneman and Twersky.

In 1931, considering popular responses to epidemic terror, he published 'The Pied Piper of Hamelin and the Coming of the Black Death' in the *Transactions and Proceedings of the Torquay Natural History Society.* The shadows drew in. With political and military brutality beginning to overwhelm Europe, Hankin began to reflect on what kind of creature mankind really was. To an extent not readily acknowledged, he thought, decidedly beastly. The content of *Nationalism and the Communal Mind*, published in 1937, belies its conventional title. Its pages are full of palaeo-cannibalism: the eating of hearts, the ritual killing of kings, the drinking of blood from human skulls, all of which the courteously mild Ernest Hankin had found in the records of remote ancestors, and not entirely mistakenly. We are still those creatures, he thought, as the shrieking got louder and the bombs began to be loaded into the bellies of big planes. There was a difference, he wrote, between 'national self-respect' (good) and 'national self-assertion' (not good); wound the former and you'll end up with the latter. But then, like so many of his generation, he rather thought Herr Hitler was a case in point and had a point. Two years later, the beast emerged from his lair very much in self-assertive mode. And Ernest Hanbury Hankin died.

After his life as a scientist took a turn for the worse, Waldemar Haffkine never saw Ross, Simpson or Hankin again. It is possible that when, in the midst of his toils, humiliated and stripped of his employment, he returned to Pasteur to deliver his lectures on the plague vaccine, he ran into his old mentor Metchnikoff. But it seems unlikely. A coolness between the two had set in a long time ago; the older man let it be known that he had serious doubts about the efficacy of the famous 'Haffkine lymph', though he never spelled those out. The two of them just missed being near neighbours. But in the last few years of Haffkine's life, someone from his deep past did reappear.

In 1927, Hillel Yaffe, the bringer of pistols to the Odessa boys in 1881, caught up with Waldemar in Paris. The occasion was the memorial *Jahrzeit* of their mutual friend Alexander Marmorek, yet another bacteriologist who crossed swords with an established institution, though in his case, and this did not help, that institution was the Pasteur itself. When Haffkine was in India, Marmorek was in Paris trying to validate his conviction that Robert Koch had been mistaken when he identified tuberculin as the lethally toxic source of tuberculosis. Instead, he argued, tuberculin was an activator – the 'key in the lock' in his formulation, but the true toxin was to be found elsewhere in the serum and he, Marmorek, had found it, cultured it and developed an anti-tuberculosis vaccine. His experiments with rabbits demonstrated its efficacy. Except they didn't, at least not conclusively. When the Institut disapproved, Marmorek went rogue, reading a paper to the Paris Biological Society, resigning from his post at the Institut and setting up his own lab in the Paris suburbs. He died early in 1923 while the jury was very much out on his TB vaccine. But he had also been an ardent Zionist, a friend of Theodor Herzl's and, in Haffkine's eyes, when he had come back to Paris, altogether a Good Jew; someone whose memory was a blessing.

Yaffe was four years younger than Haffkine, in his mid-sixties when they met up again and still sporting the trimmed beard that had been the facial uniform of radical students, some of whom went on to lead revolutionary lives and die in the prisons of tyrants. Like Haffkine, he had had a police record in Russia, had had a bruising interrogation and then moved to Geneva where he organised protests against the surveillance of students by the tsar's spies. In their Geneva days, he remembered Haffkine as conspicuously irreligious or at least coolly indifferent, to the point that his friend had no idea of the dates of the religious festivals and high holy days, much less bothered to put in an appearance at the synagogue. So in Paris, thirty years later, Yaffe was amused to see his old friend don a yarmulke and recite the *hamotzi* blessing over bread before taking a bite.

It had been Yaffe, after all, who had been the one to follow his sense of Jewish history to Palestine where he had turned his medical-biological training to ophthalmology, in a region cursed by trachoma. But in the swamps at Hadera in western Samaria (Shomron), he found his true mission: the eradication of the malaria which made no distinction between Jew and Arab. Yaffe caught up with the findings of Ronald Ross, knew that the marshy breeding ground for anopheles had to be drained and the recovered soil fixed with eucalypts. Towards the end of their lives, then, they could both reasonably feel that their days on earth had not been spent in vain.

They met again a year later in 1928. Haffkine's illness had got much worse. The pains in his chest came more often; sometimes he spat blood. He knew his time was short and he dreaded dying before he had accomplished one last goal: the creation of a charitable foundation to fund Jewish religious education in eastern Europe. He had sold his house in Boulogne-sur-Seine and the proceeds went into the founding endowment. He had moved to Lausanne, ostensibly to find some quiet by Lake Geneva, but such was his determination to put everything in legal order, he was constantly on trains, back and forth to Berlin, where the Deutsche Hilfsverein worked with the conditions of east European Jews, then to Paris and back to Switzerland. The more his heart told him to stop, the more he seemed to defy its caution. Hillel Yaffe caught up with him on one of those trips and for a moment Haffkine paused. Only a moment. By the lake they talked of their time in Geneva. We should go, Haffkine said, and they did, visiting old haunts, the labs, the cafes. Through it all, Yaffe said, he saw that Haffkine had no fear of death; it was after all a natural, inescapable thing. The only anxiety was what it might interrupt.[6]

Haffkine's last year personified his argument that the embrace of Judaism did not mean the abandonment of science. In Lausanne, through a scientist friend, he was able to find a little space in a laboratory where he turned his mind to tuberculosis, which, Marmorek's efforts notwithstanding, still defied a demonstrably effective vaccine. In between hours in this last lab, he talked with a young rabbi based in Montreux and took him, improbably, to the intensively restored Byronic Château de Chillon sitting romantically on its rock on the lake. But when he died on 25 October 1930, none of his scattered family were nearby; nor any of the true friends of his life and times. He was taken in the condition in which, as he said, he had mostly lived: solitude.

There were obituaries: mean-spirited in the case of William Bulloch; generous, of course, by William Simpson. Many of the writers commented on the quiet dignity with which Haffkine habitually carried himself. But unlike so many of the grand figures of bacteriology, theoretical and applied, his story slipped quickly into oblivion. There was a Ross Institute; a Mechnikov University; a Hillel Yaffe Medical Center. Nothing bearing the name Haffkine; except in one place.

In 1925, Surgeon-Colonel Frederick Mackie, director-in-chief of what was still called the Plague Research Laboratory at Parel, wrote to Haffkine. Mackie had been on the staff there when Haffkine was director between 1899 and 1902 and remained a passionate admirer, a rarity still in the Indian Medical Service. But then Mackie was himself remarkable: the discoverer of the insect vectors of kala azar and sleeping sickness. For two years he had run a quiet and then not-so-quiet campaign to rename the Parel laboratory the Haffkine Institute, pestering the government of Bombay until the thing was done and he could write to his old chief, who was deeply moved, and wrote back expressing gratitude and saying that the years in Bombay, fighting plague, had been the happiest and most satisfying of his entire life.

In 2021, the Haffkine Institute, together with its Maharashtra state-owned commercial development arm, the Haffkine Bio-

Pharmaceutical Corporation, and India's giant drug company, Hyderabad-based Bharat Biotech, announced that the Indian government had contracted them to produce billions of doses of Covaxin to combat COVID-19. The vaccine would first of all be meant for India, but would be exported to developing-world countries at very little cost. The World Health Organization adopted Covaxin as a crucial help in extending COVID vaccination to parts of the world where vaccination rates were low or out of reach. By early 2022, Covaxin was said to offer 80 per cent protection against infection and significant mitigation of lethal severity. But in April that year, the WHO suspended delivery of the vaccine, citing irregularities in manufacturing practices. In November, a Bharat executive admitted further disturbing discrepancies between the number of enrollees in clinical trials and numbers reported. More dismayingly, it seems that a placebo control group had been dropped entirely from stage 2 of the clinical trials.

Haffkine, who pioneered scrupulous trials in India and who always paid special heed to the necessity of a control group, would have been shocked by this. But the long rooms at Parel

Haffkine diorama, Haffkine Institute Museum, Mumbai.

where he made and stored his plague culture no longer do much of that work. Instead, the Institute now specialises in producing antidotes to snake venom, one of the original fields of research during Haffkine's tenure as director and still a serious, widespread and deadly matter in contemporary India. Visitors can observe a display of non-venomous snakes. Those who make their way through the luxuriantly planted grounds will find relics of the old place: the lion finials atop the ceremonial staircase banisters that greeted Bertie, Prince of Wales, in 1875; the Durbar Room, still in its original colour scheme of green and gold. But there are other sights too: enormously enlarged models of microbes: rabies, *E. coli*, HIV, influenza and bacteriophage virus – the most common organism on the planet. On a staircase landing is an installation of life-size figures, a translation into plaster of the photograph which for a while made Waldemar Haffkine famous: the vaccinator inoculating a small Indian girl, surrounded by the people of the Calcutta bustee, an Indian and a British assistant on the scene.

The display is no masterpiece. Whereas, inadvertently, a photograph – the last of him vaccinating – may well be.

A burning day at the coalfield of Jharia, halfway between Patna and Calcutta, in May 1908. Cholera has broken out in the mining community as it does with dreadful regularity. Half the miners together with their families have already fled; many of the European managers are preparing to leave as well. At least a hundred miners die in a single week, so that the countryside, as a witness reported, 'is covered in corpses'. But manpower is so short that there is no one to take the bodies for cremation: they lie unburied in swamps and paddy fields for vultures and pariah dogs to gnaw at. Stories have been circulating about dogs bringing human remains into houses.[7]

Waldemar Haffkine is so horrified by the misery that he volunteers to inoculate the miners and managers against cholera as he had been doing in the 1890s. But as his reputation had dimmed, so the prejudice against inoculation in the Indian Medical Service had revived and the practice abandoned in some of the most vulnerable and poorest communities. When he asks for two medical assistants to help with the urgently needed operations, he is turned down. A request to establish a centre for inoculation at the mines is described by one of the many unsympathetic sanitary officials as an act of 'characteristic audacity . . . I see Haffkine has been busy advertising himself in various papers'. You know. These People. Another warns that 'I foresee trouble brewing I'm afraid if any countenance is given to Mr Haffkine's demands'.

So Waldemar has to do it by himself. His old pupils who have stored supplies have provided vaccines, syringes and so on in this emergency, but he is still, as he says, 'badly equipped for the operation – culture tubes etc.' The nature of the vaccine meant constant replenishing with material from lab animals and he can't be sure that what he has been given will be effective, with the authorities in Calcutta making sure he is kept away from production. But the mining management – they have their reasons – are desperate and eager for him to inoculate, as are the miners themselves. So he goes ahead though with 'uncertainty and anxiety' washing through him.

Which is, I think, what we read on one of the two Haffkine faces recorded by the camera. The Inspector of Mines who is taking the shot, evidently no expert photographer, has miscalculated the shutter speed, making for a lengthy exposure during which his subject, nearing fifty now, a tad fleshier, but, as always, nattily turned out, has moved. But which of the two faces came first: the working profile or the reluctant surrender to the lens? I like to think that he conceded the pose, fraught as it was with solemn urgency, but then abandoned it to return to what, in the end, always mattered most: saving people from 'frightful mortality'. Whichever the order, what lingers with the afterburn of the image is some sort of knowledge, written in his expression, that, in the face of calamity and the wilful obtuseness of the powerful, there is only so much he can do, but do it nonetheless he must; an expression, close to sorrow, that haunts our complacency.

X

And in the end . . .

This history is not going away any time soon. On 12 December 2022, 1.7 million doses of oral cholera vaccine were delivered to Haiti were they were desperately needed to contain an erupting epidemic. For the first time in a decade, cholera cases appeared in Syria in late 2022, but the disease is no respecter of borders. It has struck Lebanon, and, in November 2022, the vibrio was identified in the Yarmuk river reservoir serving domestic and agricultural needs in northern Israel. Disaster is cholera's ally in flooded Pakistan and in parts of West Africa: Cameroon and Nigeria. Early in January 2023, Malawi was reporting a terrifying spike in cases. By the spring of 2023, the epidemic had spread to Mozambique. But global supplies of cholera vaccine are so depleted that a kind of double triage is now operating. There are attempts to prioritise the countries and regions that are in most critical need. And, in an echo of the same debate that pressed on Haffkine in the 1890s, the acute global shortage has resurrected questions as to whether a single dose might suffice for effective immunity.[1]

Such decisions can only now be managed through the kind of international collaboration envisaged by Adrien Proust. But it is exactly such agencies that have now become the bête noire of militant nationalists. It would, of course, be helpful if the two inter-connected crises of our age – the health of our bodies and the health of the earth – could break free from the distorting mirror of populist politics. But if you have got this far in the story, you will know that this is seldom, if ever, the case. The visionary advances of science, including those of virology and bacteriology, occur at an ever-accelerating pace and save lives as they do so. Hard-earned, exhaustively tested, truth, just as Thomas Nettleton, Angelo Gatti and Waldemar Haffkine hoped, always seems on the verge of overtaking error, when its exhilarating progress is sandbagged by indignation about foreign substances deviously introduced into our bodies. What could this be, it is said, other than the invasion of our veins authorised by remote 'experts' claiming a monopoly of medical wisdom, but who are, in fact, imposing clinical obedience in the name of Doing Us Good? To those for whom knowledge is conveyed by revelation, the accepted hierarchy of wisdom is all the wrong way round. Discernment begins with the judgements of God, followed by the urgings of common sense (especially pronounced by those on television and social media claiming to speak for it), and only then enlightened by science, always keeping in mind that much of what gets presented as irrefutable fact is, actually, just another set of opinions.[2]

To the most frantic alarmists, champions of vaccination are demons walking among us in lab coats, disguised as politically neutral scientists.[3] Pretending to be disinterested public servants, they embed themselves in the hardened silos of the Deep State, surfacing to position themselves close to power. Every so often they appear in public, behind or beside elected officers of state whom, through some dark art, they have convinced to act in ways that curtail liberties, shrink the space of daily life, and interfere in decisions properly belonging to individuals and families: the schooling of children; the choice of whether or not to wear protective clothing.

The Anthony Faucis of this world.

Something about inoculators, vaccinators, epidemiologists gets under the skin of public tribunes for whom nothing, certainly not epidemiology, is politics-free. Their fury swells into maddened vehemence to the point where it becomes commonplace to wish inoculators banished, imprisoned or dead. Gatti was accused of spreading the disease he claimed to fight; Lord Curzon expressed the hope that Haffkine would be hanged for what the viceroy judged to be criminal irresponsibility. But no epidemiologist has been subjected to more violent abuse than Anthony Fauci. In the minds of his adversaries, the longevity of his career – director, for thirty eight years, of the National Institute of Allergy and Infectious Diseases (NIAID); adviser to presidents of both major American political parties – is just further evidence of his creation of a fiefdom, locked off from public scrutiny. That he was awarded the Presidential Medal of Freedom by George W. Bush, not least for his work in developing a global programme to fight HIV/AIDS estimated to have saved more than 20 million lives, becomes a story of how he managed to pull the wool over the eyes of Republicans who should have known better. Scientific pre-eminence, years of research, are brushed aside in the demonising caricatures so that for Tucker Carlson, once the most-watched anchor of Fox News' opinion show in the United States, Fauci is 'a hypocritical buffoon'; and, as if height somehow was a sure sign of evil intentions, 'an even shorter version of Benito Mussolini', 'a midget Stalinist' and 'a tinier version of the Dalai Lama'. Do not, then, be deceived by the forthright manner of elfin scientists, since, according to the 6-foot Carlson, Fauci is 'a dangerous fraud who has done things that in most countries at most times in history would be understood perfectly clearly to be very serious crimes'.[4] In one of the *obiter tweeta* he periodically delivers to the world (in a misunderstanding of the parts of speech as well as science), Elon Musk let it be known that his 'personal pronouns' were 'prosecute Fauci'.

This arch-malefactor is, apparently, not just your regular criminal, but a genocidal *war* criminal. In November 2021, Lara Logan, one of Carlson's colleagues at Fox News, accused Fauci of causing 'immense suffering' and declared that he 'does not represent science; what he represents is Josef Mengele', the Nazi extermination camp doctor who performed experiments on Jews in Auschwitz.[5] A month later, Jesse Watters, also at Fox News, urged 'activists' to ambush Fauci with questions that would be 'the kill shot . . . Boom boom!

He's dead! He's gone.' Fox described this as 'metaphor', but at least some of those who think Fauci the enemy aren't that subtle. One of seven emails sent to Fauci anonymously but in fact written by Thomas Patrick Connally, in April 2021, expressed the ardent hope 'that you and your entire family will be dragged into the street and beaten to death and set on fire'. Connally was subsequently prosecuted and convicted, but another deranged man set off from Sacramento, California, with an AK-15 in his car and a list of people he would kill in Washington, among whom, of course, Fauci prominently featured. The death threats have been persistent enough for Fauci to need round-the-clock security.

The obsession with Fauci has become a tic, twitching on the angry face of populist conservatism. Its rage can't afford to do without him. Retirement from the directorship of NIAID at the end of 2022 has not spared him from investigation by committees of the Republican-majority House of Representatives, seeking, so their spokespersons claim, to throw light on 'the origins of COVID-19'. The House committees assigned to pursuing that line of inquiry are the Oversight Committee and the Energy and Commerce Committee.

The demonisation of Anthony Fauci is the result of conservative media framing him not just as a 'fraud' but as a personification of the 'medical deep state', as another Fox presenter, Laura Ingraham, tendentiously coined it. What could possibly be less American than an epidemiologist out to rob Americans of their liberty of movement; the sovereignty of their physical person; the right, whenever and wherever they choose, to bare their face? In the most feverish conspiratorial minds, juiced by pulp sci-fi, the very act of vaccination, involving as it does the puncturing of the defensive casing of skin, becomes a pathway by which foreign bodies – a nanochip invented by Bill Gates, for example – can be surreptitiously introduced to enslave free citizens, without them even being aware of their captivity. But resistance to vaccine mandates, the guidance of government public health institutions and the offerings of pharmaceutical corporations has now gone well beyond conspiracy theory communities. Ominously, the rhetorical coinage of science as the enemy of liberty has generated populist dividends for those seeking power. The governor of Florida, Ron de Santis, widely assumed to be seeking the 2024 Republican nomination for president, has spoken of 'biomedical security state'.[6] In February 2023, the Lee County Republican Party passed a 'ban the jab' resolution 'to stop the genocide [*sic*] because we have foreign non-governmental entities that are unleashing biological weapons on the American people'. A 'Health Freedom Movement' adamant that decisions affecting health and survival should rest exclusively with the 'sovereign rights of the individual' has taken root in the culture of libertarianism and conservatism, and not just in the United States, where it is most powerfully established. The arbitrariness and unaccountability of public health decisions in authoritarian states like China, abruptly shifting from 'Zero COVID' coerced confinement to the opposite, has only reinforced suspicions that quarantines and the like were imposed in the interests of state power, not the safety of citizens.

The libertarian rhetoric of Health Freedom requires an unmasking of public officials who in its view have abused their purported scientific authority to violate the natural rights of individuals and families, but also a gallery of populist medical heroes determined to challenge that usurped authority. Predictably, then, it was not long before a search was on in the United States for an anti-Fauci, a people's physician who swore by a drug, or a mixed cocktail of them, that would obviate the need for vaccines. Enter Dr Vladimir Zelenko, characterised as 'a simple country doctor' but in fact a suburban physician whose practice was located in the ultra-orthodox Jewish community of Kiryas Joel in New York State. In the last week of March 2020, Zelenko wrote and published an open letter to President Donald Trump claiming that a cocktail of hydroxychloroquine, a drug used against immune system disorders including lupus, plus azithromycin and zinc, prevented the onset of any serious, life-threatening disorders in COVID-19-infected patients. Though Zelenko claimed that among the 600-odd patients using the prescription there had been 'zero' cases of intubation, ventilation or death, he had conducted no comparative trials, nor published any of his 'research'. His entirely anecdotal promotion did not, however, prevent the adoption of hydroxychloroquine as the prophylaxis and therapy of choice for hard-right media personalities, including Ingraham and Joe Rogan, who pushed it relentlessly as the free Americans' COVID antidote. For Trump, the drug he described as 'the greatest game-changer in medical history' would short-cut the exasperating wait for vaccines that might compromise his chances of re-election. Though irrational, there was also something about an oral medication that somehow seemed less invasive than being punctured by a needle. On 18 May 2020, at a press conference Trump announced that he had been taking hydroxychloroquine preventively. 'All I can tell you is so far I seem to be OK . . . I get a tremendous amount of positive news on the hydroxy . . . what do you have to lose?'[7] If 'you' included people with immune disorders or persistent malarial symptoms, the answer was plenty, as Trump's and Fox News' promotion of hydroxychloroquine generated a run on the drug, and an acute shortage for those vulnerable people who actually depended on it. In August 2020, a scientifically rigorous trial of the drug by infectious disease researchers at the University of Minnesota concluded that it had no measurable effect in preventing infection by anyone exposed, nor in mitigating severity of symptoms. But this did little to damp down the craze, nor to stop Fox News hosts from continuing to promote its virtues.[8]

The airwaves and hustings lit up with the enthusiasms of digital mountebanks. Right-wing talk radio hosts and Trump champions presented the scorning of masks as an act of courageous libertarian defiance. Conspicuous attendance at the mass indoor gatherings they had been warned to avoid became gestures of patriotic defiance, even when some of the danger-deniers like Herman Cain – Tea Party activist, CEO of Godfather's Pizza and former presidential candidate – died from COVID-19 after going to a Trump rally in Tulsa. In August and September, five popular right-wing talk radio hosts, including Phil Valentine, who released a single called 'Mr Vaxman'; Marc Bernier ('Mr Anti-Vax'); Jimmy de Young, who insisted that vaccination was 'a form of government control'; and Bob Enyart, who believed vaccines were 'immorally developed', died, one after the other, from COVID-19.

None of this pre-empted the spectacular popularity of yet another vaccine-alternative: ivermectin, an anti-parasitic and anti-inflammatory drug most commonly used to rid livestock and horses of nematode-borne disease, but occasionally prescribed for head lice and rosacea. In December 2020, Dr Pierre Kory, a lung specialist at a Milwaukee hospital with painful immediate experience of severe COVID patients, gave the first of two testimonies about ivermectin to the Homeland Security Committee of the United States Senate. Kory was adamant that despite 'the apparent success' of vaccine development, ivermectin, 'the penicillin of COVID' as he called it, had been shown to be dramatically effective at both preventing and treating the virus. One of the many papers he cited (several of which were subsequently retracted after doubts about method and evidence were raised) was an Australian study reporting the inhibition of the virus's capacity to replicate when grown in vitro.[9] What was not reported, at least by Kory, was that the concentration of the drug needed to achieve that result in a petri dish was incomparably higher than anything that could be safely tolerated by humans. But ivermectin – especially in 2021 when, after a respite, COVID returned in its Omicron variant – became the next populist panacea. It was cheap, available over the counter, and required no needle shots. Right-wing talk radio hosts embraced it as an alternative to vaccination, especially when the latter was a required condition of employment. In short order, ivermectin became the wonder drug of choice for organisations of doctors suspicious of, or hostile to, vaccination, including Kory's Front Line COVID-19 Critical Care Alliance, the British Ivermectin Recommendation Group (BIRD) and America's Frontline Doctors.[10] This happened despite the lack of any conclusive trials or any evidence of its success, either in prevention or in therapeutic treatment. To the dismay of government health agencies trying to persuade the population to receive boosters, the endorsement by those professional-sounding organisations made ivermectin ideal for anyone harbouring suspicions about Deep State medicine. In August 2021, 88,000 doses of the anti-parasitic drug were sold, compared to 3,000 two years before. One of its most irrepressible enthusiasts, Joe Rogan, who had been taking ivermectin weekly, went down with a nasty dose of COVID, as did Dr Pierre Kory himself, eight months after his testimony to the Senate in December 2020. As use of ivermectin shot up in the summer of 2021, so did hospital admissions for ivermectin poisonings, with patients reporting vomiting, diarrhoea, loss of balance, dizzy spells and, in some cases, serious cerebral disorders. A randomised comparative trial organised by scientists at McMaster University in Canada of 1,358 people, half of whom were given the drug and half a placebo, with subsequent tracking of their medical history, concluded that ivermectin made no difference whatsoever either to the chances of infection or to the mitigation of symptoms. But then earlier that same year, and to no avail, Merck, the manufacturer of the drug, had already warned that there was no evidence that it was appropriate or effective for COVID-19.

The mass appeal of ivermectin, like hydroxychloroquine, had nothing to do with reliable science, much less data derived from clinical trials. It was what it was *not* that accounted for its intensive promotion by populist media and vaccine-sceptical physicians. It was not anything ordered up by impersonal institutions of the Deep State and its willing tool, Anthony Fauci. It was not an administered puncture to which one meekly submitted. Somehow, ivermectin was more . . . American; freely chosen, over the counter, self-dosed. Even its associations with the guts of animals made it somehow more authentically homestead. Likewise, its champions such as Kory presented themselves as practical doctors, working at the brutal ICU rockface of the disease; hardened by hands-on bedside experience, unlike remote, theoretically driven virologists, concocting vaccines of dubious effectiveness.

Fauci's most zealous prosecutors in the United States Congress represent him – and vaccines – as instrumental in a design to rob citizens of their God-given liberty. But they have also challenged his claim to personify the authority of science, claiming (without the credentials to do so) that he is himself guilty of playing fast and loose with the integrity of research. Senator Ted Cruz of Texas characterised Fauci as 'an unelected technocrat [who] distorted science and facts *in order* [my emphasis] to exercise authoritarian control over millions of lives'.[11] The most zealous of all Fauci's self-appointed Senatorial prosecutors, Rand Paul of Kentucky, the first in the Senate to go down with COVID, has made it a personal mission to devalue what he considers the overestimation of vaccine protection, insisting that post-infection immunity is 'nature's vaccine'. Specifically, Paul claimed, on the strength of two pre-prints (neither published nor peer-reviewed), that the immunity conferred by infection was at least as effective as vaccination, if not more so, in preventing reinfection and stated that he himself would not be getting vaccinated. Any other view, especially Fauci's urging vaccination and boosters on the previously infected, Paul insisted, was 'unscientific'. Paul took special umbrage at Fauci's implication that his own views were coloured by hostility to science. How could that be true, he asked, since he was, after all, a trained ophthalmologist? But Rand Paul's confidence in naturally acquired immunity took no account, first, of questions over the duration of that immunity and, more critically, of whether or not immunity acquired against, for instance, the Delta version of the virus would work against Omicron or other emerging variants. Paul publicly rejoiced that his opinions would deter previously infected people from getting vaccinated or boosted.

The hounding of Fauci has gone much further. During the spring and summer of 2021, Paul submitted a bill to Congress that would have eliminated NIAID altogether, along with Fauci's position as director (or 'dictator-in chief' as he once called him), to be replaced by three separate institutes, with their heads all subject to Senate confirmation. He then urged the Department of Justice to initiate criminal proceedings against Fauci for lying to Congress. The perjury, so Paul claimed, had been committed during a fiery sparring match between the two men in May of that year in the Senate, when Paul had put a sinister interpretation on NIAID's funding, in 2014–15, of a research collaboration between EcoHealth Alliance and the Wuhan Institute of Virology (WIV). The context of those collaborations was the creation by the Obama administration, following the successful containment of an Ebola epidemic in west Africa, of an Emergent Pandemic Threat (EPT) programme to predict new infections resulting from zoonotic spillovers. EcoHealth Alliance was at the centre of that work, and since the WIV already had experience of studying the sources of SARS-1, MERS and avian influenza, further collaboration seemed not only sensible but urgent.

None of this prevented a sinister imputation being placed on a chimeric virus made in the WIV from a spike protein taken from horseshoe bats, a mammal known to be the reservoir of previous SARS viruses. This virus was capable of binding to ACE-2 receptors in mice modified to be closer to human susceptibility. It is true that the terms of the grant required EcoHealth Alliance and its sub-grantee the WIV to report any enhancement or 'gain of function' likely to affect transmissibility to humans. It is also true that the grantees had not done this, though their defence – a critical one – was that the chimeric virus was genetically remote from SARS-CoV-2. They were, in fact, in a bind. On the one hand, the whole point of the collaboration was to study the likelihood of coronaviruses jumping, zoonotically, from wild mammals like horseshoe bats to humans; and this could only be done by modifying the virus so that it would engender illness in laboratory mice. On the other hand, the closer they got to realising exactly these potentially predictive experiments, the more they risked crossing the red line of the National Institutes of Health (NIH). In 2011, avian flu virus, thought to present no danger to mammals, genetically modified in the lab of Ralph Baric at the University of North Carolina, was shown to be transmissible between ferrets. Red lights flashed; rules were instituted for reporting any such experiments that could be deemed potentially dangerous to human populations and for three years a moratorium was imposed on any such 'gain of function' experiments. Though the ban was lifted in 2017 as further waves of diseases hit with ever-increasing frequency, needing scientists to 'see round the corner' at what the immediate future might hold, 'gain of function' had become a byword in political as well as some scientific circles for unacceptable risk.

In 2020, the refusal of the WIV to provide EcoHealth Alliance and the NIH with records of experimental protocols, and its deletion of genomic sequences, played into the hands of those insisting that the origins of COVID-19 were to be found in a deliberately engineered chimeric virus in Wuhan that had escaped in a laboratory leak. Even if that was decidedly unproven, WIV resistance to sharing lab records made an accusation of, at the very least, reckless experimentalism inevitable. In the light of its inability to see that documentation, in August 2022, the NIH terminated part of its grant to EcoHealth Alliance. Appeals were made to the NIH – by, among others, seventy seven Nobel laureates – to reconsider its decision, but to no avail. Some of the scientists best placed to make reliable predictions about emerging coronavirus threats, based on experimental work done in the very place where such threats arose, now had their hands tied.

Rand Paul was in no doubt that NIAID's funding to the EcoHealth–WIV research amounted to an act of criminal responsibility. This stopped short (but not by much) of Tucker Carlson's claim on Fox News that 'the guy in charge of the American response to COVID turns out to be the guy who funded the creation of COVID'[12] or Donald Trump's view (repeated in November 2022, during the speech announcing his presidential candidacy) that the 'China virus' had been cooked up as a biological weapon, not least to damage his re-election.

During the Senate hearings, Fauci insisted that the WIV chimeric virus did not enhance the ability to cause disease in humans, the criterion for determining whether gain-of-function work should be supported or suspended. But fine points of methodology were lost in bitter arguments over definitions about what did or did not make for GOFROC (gain-of-function research of concern). The back-

and-forth arguments were, in any case, somewhat beside the point. There was, and is, in fact nothing inherently alarming about the genetically manipulated creation of new viruses. Ralph Baric has always maintained that this is the only way to test the likelihood of a novel virus's emergence as potentially a lethal pathogen in the human population. In an admirable attempt to defuse the myths and fears attached by populist media and politicians to anything called 'gain of function' research, the authors of 'Virology Under the Microscope' in the *Journal of Virology* have explained that any number of indisputably benign advances could not have happened without it. These include melanoma and solid tumour therapy; the repair of cardiac pacemakers; the combatting of citrus greening disease; rabbit pest control; the treatment of bacterial diseases; increased nitrogen fixation to reduce the use of fertilisers; the understanding that viruses like H5N1 bird flu are capable of transmission to mammals including humans; and, not least, adenovirus vaccines for dengue fever and COVID-19.[13] For that matter, Baric and Stanley Perlman at the University of Iowa had used 'passaging' – the introduction into a healthy animal of viral material collected from an infected animal – followed by successive repetitions until the required strain of enhanced virulence to test immune response was achieved.

Does this sound familiar? It is, in fact, the procedure used at the Institut Pasteur to create rabies and diphtheria vaccines and then by Waldemar Haffkine (thirty nine passages through guinea pigs) to make his cholera vaccine in the summer of 1892.

Much of vaccine history presupposes an unbroken chain of animal and human biology. The inter-connectedness is not always mutually benign. Rats and fleas have scuttled and gorged their way through the centuries of this story and still do. Human appetites have industrialised livestock farming to the point where massed animals have become breeding grounds for diseases, sometimes, but not always, held back by antibiotics. The same appetite has degraded the barriers between wild species and human habitats so that zoonotic diseases are coming at us in ever shorter intervals.

M. W. Sharp, portrait of Benjamin Jesty, 1805.

The feedback loop is not invariably ominous. The involuntary gifts of animals are written into the etymology of the word 'vaccine': from cows. In 1774, Benjamin Jesty, a farmer living in the Dorset village of Yetminster, took his wife and two sons to the neighbouring hamlet of Chetnole where a farmer had a cow showing lesions of cowpox on its udder. Jesty himself and two of his milkmaids had contracted the pox some years earlier with minimal symptoms. Nine years before, two Gloucestershire doctors had already reported this revelation; one of them, John Fewster, had sent a report to the London Medical Society suggesting vaccination rather than inoculation, not least because the subjects seemed not to be at all contagious after their treatment. Now, Farmer Jesty took a darning needle and pricked the arms of his wife and two sons, thus vaccinating twenty two years before Edward Jenner. But his pioneering remained unpublicised for decades. In 1805, after Jenner had taken much of the credit for the breakthrough, Jesty came to London where his son Robert allowed himself to be demonstratively vaccinated a second time. Jesty himself appears to us in all his rotund, rosy-faced country bluffness in a portrait painted by Michael William Sharp.

Perhaps it's the nourishing, domestic character of cattle, their ubiquity in the lactic flow of life as ungulate second mothers, that makes their good pox, the matter of infection protection, feel like a 'natural gift'. Wildness, on the other hand, whether introduced into the densely populated human world by the traffic of false medicine or bush meat, or from its migration into the spaces of human occupation, can seem the bearer of dangerous ills. In medieval folk traditions, iron horseshoes, hung on a wall, were thought to protect against visits from the Devil and thus became tokens of good fortune. Horseshoe bats, especially in the region of south-west China bordering Myanmar, Cambodia and Vietnam, where the fifth pandemic of bubonic plague arose, have been identified, since 2005, as carriers of novel coronaviruses, including SARS.[14]

But then there are horseshoe crabs: a wild creature whose defence against foreign bodies, developed over millions of years, has become ours too. Thus, a survivor from deep, archaic time has yielded a modern necessity.

Atlantic horseshoe crabs coming ashore to spawn on the beach, Delaware Bay, Delaware.

In the dog days of summer 2020, when the glowing light of August fooled us into believing the virus was being held at bay, I threaded a path between the remains of horseshoe crabs. Hundreds of them, lying this way and that, patterned the Cape Cod beach at low tide; dozens more lay in the grooved mud. Most were prone, bronze carapaces like so much armour and helmet gear fallen from the losing side in some ancient sea battle. Eyelets for their pair of compound eyes pierce the frontal plate, the prosoma, much where one would expect them. Eight other simpler eyes are distributed around their body. The spectacle was not a beachy graveyard; many of the shells were washed-up exoskeletons from one of the eighteen moults horseshoe crabs make during their growth to spawning adulthood, and so were evidence of teeming life out there in the deep Atlantic. But there were some full carcasses lying supine, tumbled over by the force of incoming waves. Horseshoe crabs can live to twenty five, a nice old age, but the slower veterans are often too feeble to use their spiky telson tail to right themselves after taking a hit in the tide. Those upside-downers exposed the neatly packed array of working parts within the bowl of their shell: stacked rows of book gills that manage the gaseous exchange keeping them alive; five pairs of crawling legs; a pair of small appendages armed with bristles that rub and grind up razor clams and mussels before guiding the food into the horseshoe's mouth. The telson, extending out from the posterior plate, acts like a directional rudder, especially for comparatively weak juveniles, while they are swimming, but was imagined by romantically minded nineteenth-century naturalists to be a weapon, possibly with venom, suggesting the nickname 'sword crab'. But although they move forward from the discarded exo-skeleton as they moult (rather than reversing like true crabs), horseshoe crabs lack the antennae and mandibles of crustacea. They belong instead to the class of merostomata (meaning, literally, 'legs attached to the mouth'), making them anatomically closer to spiders and long-extinct eurypterids, sea scorpions, the most impressive of which could grow to be 2½ metres long.

Horseshoe crabs are prodigious survivors. Palaeontologists cringe when glossy natural history magazines call them 'living fossils'. But the larvae and juveniles, especially, do resemble trilobites and other creatures of the Cambrian period. With their clanky armour plating and hairy ganglia, adult horseshoe crabs look like a hold-over from an era when the marine world was primarily populated by complex, but invertebrate, life forms. This makes *Limulus polyphemus* and its three Asian cousin species the most stupendously preserved of all time travellers. Horseshoe crabs as we now see them first appeared 200 million years ago, but fossil evidence has revealed a smaller but metabolically ancestral form twice as ancient. Those proto-horseshoe crabs first arose in the Palaeozoic oceans 400 million years distant, when most of the earth's land was still massed in the super-continent Pangaea, and were among the earliest arthropods to creep momentously towards solid terrain. The great Ordovician extinction of 440 million years ago carried off a quarter of all marine species but not the horseshoe crabs. An astonishing fossilised specimen in the Mazon Creek site in Illinois dated to 310 million years ago shows a central neural structure essentially identical with that of the modern horseshoe crab.[15] They went on to survive the Triassic–Jurassic extinction 201 million years ago that did away with 70 per cent of all terrestrial and marine life, and the Cretaceous extinction that wiped out three quarters of sea creatures excepting horseshoe crabs and those johnnies-come-lately the eel, crocodile, turtle and shark. Around 67 million years ago, the Laurentian continent, which would become North America, broke from the Eurasian land mass, wrenching open a vast rift soon flooded by a new ocean: the Atlantic. Horseshoe crabs were then separated into occidental and oriental varieties. *Tachypleus gigas* is native to the waters of Indonesia, especially along the Sumatran coastline and southern China; *Tachypleus tridentatus*, the three-spine horseshoe, much the biggest of the species, lives in the waters of Japan, Vietnam and Korea, as well as China; while *Carcinoscorpius rotundicauda*, the mangrove horseshoe, exists (but barely) in brackish waters in India, Bangladesh and Malaysia. These three Asian varieties are all now in critical decline, overharvested as bait for the eel and whelk varieties that end up in sushi bars. They are also taken in Asia for remedies that are as spurious as those attributed to the scales of the pangolin. Their Atlantic cousins can still be found in substantial but dramatically thinning populations from the Gulf of Mexico, through the Carolinas, Chesapeake and Delaware, and as far north as New Hampshire and Cape Cod where I meandered through their beachy cemetery. They too are facing a crisis, though not as extreme as the Asian horseshoe crabs. But it is one that matters, since it turns out that at this moment of reckoning, the state of their health is fatefully linked to our own.

In the summer of 1963, Jack Levin, a bright young haematologist from the Johns Hopkins University medical school, arrived at the lab of a senior colleague from the same university, Frederick Bang, in Woods Hole, Massachusetts. Levin had never seen a horseshoe crab before, but Bang plunged his hand into the tank and brought out one of the marine wonders; marvellous not just because of its species antiquity, but because of the colour and, more importantly for the scientists, a unique property of its blood. Horseshoe crab blood is a spectacularly beautiful blue; milky in some lights, deep cobalt in others. Its hue is determined by the copper-based haemocytes which manage the exchange of gases in much the same way as our iron-based red haemoglobin. Ten years before, in 1953, working on the vascular system of the animal at Johns Hopkins where he was chair of the Department of Parasitology, Bang had noticed something remarkable. Horseshoe crabs have an open circulatory system, which makes them highly vulnerable to bacterial infections. But on contact with the slightest presence of some types of marine bacteria, their blood clots into a dense, knotty mass or a viscous gel, which, Bang concluded, rendered the toxin inert.[16] But when Bang worked with Levin, they saw clearly that it was endotoxins that triggered the coagulation mechanism. Endotoxins form part of the outer membrane wall of a particular type of bacteria which are capable of engendering lethal threats to higher organisms including anaphylactic shock, sepsis, and violent inflammations and fevers. Should endotoxins be present, even minutely, in drugs and implants – and vaccines – the consequences could be lethal. Up to that point, testing for endotoxins had been largely done with rabbits whose temperatures were measured every few hours over a period of days. But the method was laborious and, because of the possibility of independent causes of fever, unreliable. It occurred to Frederick Bang that since horseshoe blood reaction was instantaneous, a blue-blood test could be fast and faultless.

'Warp speed' was not an option in the 1960s. It took until 1972 for pharmaceutical-makers to realise the potential of horseshoe blood for their products and another five years before the Federal Drug Agency finally approved the production of lysate material extracted from horseshoe crab blood for the safety testing of therapeutic drugs and vaccines. Every year now, hundreds of thousands of the animals are taken to pharmaceutical facilities in South Carolina, Delaware and Massachusetts for copious bleeding. Conveniently for attachment to the retaining calipers and the receiving vessels, horseshoe crabs can be slightly bent at the hinge dividing their anterior (prosoma) from the posterior (opisthosoma) sections of their body. A hypodermic needle punctures the pericardial membrane allowing the collection of between 50 and 400 millilitres of blood, depending on the animal's age and condition, amounting to between 25 per cent and 40 per cent of its body weight. The targets of the procedure are amoebocytes carrying the clotting function. Extracted and compacted by centrifuge, the addition of water to the concentrate then triggers a bursting or 'lysing', releasing the coagulating protein.

Before the coronavirus pandemic hit, 70 million tests using LAL (limulus amoebocyte lysate) were performed every year on a variety of drugs, vaccines and implants, raising the going rate to $60,000 a gallon. Now, with the prospect of millions of doses of newly developed vaccines needing to be tested for endotoxins, the sky is likely to be the limit: a bonanza for the four American facilities producing the testing material.[17] Synthetic recombinants, made in Switzerland, have been available for some years and have been shown to be effective in signalling endotoxin presence. But in June 2020, concerned about the slightest possibility of dangerous contaminants, the United States Food and Drug Administration cancelled provisional approval for recombinants, restricting it instead to the horseshoe crab-derived lysate, upheld as the gold standard.

This has come at a cost. Steeply rising demand (before the pandemic) has already resulted in damage to the stability of the ecosystem in which *Limulus polyphemus* is embedded. For millions of years, horseshoe crabs have followed the same life cycle, swimming from ocean depths of between 20 and 60 metres, drawn by high tides during full and new moons to spawn on the Atlantic beaches. As they approach land, most of the smaller males use claspers to fasten themselves to the backs of the females, who then lug them on to the shore, so that they can get on with inseminating thousands of eggs laid in shallow circular depressions dug in the sand. As you would expect, there are some free-roaming laddish 'satellite' males that cruise the beaches condescending to do a bit of fertilisation as the mood takes them or not. Females will dig twenty such nests in a single spawning, amounting to between 80,000 and 90,000 eggs: enough to survive the appetite of the predators who have been waiting for just this moment. They are, in their own right, a natural wonder, none more so than the red knots who make a stop off on their 19,000-mile epic pan-American migration, from Tierra del Fuego to their breeding ground in the Arctic Circle, to feast on horseshoe eggs, rich in fatty acids. A host of other shorebirds feed off the same diet: sanderlings, palmated plovers, black-backed and laughing gulls. Larvae that have escaped the avian swarm then have a perilous six days' crawl to get to the shallows, where they are in danger of being eaten by another group of predators, including green crabs, loggerhead turtles and leopard sharks. Survivors of the gauntlet of the shallows will winter in place in the ocean before swimming deeper for the first of their eighteen moults to full adulthood. Five years pass before they are sexually mature enough to make a final moult before returning to the spawning grounds on the beach.

You walk along the shore, watch the gently rhythmic push and pull of the tide, sanderlings dashing pointlessly up-beach to get out of the way of incoming surf, and it all seems perennially imperturbable. But history's first lesson is that nothing is, certainly not the multi-million-year epic of the horseshoe crabs. Their defence against foreign bodies is being imperilled by ours. Though biomedical producers are required to return the animals to the sea within seventy two hours of blood collection, many perish from the rigours of (careless) handling. Recent research has suggested that casual stacking of the captured animals in plastic bins, where their spiny telsons can pierce the bodies of those heaped on them, or crushing in the overcrowded containers has increased mortality rates, once thought to be as low as 3 per cent but recently estimated at 20 or even 30 per cent. Abrupt upward temperature changes en route to the biomedical facilities have impaired, often fatally, the ability of the cold-blooded animals' gills to expel carbon dioxide from their bodies. The loss of as much as nearly half their blood, along with wound trauma from the milking puncture to the pericardial membrane, has been shown to affect normal heart function and in manifold ways damages the horseshoe crabs' ability to repair and restore their normal metabolism. It has been observed that, after blood-letting, many of the animals no longer walk and crawl as they should. And the capture itself during spawning, the moment when the horseshoe crabs are most prolific, has adversely affected the regularity of their reproductive cycle. In the spring and summer of 2022, observers reported the lowest number of spawning crabs in twenty years. The population of shorebirds especially has been correspondingly depleted.[18] A precious ecosystem is breaking.

At which point you would expect the law of diminishing returns to kick in – and, to some extent, it has. Pharmaceutical companies minting money from FDA LAL tests are now scrambling to produce recombinant alternatives that have been available for over two decades and have now been approved in other countries including Japan. But LAL producers, defending their patch, have insisted that threat to horseshoe sustainability has been exaggerated, and the US Pharmacopeia has listened to them. Predictably, the habitually depressing trade-off between immediate profitability and ecological preservation has come to determine the fate of the horseshoe crabs. It is a bitter irony that this improbable connection across vast measures of time and tide should have turned into a zero-sum calculation about comparative damage.

It may yet turn out well, or at least not disastrously for us, or for *Limulus polyphemus*, our arthropod defenders against tiny toxins. We must hope so, for, even as paranoia about borders and frontiers continues to dominate populist rhetoric, the inseparability of humans – and, for that matter, the indivisibility of humankind and nature – remains the saving imperative of our beleaguered time. That has been my story and it remains my faith. Contrary to what you'll read in tabloid headlines, or hear in the hoots and yells of social media, in our present historical extremity, there are no foreigners, only familiars: a single precious chain of connection that we snap at our utmost peril.

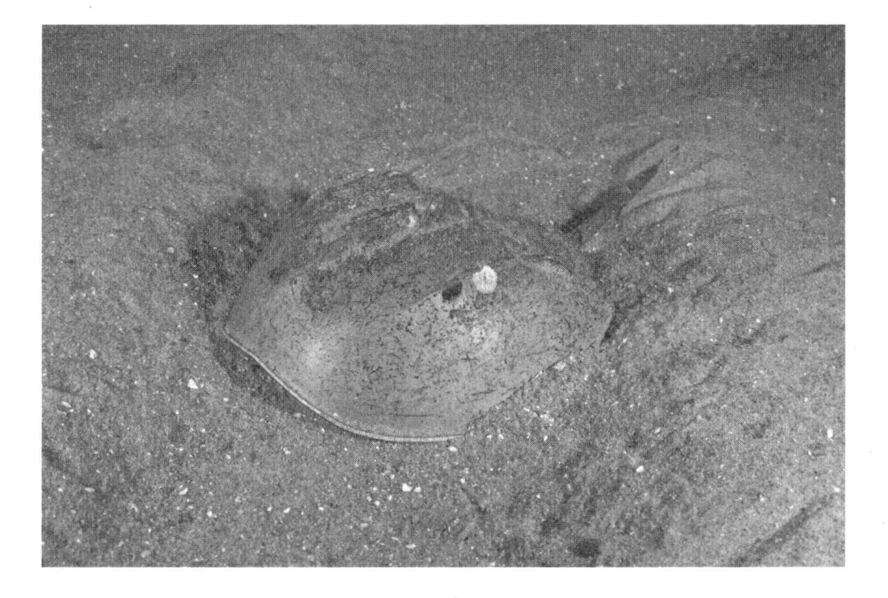

ACKNOWLEDGEMENTS

It will surprise no one to learn that *Foreign Bodies* is a book born of the COVID-19 pandemic, during which I swerved sharply from writing another book entirely, the one my publishers happened to be expecting. So I owe an even greater debt than usual to my agent and dear friend Caroline Michel for her passionate belief in the new project, an enthusiasm so infectious that scepticism had no option but to melt before its warmth. All the same I must also thank Ian Chapman at Simon & Schuster UK for his open-mindedness to the unanticipated subject, followed by generous and enthusiastic commitment. As always, Caroline's critical insights and perceptiveness have been my indispensable guide as the book took shape.

Katherine Ailes at Simon & Schuster has been the most inspirational editor and collaborator imaginable: sympathetically immersed in the text, critically eagle-eyed for its many imperfections, inexhaustible at the task in hand, and (unlike the author) calmly assured when deadlines loomed. I must also thank Alex Eccles for his devoted editorial work unknotting tangles, spotting snafus and a great deal else. Also at Simon & Schuster: Suzanne Baboneau, Polly Osborn, Holly Harris, Justine Gold, Matt Johnson, Paul Gooney, Karin Seifried, Jonathan Wadman, Liz Marvin, Alex Newby, Nige Tassell. Out of house, my thanks to Doug Kerr for his work on the beautiful cover design and to the brilliant Emma Harrow for the publicity campaign.

At my agents, Peters, Fraser and Dunlop, thanks (especially) to

Kieron Fairweather, and to Lucy Berry, Laurie Robertson and Becky Wearmouth.

The Waldemar Haffkine papers are preserved in the Archive Department of the National Library of Israel in Jerusalem. My deepest thanks to those who made my research there a time of joyful illumination and discovery, in particular Drs Rachel Misrati and Stefan Litt, and those who helped me 'hands-on' negotiate the vast trove of Haffkine's documents: especially Meytal Solomon, and Mona Kepach, Lana Shashar, Hallel Sasposnik, Alexander Gordon and Zmira Reuveni. David Langrish at the library of the Wellcome Collection was extraordinary helpful with the Bombay Plague Visitation Album and I am also grateful to Nicole Ioffredi at the British Library, the custodian of another copy of that precious and peculiar document of the mind of the medical Raj.

The literary ship would have foundered (and its skipper-author been entirely lost to the waves) without the extraordinary support and critical help of Marta Enrile-Hamilton and Naomi Sanyang at almost every stage of writing. That their indefatigable resourcefulness and exactness has been combined with patience and warmth has been a boon beyond measure. I must also thank my assistant at Columbia University, Jennifer Sonntag, for her lightning responses in unearthing obscure documents of medical history and debate and much else.

Acutely conscious that I am a newcomer to medical and scientific history, I am immensely grateful to the brilliant science writer and former Consultant Editor of *Nature*, Philip Ball, for his kindness in reading the typescript, the better to save me from hapless solecisms and confusions, and for suggestions where I might have gone astray. Thanks also to my dear friend Stella Tillyard for reading a draft and for her own editorial suggestions, helpful criticisms and generous enthusiasm.

When they have got over their surprise at learning about the subject of *Foreign Bodies*, and the part in it played by Waldemar Haffkine, many good friends have reassured me about the importance, both contemporary and historical, of the stories I have thought to tell, especially Dr Barbara Alpert, Anthony and Beverley Silverstone, Jan Dalley, Chloe Aridjis, Suzannah Lipscomb, Whitney McVeigh and Kavita Puri. Alice Sherwood has listened, many times, to the dizzying twists and turns this narrative has taken, and has responded with her unfailing generosity and wisdom.

But there is a peerless guide for this kind of work who was a close and much loved friend in our days together as undergraduates in Cambridge and young scholars whom, had his life not been cut brutally short, I would have consulted for constructive criticism and reassurance: Roy Porter, rightly called 'the greatest medical historian of his generation' and the author of countless works of intellectual and medical history, above all, perhaps, in this context, *The Greatest Benefit to Mankind: A Medical History of Humanity* (New York and London, 1999). As it is I have had to make do with the shade of Roy, imagining what he might have said about the ambition, not to say chutzpah, of my working in his field, but also knowing his broad, sunny and knowing smile would have set me on my way.

However gripped I was by these stories, I don't believe I would have had the presumption to commit them to writing had it not been for the constant encouragement of my wife, the developmental biologist Ginny Papaioannou, my guide over many years in matters scientific. She has been kind enough to read the typescript for errors and misjudgements, and indeed for literary missteps and repetitions. Most of all, when my faith in the project flagged, she restored it, persuading me that both for the causes of both history and science, the past and the contemporary, it was a worthwhile and valuable endeavour. For all this, and for so much more, this book is lovingly dedicated to her.

NOTES

A note on place names

I have used place names as they were known at the time of the events I describe and as they appear in literature and archival sources rather than in post-imperial nomenclature; thus Bombay and Canton rather than Mumbai and Guanzhou. Using 'Odessa' rather than the Ukrainian transliteration 'Odesa' does not in any way imply subscribing to Russian claims claims of sovereignty over Ukrainian cities and towns.

All HA references are to the Haffkine Archive in the National Library of Israel: Arc. Ms. Var. Haffkine.

CHAPTER ONE: Et in suburbia ego

1 Jack R. Layne Jr. and Richard E. Lee, 'Adaptation of Frogs to Survive Freezing', *Climate Research,* 5 (1), 1995, 53–59
2 Lyme disease is now also common across Europe, especially in two regions: the Scandinavian and Baltic countries from Sweden to Estonia and Lithuania, and in central Europe from Austria to Slovenia. As of 2020, around 200,000 cases have been reported in Europe and around 400,000 in the USA. In the spring of 2022, Pfizer, in partnership with the pharmaceutical company Valneva, announced the beginnings of phase 3 trials of the VLA15 vaccine for Lyme in the hope that a submission to American and European CDC may happen in 2025. On the European spread, Adriana Marques et al, 'Comparison of Lyme Disease in the United States and Europe', *Emerging Infectious Diseases,* 27 (8), August 2021, 2017–2024. In mid-March 2023, the Centers for Disease Control in the United States reported a doubling of cases, between 2011 and 2019 in north-eastern states cases of a previously rare disease, babesiosis, transmitted by deer ticks, with symptoms and aetiology similar to Lyme. In the 16 March 2023 *New York Times* story, Dr Peter Krause of the Yale School of Public Health comments that the construction of new housing in tick-abundant areas may well be a factor in the rise of babesiosis.
3 Tom Perkins, 'Hold the Beef', *The Guardian,* 10 December 2021

4 Dom Phillips et al, 'Revealed: rampant deforestation of Amazon driven by global greed for meat', *The Guardian*, 2 July 2019

5 This is a point well made by Professor Delia Grace, epidemiologist and veterinarian, who leads research at the International Livestock Research Institute in Nairobi, Kenya and the Consultative International Group for Agricultural Research programme on Agriculture for Human Nutrition and Health. Delia Grace et al, 'The Multiple Burdens of Zoonotic Disease and an Ecohealth Approach to their Assessment', *Tropical Animal Health and Production*, 44, 2012, 67–73

6 George III's pangolin-scale coat is preserved in the Leeds Armory Museum.

7 Ping Liu, Jingpin Chen et al, 'Are Pangolins the Intermediate Host of 2019 Novel Coronavirus (SARS–CoV-2)?', *PLOS Pathogens*, 16 (5), 14 May 2020; see also Songyi Ning et al, 'Novel Putative Pathogenic Viruses Identified in Pangolins by Mining Metagenomic Data', *Journal of Medical Virology*, 94 (6), 2500–2509, 3 January 2022

8 There is now a substantial scientific literature on the relationship between environmental degradation and recurring waves of lethally infectious diseases. In particular, Bryony Jones, Delia Grace et al, 'Zoonosis Emergence Linked to Agricultural Intensification and Environmental Change', *Proceedings of the National Academy of Science*, 110 (21), 21 May 2013, 8399–8404; Shahid Jameel, 'On Ecology and Environment as Drivers of Human Disease and Pandemics', *ORF Issue Brief*, 388, July 2020; Kate E. Jones, Nikkata G. Patel, Marc A. Levy, Adam Storeygard, Deborah Balk, John A. Gotteman and Peter Daszak, 'Global Trends in Emerging Infectious Diseases', *Nature*, 451, 21 February 2008, 990–993; Jeff Tollefsen, 'Why Deforestation and Extinctions Make Pandemics More Likely', *Nature*, 584, 7 August 2020, online

9 Chengxiang Gu, Ming Wang et al, 'Saline lakes on the Qinghai-Tibet Plateau harbor unique microbial assemblages mediating microbial environmental adaptation', *i Science*, 12, 17 December 2021

10 David Quammen, *Spillover: Animal Infections and the Next Human Pandemic* (New York and London, 2013) is a gripping account of this history and, published ten years ago, chillingly prescient. The most powerful and persuasive research on zoonotic diseases has been by Edward Holmes and his colleagues and students; Edward Holmes, *The Evolution and Emergence of RNA Viruses* (Oxford, 2009); Holmes et al, 'The Origins of SARS COV-2: A Critical Review', *Cell*, 184, 16 September 2021

11 Apoorva Mandavalli, 'Scientists zero in on the origins of monkeypox outbreak', *New York Times*, 23 June 2022; Eskild Petersen et al, 'Human Monkeypox: Epidemiologic and Clinical Characteristics, Diagnosis and Prevention', *Infectious Diseases of North America*, 33 (4), December 2019, 1027–1043

12 Phi-Yen Nguyen et al, 'Re-emergence of Human Monkey-pox and Declining Population Immunity in the Context of Urbanization in Nigeria, 2017–2020', *CDC Control and Prevention*, 27 (4), April 2021, online

13 Kristian Andersen et al, 'The Proximal Origin of SARS-CoV-2', *Nature Medicine*, 26, 17 March 2020, online

14 Katherine J. Wu, 'The strongest evidence yet that an animal started the pandemic', *The Atlantic*, 16 March 2023
15 Felicia Goodrum et al, 'Virology Under the Microscope – A Call for Rational Discourse', *Journal of Virology*, 97 (2), commentary, 26 January 2023
16 Nicoletta Lanese, 'Omicron variant may have evolved in rats', *Live Science*, 2 December 2021; in the same article, the evolutionary biologist Michael Worobey of the University of Arizona is quoted speculating that the development of the Omicron variant might have first been in animal species such as rats.
17 Dennis Carroll et al, 'The Global Virome Project', *Science*, 359 (6378), 23 February 2018, 872–874
18 David Adam, 'COVID's true death toll: much higher than official records', *Nature*, 10 March 2022, online
19 For discussions on state approaches to COVID-19, Peter Baldwin, *Fighting the First Wave: Why the Coronavirus Was Tackled So Differently Across the Globe* (Cambridge, 2021); Adam Tooze, *Shutdown: How Covid Shook the World Economy* (New York, 2021)
20 One of the first acts of the Biden administration in late January 2021 was to rejoin the World Health Organization.
21 On the history of stigmatising cholera as Asiatic, Professor Sir Richard Evans, 'The Great Plagues: Epidemics in History from the Middle Ages to the Present Day', Gresham College Lectures, April 2013, online. On the demonisation of Chinese people in Italy during the pandemic, K. Y. C. Adja, D. Golinelli et al, 'Pandemics and Social Stigma: Who's Next? Italy's Experience with COVID-19', *Public Health*, 185, 2020, 39–41. Samuel K. Cohn Jr, in his important *Epidemics: Hate and Compassion from the Plague of Athens to AIDS* (Oxford, 2018), argues from a rich trove of sources that, historically, there has been nothing inevitable about the stigmatisation of groups thought to be responsible for epidemics. My concern in this work, rather, is with how inoculators and vaccinators have been the targets of suspicion. But current political rhetoric in the United States from supporters, both in Congress and the right-wing media, of 'lab leak' theories of the origin of SARS-CoVID-2 is charged with Sinophobic assumptions.
22 On this kind of default language, Patrick Wallis and Brigitte Nerlich, 'Disease Metaphors in New Epidemics: The UK Media Framing of the 2003 SARS Epidemic', *Social Science and Medicine*, 60 (11), June 2005, 2629–2639; Jacques Tassin et al, 'Devising Other Metaphors for Bio Invasion', *Nature, Sciences, Sociétés*, 20 (4), 2012, 404–414; Laura N. H. Verbrugge et al, 'Metaphors in Invasive Biology', *Ethics, Policy & Environment*, 19 (3), 2 September 2016
23 Conor Stewart, 'Coronavirus Deaths from COV-19 in EU and UK', *Statista*, 3 April 2022, online
24 Michelle Ye Hee Lee and Min Joo Kim, 'As world reopens, North Korea is one of two countries without vaccines', *Washington Post*, 24 April 2022
25 P. M. Rabinowitz, M. Pappaioanou et al, 'A planetary vision for one health', *National Library of Medicine*, 3 (5), 2 October 2018. There is, however, a current effort to 'reclaim' the *meaning* of 'global health' from a top-down

exercise in better-off nations coming to the rescue of less-well-off regions, towards a less patronising involvement in addressing social and economic inequities as intrinsic to vulnerability. Liam Smeet and Katherine Kyobutungi, 'Reclaiming Global Health', *The Lancet*, 16 February 2023, online

CHAPTER TWO: 'The fresh and kindly pock'

1 Voltaire to Louis-Nicolas le Tonnelier, Baron de Breteuil, 5 December 1723, *Oeuvres Complètes, Correspondence* (Paris, 1880), vol 1, 100–104. On Voltaire's medical beliefs, Margaret Sherwood Libby, *The Attitude of Voltaire to Magic and the Sciences* (New York, 1935), 240–268
2 *Mariamne* had a rough ride to acceptance. It was a disastrous flop when first performed at La Comédie Française, then much revised and more warmly received.
3 Two years later, in 1725, Rohan and Voltaire exchanged public insults, serious enough for Voltaire to be sent to the Bastille a second time, followed by the exile which took him to England in 1726.
4 Voltaire, 'Aux mânes de M. Genonville, conseiller au Parlement et ami intime de l'auteur', (Paris, 1728). 'Nous nous aimions tous trois. La raison, la folie, l'amour, l'enchantement de plus tendres erreurs.'
5 Voltaire to Breteuil, ibid, 103
6 Thomas Sydenham, *Observationes Medicae Circa Morborum Acutorum Historiam Curationem* (London, 1676)
7 *The Works of Thomas Sydenham, M.D.*, translated from the Latin edition of *Dr. Greenhill with a Life of the Author* by R. G. Latham, M.D. (London, 1843), 133
8 Miguel Quirsch et al, 'Hazards of the Cytokine Storm and Cytokine-Targeted Therapy in Patients with COVID 19', *Journal of Internet Medical Quarterly*, 22 (8), 13 August 2020, online
9 Voltaire to Breteuil: 'La petite vérole, par elle meme, dépouillée de toute circonstance étrangère n'est qu'une épuration du sang, favorable à la nature et qui en nettoyant de ce qu'il a d'impur lui prépare une santé vigoureuse.'
10 Michelle Dimeo and Joanna Ware, 'The Countess of Kent's Powder: a 17th-century cure-all', *The Recipes Project*, 11 December 2014, online
11 Jean Delacoste, *Lettre sur l'inoculation de la petite vérole comme elle se pratique en Turquie et Angleterre addressée à M. Dodart Conseiller d'Etat & Prenier Medecin du Roy* (Paris, 1723); see also Genevieve Miller, *The Adoption of Inoculation for Smallpox in England and France* (Philadelphia, 1957)
12 Voltaire insisted a number of times that he had written much of the *Letters* while in England in 1728, which raises the fascinating possibility that he wrote at least some of the work in English at which he had become impressively fluent, writing, for instance, to correspondents such as Alexander Pope in that language.
13 An edition in the original French had already appeared with the imprimatur of Basle, but was in fact printed in Amsterdam.

14 Voltaire to Thieriot, 12 August 1726, *Correspondence,* vol 85, 303

15 Norma Perry, 'Sir Everard Fawkener, Friend and Correspondent of Voltaire', in Thedore Besterman (ed), *Studies on Voltaire and the Eighteenth Century* (Geneva, 1975)

16 Voltaire, *Letters Concerning the English Nation* (Oxford, 1994, ed Nicholas Cronk), following the text of 1733 (London and Lyon), 44

17 *A. de la Mottraye's Travels through Europe, Asia and into Parts of Africa; With Proper Cutts and Maps etc* (London 1723–24), a French edition, printed in The Hague in 1727. Hogarth's illustrations were largely drawn from Jean Scotin's *Recueil de cent estampes representant différentes nations du Levant . . .* after Jean Baptiste Vanmour (1712–13).

18 Ibid, vol 2, 74–75

19 Voltaire, *Letters,* 45

20 Mottraye, *Travels,* vol 2, 58

21 There is a rich biographical literature on Mary Wortley Montagu to whom I was first introduced in the 1960s in Cambridge and New York by the late Robert Halsband, a pioneer in studies of this extraordinary woman. R. Halsband, *The Life of Lady Mary Wortley Montagu* (Oxford, 1956); more recently, Isobel Grundy, *Lady Mary Wortley Montagu* (Oxford, 1999) and Jo Willett, *The Pioneering Life of Mary Wortley Montagu: Scientist and Feminist* (London, 2021)

22 Voltaire, *Letters*

23 Michael McCormick, 'Gregory of Tours on 6th Century Plague and Other Epidemics', *Speculum,* 96 (1), January 2021, 55–56

24 P. S. Codellas, 'The Case of Smallpox of Theodorus Prodromus (XII Cent. AD)', *Bulletin of the History of Medicine,* July 1946, 207–215; John Lascaratos and Constantine Siamis, 'Two cases of smallpox in Byzantium', *International Journal of Dermatology,* 41 (11), December 2002, 792–795

25 Ann G. Carmichael and Arthur M. Silverstein, 'Smallpox in Europe before the Seventeenth Century: Virulent Killer or Benign Disease?', *Journal of the History of Medicine and Allied Sciences,* 42 (2), April 1987, 147–168

26 Beatriz Puente-Ballesteros, 'F.-X. d'Entrecolles and Chinese Medicine: A Jesuit's Insights in the French "Controversy Surrounding Smallpox Inoculation', *Revista de Cultura/Review of Culture,* 18, 2006, 89–98. On Chinese practices, Angela K. Che Leung, "Variolation" and vaccination in late Imperial China 1570–1911', in S. Plotkin and B. Fantini (eds), *Vaccinia, vaccination, and vaccinology: Jenner, Pasteur and their successors* (Paris and London, 1996), 65–71

27 Donato Paolo Mancini, 'Nasal version of Oxford/AstraZeneca Covid vaccine fails in trial', *Financial Times,* 11 October 2022

28 Not entirely, for hard-paste porcelain was beginning to be made in Meissen in Saxony. But see *Lettre du Père d'Entrecolles . . . au Père d'Orry* (1712), idem, 1722, which formed the basis for the extended account of (Père) Jean-Baptiste du Halde, the Jesuits' armchair historian of China whose account in 1736 was praised by Voltaire: *Description Géographique,*

Historique, Chronologique, Politique et Physique de l'Empire de la Chine et la Tartarie (Paris, 1736)

29 For Holwell and the Black Hole, Jan Dalley, *The Black Hole: Money, Myth and Empire* (London, 2006)

30 John Zephaniah Holwell, *An Account of the Manner of inoculation for Small Pox in the East Indies with some Observations of the Practice and Mode of Treating that Disease in those Parts* (London, 1767)

31 Margaret DeLacy, *The Germ of an Idea: Contagionism, Religion and Society in Britain, 1660–1730* (Basingstoke, 2016), 127 ff

32 Perrot Williams, M.D., *Part of Two Letters concerning a method of procuring small pox in South Wales . . . from Perrot Williams, MD Physician at Haverford West to Dr Samuel Brady, Physician to the Garrison at Portsmouth* (London, 1723)

33 Arnold C. Klebs, *Die Variolation in achtzehnter jahrhundert: ein historische Beitrag zur Immunitatsforschung* (Giessen, 1914), 7

34 Marie de Testa and Antoine Gautier, 'Unde grande famille latine de l'empire Ottomane. Les Timonis: medecins, drogmans et hommes d'église', in *Drogmans et Diplomates Europeens Aupres de la Porte Ottomane* (Istanbul, 2003), 235–255; see also the note by Randoph P. Stearns, 'Fellows of the Royal Society in North Africa and the Levant, 1662–1800', *Notes and Records of the Royal Society of London,* 1954, 11 (1), January 1954, 77–78. On the broader significance of translation, see the very suggestive article by Anne Eriksen, 'Smallpox Inoculation: Translation, Transference and Transformation', *Palgrave Communications,* 6 (52), 2020, online

35 The Peace of Carlowitz had, in fact, transferred the whole of the Peloponnese from Ottoman to Venetian sovereignty, but was returned to Turkish control after further hostilities in 1715.

36 Sophie Vasset, 'Medical Laughter and Medical Polemics: The Woodward-Mead Quarrel and Medical Satire', *Revue de la Société d'Etudes Anglo-Américaines des XVIIe et XVIIIe Siècles*, 70, 2013, 109–133; Joseph Levine, *Dr Woodward's Shield: History, Science and Satire in Augustan England* (Berkeley, 1977)

37 Emmanuel Timoni and John Woodward, 'An Account or History of Procuring the Smallpox by Incision or Inoculation as it has for some time been practis'd in Constantinople, being the Extract of a Letter of Emmanuel Timonius, Oxon et Padova, dated at Constantinople December 1713', *Philosophical Transactions of the Royal Society* (1714–16), 29 (339), 72–82. On Timoni (albeit with a strenuous effort to reclaim him from any Italian cultural identity), Effie Poulakou-Rebelakou and John Lascaratos, 'Emmanuel Timonius, Jacobus Pylarinus and Inoculation', *Journal of Medical Biography,* 11 (3), August 2003, 181–182; John Lascaratos, 'Emmanuel Timonis, Biography and Ergography. The Famous Physician from the Island of Chios. Emmanuel Timonis and his Era', *Medical Association of Chios,* 2000, 31–42; S. Bartsokas and S. G. Marketos, 'Emmanuel Timonis, Jakovos Pylarinos and smallpox inoculation', *Journal of Medical Biography,* 4, 1996, 129–136

38 J. Pylarinos, *Nova et tuta Variola: Excitandi per Transplantionem Morbidus; Nuper inventa et in unum tracta* (Venice, 1715; London, 1716; Leiden, 1721); C. N. Alisivatos and G. K. Pournaropoulos, 'The work of the Greek physician Jakovos Pylarinus on "variolation"', *Proceedings of the Athens Academy,* 1952

39 Giacomo Pylarini, 'Nova & tuta vaiolas excitandi per transplantationem methodus nuper inventa & in usum tracts', *Transactions of the Royal Society*, 29 (347), 31 March 1716, 396–7

40 Teresa Heffernan and Daniel O'Quinn (eds), *The Turkish Embassy Letters: Lady Mary Wortley Montagu* (Peterborough, Ontario, 2013), 126

41 Alicia Grant, *Globalisation of Variolation: The Overlooked Origins of Immunity for Smallpox in the 18th Century* (London, 2019), 61 ff

42 The Circassians were subject to an Islamic conversion campaign enforced by Ahmed III at exactly this period.

43 J. G. Scheuchzer, 'An Act of Success of Inoculation', *Philosophical Transactions of the Royal Society* (London, 1729)

44 'An Account of the Practice of Inoculation in Arabia in a Letter from Dr. Patrick Russell, Physician at Aleppo to Alexander Russell, MD., FRS', *Philosophical Transactions,* 1768, 142–150

45 Peter Kennedy, *An Essay on External Remedies Wherein is Considered Whether all the Curable Distempers Incident to Human Bodies May not be Cured by Outward Means* (London, 1715), 153 ff

46 A poor choice of comparison perhaps since the 'itch' was usually the condition of genital scabies.

47 Robert Halsband (ed), *The Complete Letters of Lady Mary Wortley Montagu,* vol 1, 1708–1720 (Oxford, 1967), online. For her time in Turkey, see Grundy, 134–166

48 Robert Halsband, *The Life of Lady Mary Wortley Montagu* (Oxford, 1956), 71–72

49 Lady Mary Wortley Montagu, *Letters of the Right Honorable Lady M... W...M..., Written during Her Travels in Europe, Asia and Africa to Persons of Distinction, Men of Letters &c... which contain, among other Curious Relations Accounts of the Policies and Manners of the Turks* (3 volumes, London, 1763), 3

50 As reported by the Smyrna consul William Sherard who was no admirer of Timoni. See Stearns, 'Fellows', 88

51 On the cultural implications of Maitland's account, see Eriksen, op cit

52 Halsband, *Life,* 81

53 Art Boylston, 'The Newgate Guinea Pigs', *London Historians,* September 2012

54 John Arbuthnot, *Mr Maitland's Account of Inoculating the Smallpox Vindicated from Dr. Wagstaffe's Misrepresentations of the Practice with Some Remarks on Mr. Massey's Sermon* (London, 1722)

55 Jacob de Castro Sarmento, *Dissertatio in Novam, Tutam, ac Utilem Methodum Inoculationis seu Transplantationis Variolorum* (London, 1721; German translation, Hamburg, 1722). On Sarmento, R. Bennett, 'Sephardim

and Medical Practice in 18th century London', *Transactions of the Jewish Historical Society of England*, 1878, 84–114; Matt Goldish, 'Newtonian *Converso* and Deist: The lives of Jacob (Henrique) de Castro Sarmento', *Science in Context*, 10 (4), Autumn 1997, 651–675. For engraved portraits of Sarmento, Alfred Rubens, *Anglo-Jewish Portraits* (London, 1935), 265–6

56 Edmund Massey, *A sermon against the dangerous and sinful practice of inoculation* (London, 1722)

57 On Wagstaffe as Tory satirist and his relationship with Arbuthnot, Alasdair Raffe, 'John Bull, Sister Peg and Anglo-Scottish Relations in the Eighteenth Century', in Gerard Carruthers and Colin Kidd (eds), *Literature and Union: Scottish Texts, British Contexts* (Oxford, 2018), 44

58 William Wagstaffe, *A Letter to Dr. Freind Shewing the Danger and Uncertainty of Inoculating the Small Pox*, 12 June 1722, 5

59 Legard Sparham, *Reasons against the practicing of inoculating the small pox. As also a brief account of this poison infused after this manner into a wound* (London, 1722). See also the Lincolnshire antiquarian Francis Howgrave, *Reasons against the Inoculation of the Small-pox in a Letter to Dr Jurin* (Stamford, 1724)

60 John Arbuthnot, *Mr Maitland's Method of inoculating the small pox vindicated from Dr. Wagstaffe's misrepresentation of the practice, with some remarks on Mr Massey's sermon* (London, 1722)

61 Shawn Buhr, 'To Inoculate or Not to Inoculate: The Debate and the Smallpox Epidemic of Boston in 1721', *Constructing the Past*, 1 (1), 2000

62 Kathryn S. Koo, 'Strangers in the House of God: Cotton Mather, Onesimus and an Experiment in Christian Slaveholding', *Proceedings of the American Antiquarian Society*, 2007, 143–175

63 Zabdiel Boylston, *An historical account of the Small Pox inoculated in New England with some account of the nature of the infection in the natural and inoculated way and their different effects on human bodies* (London, 1726). Boylston dedicated the publication to Caroline, still then Princess of Wales; Arthur William Boylston, *Defying Providence: Smallpox and the Forgotten 18th Century Medical Revolution* (North Charleston, 2012), 25–70

64 Boylston, *Defying Providence*, 29

65 Buhr, op cit. In the 7–14 August 1721 issue of *The New England Courant*, Douglass sarcastically suggested inoculating native Americans and offered to pay five pounds for every Indian dying of inoculation.

66 Boylston, *Defying Providence*, 48

67 [Lady Mary Wortley Montagu], 'A Plain Account of the Inoculation of the Small Pox by a Turky [sic] Merchant', *Flying-Post*, 13 September 1722, in Heffernan and O'Quinn (eds), *The Turkish Embassy Letters*, 256–257

CHAPTER THREE: SAFELY! QUICKLY! PLEASANTLY!

1 T. Nettleton, 'A Letter from Dr Nettleton, Physician at Halifax in Yorkshire to Dr. Whitaker, concerning the Inoculation of the Small Pox', *Philosophical*

Transactions of the Royal Society, 32 (370), 31 March 1722, 35–48. The letter was also commercially and more confidently published in April 1722 as *An Account of the Success of inoculating the small pox in a Letter to Dr. William Whitaker* (London, 3 April 1722). For Nettleton's contribution to quantifying results, Arthur Boylston, M.D., 'Thomas Nettleton and the Dawn of Quantitative Assessment of the Effect of Medical Intervention', *Journal of the Royal Society of Medicine*, August 2010, 334–339

2 'A Letter from the Same Learned and Ingenious Gentleman concerning his Farther Progress in Inoculating the Small Pox to Dr. Jurin, R.S., June 16th, 1722', *Philosophical Transactions of the Royal Society*, 32, 1722, 49–52

3 Ibid, 49–50

4 An excellent account of this relationship is Andrea A. Rusnock, *Vital Accounts: Quantifying Health and Population in 18th century England and France* (Cambridge, 2002); idem, *The Correspondence of James Jurin 1684–1750* (1996); idem, 'The Weight of Evidence and the Burden of Authority: Case Histories, Medical Statistics and Smallpox Inoculation' in Roy Porter (ed), *Medicine in the Enlightenment* (London, 1995), 289–315

5 Miller, *The Adoption of Inoculation*, 111–116

6 Rusnock, 63

7 James Jurin, *An Account of the success in inoculating the small-pox in Great Britain for the year 1725 with an Account of the miscarriages in that practice and the mortality of the natural small pox* (London, 1726), 55–57

8 *Part of Two Letters Concerning a method of procuring the small pox in South Wales. From Perrot Williams MD, Physician at Haverford West to Dr Samuel Brady, Physician to the Garrison at Portsmouth* (London, 1723)

9 James Kirkpatrick (Kilpatrick), *The Analysis of Inoculation Comprizing the History, Theory and Practice of It, With a Consideration of the Appearances in the Small Pox* (London, 1754)

10 James Jurin, 'A letter to the learned Dr Caleb Cotesworth FRS of the College of Physicians, London and Physician to St Thomas's Hospital containing a comparison of the danger of the natural Small Pox and of that given by inoculation', *Transactions*, 31 December 1722

11 John Gasper Scheuchzer, M.D., *An Account of the Success of Inoculating the Small Pox in Great Britain for the years 1727 and 1728* (London, 1729)

12 Isaac Massey, *A Short and Plain Account of Inoculation with Some Remarks on the Main Arguments Made Use of to Recommend that Practice by Mr Maitland and others* (London 1722)

13 Jurin, 'letter', op cit

14 Thomas Fuller, *Exanthematologia or an Attempt to Give a Rational Account of Eruptive Fevers, Especially Measles and Small Pox* (London, 1730); James Kirkpatrick (Kilpatrick), *The Analysis of Inoculation* (London, 1754), 34–8

15 David van Zwanenberg, 'The Suttons and the Business of Inoculation', *Medical History*, 22 (1), 1978, 73; Gavin Weightman, *The Great Inoculator: The Untold Story of Daniel Sutton and his Medical Revolution* (Yale, 2020),

34–38; Arthur Boylston, 'Daniel Sutton: A forgotten eighteenth century clinician and scientist', *Journal of the Royal Society of Medicine*, 105 (2), February 2012, 85–87; Stanley Williamson, *The Vaccination Controversy: The Rise, Reign and Fall of Compulsory Vaccination* (Liverpool, 2007), 48–73

16 Lucy Ward, *The Empress and the English Doctor: How Catherine the Great Defied a Deadly Virus* (London, 2022)

17 Kirkpatrick, *Analysis*, 230 ff

18 Thomas Dimsdale, *The Present Method of Inoculating for the Small-Pox* (London, 1768)

19 Surprisingly, and though he appears fleetingly in numerous histories of inoculation, the only dedicated biography of Gatti is the brief but useful study by Veronica Massai, *Angelo Gatti: Un medico toscano in terra di Francia* (Florence, 2008)

20 Yasmine Marcil, 'Entre voyage savant et campagne medicale: le séjour en Italie de la Condamine', *Diciottesimo Secolo*, vol 3, 2018, 23–46; idem, 'Entre France et Italie, le mémoire en faveur de l'inoculation de la Condamine', online, 2018. For context, R. Pasta, 'Scienza e instituzioni nell'età leopoldine: Riflessioni e comparazione' in G. Barsanti, V. Beccagli and R. Pasta (eds), *La politico della scienza Toscana e i strati Italiani nel tardo settecento* (Florence, 1994), 1–34

21 *Journal du Voyage fait par order du Roi a l'Equateur á la mesure des Trois Premiers Degrès du Meridien* (Paris, 1751), 195. Condamine also notes this in his *Memoire sur l'Inoculation de la Petite Vérole* (Paris, 1754), 59, and the mass deaths of indigenous Americans from smallpox became a feature of eighteenth-century enlightenment critiques of European imperialism in the work, for example, of the Abbé Raynal's *Histoire des Deux Indes* (Amsterdam, 1770)

22 Condamine, Memoire, 58

23 Massai, 23–25. Most accounts of Gatti have Baron d'Holbach issuing the invitation; the two materialist philosophers were close, so both may have been involved in Gatti's move to Paris.

24 Arnold H. Rowbotham, 'The *Philosophes* and the Propaganda for Inoculation of Smallpox in Eighteenth Century France', *University of California Publications in Philology*, 18, 1935, 265–290

25 Angelo Gatti, *Réflexions sur les prejugés qui s'opposent aux progrès & la perfection de l'inoculation* (Brussels [Paris], 1764)

26 Ibid, 2

27 Angelo Gatti, *Nouvelles Réflexions sur la pratique de l'inoculation* (Brussels [Paris], 1767), 12–13. The emotionally diluted (but still very valuable) English edition, translated by Mathieu Maty, is *New Observations on Inoculation*. The imprimatur is Dublin but the date is carelessly given as 1758 – before Gatti was actually in France. In a prefatory note, Maty describes Gatti's 'offspring' – the 'Brussels' edition – as coming to him 'in rich French dress' being 'sent back' 'in a plain English frock' and hopes

the author will not be offended 'if, in taking off some of the trimmings, I should here and there have made rents in the cloth'.

28 William Watson, M.D., New Observations on Inoculation. An account of a Series of Experiments Instituted with a View to Ascertaining the most Successful Method of Inoculating the Small Pox (Dublin, 1768); Arthur Boylston, 'William Watson's Use of Controlled Clinical Experiments in 1767', Journal of the Royal Society of Medicine, 107 (6), June 2014, 246. Interestingly, the results of Watson's experiments at the Foundling Hospital were added as an appendix to Maty's English translation of Gatti's New Observations by way of empirical support for his contention that 'preparations' and elaborately therapeutic post-inoculation regimes were gratuitous, as well as the desirability and safety of inoculating children.

29 Watson, *New Observations,* 85 (in the edition bound with the Maty/Gatti *New Observations*).

30 Gatti, Nouvelles Reflexions, 84 ff

31 Elizabeth A. Fenn, Pox Americana: The Great Smallpox Epidemic of 1775–1781 (New York, 2001)

CHAPTER FOUR: Proust's travels

1 Robert le Masle, *Le Professeur Adrien Proust 1834–1903* (Paris, 1935)

2 Christian Péchenard, *Proust et Son Père* (Paris, 1993), 111–112

3 The indispensable guide to the history and deliberations of the International Sanitary Conferences is Norman Howard-Jones, *The Scientific Background to the International Sanitary Conferences, 1851–1938* (WHO, Geneva, 1975) and online by Cambridge University Press, 2012. The full *procès-verbaux* of each conference are also available online at http://nrs.harvard.edu/urn-3:hul.eresource:contagio?utm_source=library.harvard, thanks to Harvard University Library's superbly helpful primary source collection on *Contagion: Historical Views of Diseases and Epidemics.*

4 The only biography of Adrien Proust in his own right (rather than the relationship with Marcel) is Daniel Panzac, *Le Docteur Adrien Proust: Pere Méconnu, Précurseur Oublié* (Paris, 2003); see also Donatella Lippi et al, 'Adrien Proust (1834–1903): An Almost Forgotten Public Health Pioneer', *Vaccines,* (10) 5, 20 April 2022, 644; the oddly titled short essay, B. Straus, 'Achille-Adrien Proust, doctor to river basins', *Bulletin of the New York Academy of Medicine,* 50, July–August 1974, 833–838

5 Panzac, *Le Docteur Adrien Proust,* 227–9

6 Péchenard, 23 ff, begins his deeply evocative book with an account of the funeral.

7 Jean-Yves Tadié, *Proust: A Life* (trans. Euan Cameron; London and New York, 2000), 47–49

8 Ibid

9 Péchenard, 156–171

10 Ibid

11 The definitive history of official debates and approaches to cholera, in particular the debate between 'sanitationists' and 'contagionists', is Peter Baldwin, *Contagion and the State in Europe 1830–1930* (Cambridge, 2005), especially 123–243
12 Alain Ségal and Bernard Hillemand, 'L'hygiéniste Adrien Proust, son univers, la peste et ses idés politiques sur sanitaire internationale', *Histoire des sciences médicale,* I, XLV, 1, 2011, 63–69; Bernard Hillemand, 'Rénovation de la prévention des epidémies du XIXe siècle: Role majeure des pionniers et novateurs de l'Académie de Médicine injustement oubliés', *Bulletin de l'Académie Nationale de la Médicine,* 195, 2011, 755–772; Baldwin, 123–243
13 On the economic importance of the Parisian human manure trade well into the nineteenth century, S. Barles, 'Urban metabolism and river systems: an historical perspective, Paris 1790–1970', *Hydrology and Earth Systems,* 2007, 1757–1769
14 *Essai sur l'Hygiene Internationale, ses applications contre la peste, la fièvre jaune et le choléra asiatique...* (Paris, 1873), 301. This was Proust's core conviction: that the disease 'attached itself to the feet of the traveller', *Essai,* 287 and passim
15 Filippo Pacini, 'Osservazione microscopiche e deduzione patologiche sui cholera asiatico', *Gazzetta Medica Italiano-Toscano* (2e serie), 4 (50), 397–340, 405–412
16 Gian Piero Carboni, 'The enigma of Pacini's *Vibrio cholerae* discovery', *Journal of Medical Microbiology,* 70 (11), November 2021, online; Filippo Pacini, *Sulla causa specifica del colera* (Firenze, 1865)
17 Howard-Jones, 27–28
18 For a full account of the proceedings at Constantinople, Howard-Jones, 27–34; *Procès-Verbaux de la Conférence Sanitaire Internationale, le 13 février, 1866* (Constantinople, 1866)
19 Howard-Jones, 29
20 Darcy Grimaldo Grigsby, 'Rumor, Contagion and Colonization in Gros's Plague-Stricken Jaffa', *Representations*, Summer 1995, 46
21 Fauvel's first extensive medical analysis was *De l'influence de la connaissance des causes et des traitement des maladies* (Paris, 1844)
22 Adrien Proust, 'Rapport sur un mission sanitaire en Russie et en Perse en 1869', *Journal Officiel de l'Empire,* 10 July 1870, 1–31; Panzac, 31–55. The impressively detailed account of the cholera hot zone along the Russian–Persian frontier is based on his own exhaustive geo-epidemiological observations, recorded in the *Essai,* 366–385
23 Abbas Amanat, *Pivot of the Universe: Nasir Al-Din Shah Qajar and the Iranian Monarchy 1831–1896* (Los Angeles and Berkeley, 1997)
24 Proust noted that 'in Persia there exist a certain number of pools, elegantly lined in marble with constantly refreshed clear and limpid water but the reservoir is in the middle of the delightful gardens that are the exclusive preserve of the Shah's palace'. *Essai,* 372

25 *Essai,* 308. Proust was dismayed to see women washing clothes in the same ditches and canals contaminated by human evacuation.
26 Claude Francis and Fernande Gontier, *Proust et les Siens* (Paris, 1981), 48
27 Panzac, 45
28 Proust's rule was that 'the speed of the progress of epidemics has always corresponded to the rapid growth of communications'. *Essai,* 301
29 David Arnold, 'Cholera and Colonialism in British India', *Past and Present,* 113, November 1986, 118–151
30 J. D. Isaacs, 'D. D. Cunningham and the aetiology of cholera in British India', *Medical History,* 42 (3), 1998, 279–305
31 J. M. Cunningham, *Cholera: What Can the State do to Prevent it?* (Calcutta, 1884), vi
32 *Cholera Inquiry by Doctors Klein and Gibbes, and Transactions of a Committee Convened by the Secretary of State for India* (1885); Baldwin, 183–184
33 Howard-Jones, 58–65

CHAPTER FIVE: Sans frontières

1 On the friendship between Haffkine and the Aclands, and the photographic record of it in 1899, Giles Hudson, 'Epidemic Encounters', *Inside HSM: Stories from the History of Science Museum, University of Oxford,* 2021, online
2 The many portraits and studies of Haffkine at work collected (by him) – and now in the Waldemar Mordecai Wolff Haffkine Archive (hereafter HA) at the National Library of Israel in Jerusalem – reveal a bachelor's choice wardrobe. Wherever he was, Haffkine was almost always elegantly dressed for both social and professional occasions. And he kept multiple copies of his portraits, including commissioned engravings.
3 HA 325.01.105; Paris notes
4 HA 325.01.87; Haffkine's departure in 1904
5 Leonard Rogers, cited in Joel Hanhart, *Waldemar Mordekhai Haffkine (1860–1930): Biographie Intellectuelle* (Paris, 2016), 132; see also Wellcome Collection, Rogers Papers, PP/ROG/A55/106
6 The full report of the Commission was not published until 1902, an ominous year for Haffkine, but the informal circulation of its vaccine scepticism was already damaging his authority in 1899.
7 'It grieved me very much,' Haffkine wrote to Nightingale, 'to see how unwell you are.' HA 325.01.105; Paris notes
8 W. M. Haffkine, 'A Lecture on Vaccination against Cholera', delivered in the Examination Hall of the Conjoint Board of the Royal Colleges of Physicians and Surgeons on 18 December 1895
9 HA 325.01.105; Paris notes
10 On Jewish Odessa, see Steven J. Zipperstein, *The Jews of Odessa, 1794–*

1881 (Stanford, 1986); Simon Schama, *Belonging: The Story of the Jews 1492–1900* (London, 2017), 604–621

11 Lorraine de Meaux, *The Gunzburgs: A Family Biography* (London, 2019)

12 Joel Hanhart, 27

13 Luba Vikhanski, *Immunity: How Elie Metchnikoff Changed the Course of Modern Medicine* (Chicago, 2016), 110

14 Mark Popovsky, *The Story of Dr. Haffkine,* trans. M. Vezey (Moscow, 1963), 12–28

15 Hanhart, however, thinks the accusation of Haffkine's membership of Narodnaya Volya was intended to make the treason charges against him in February 1882 more weighty. Edyth Lutzker, on the other hand, stood by the Alexander Popovsky version of Haffkine's membership of the revolutionary organisation. Edyth Lutzker, 'Waldemar Haffkine CIE', in *Haffkine Institute Platinum Jubilee Commemorative Volume, 1899–1974* (Bombay, 1974)

16 Haffkine's (then spelled Khawkine) letter of appointment made it clear that he was to be a 'laboratory assistant'. HA 325.01.05; Paris notes

17 HA 325.01.151; sections from articles and belles-lettres

18 Kendall A. Smith, 'Louis Pasteur, the father of immunology?', *Frontiers in Immunology,* 3 (68), April 2012, online

19 W. M. Haffkine, 'Maladies infectueux des paramécies', *Annales de l'Institut Pasteur*, 4, 1890, 168–192

20 W. M. Haffkine, 'Recherches sur l'adaptation au milieu chez les infusoires et les Bacteries', *Annales de L'Institute Pasteur*, 4, 1890, 363–379

21 HA 325.03.326; Pasteur Institute lectures

22 George Bornside, 'Jaime Ferran and preventive inoculation against cholera', *Bulletin of the History of Medicine,* 1981, 516–532

23 An interesting later correspondence between Haffkine and Roux on the Ferran vaccine is in HA 325.01.32

24 This was partly because the operation of the immune system was not yet fully integrated into the working principles of Pasteurian bacteriology, so that many of the senior scientists, including Roux, followed their leader's assumption that the introduction of attenuated microbes somehow depleted the trace elements necessary for the infection to grow, rather than being stalled by a cellular immune response.

25 Quoted in Ilana Löwy, 'Guinea Pigs to Man: The Development of Haffkine's Anticholera Vaccine', *Journal of the History of Medicine and Allied Sciences,* 47 (3), July 1992, 279. On this dramatic moment, see also Barbara J. Hawgood, 'Waldemar Mordecai Haffkine CIE (1860–1930): Prophylactic Vaccine Against Cholera and Bubonic Plague in British India', *Journal of Medical Biography,* February 2007, 10–11

26 W. M. Haffkine, 'Le choléra asiatique chez le cobaye', *Comptes Rendus de la Société Biologique,* 9, 1892, 635–637

27 W. M. Haffkine, 'Sur le choléra asiatique chez le lapin et le pigéon', *Comptes rendus de la Societe Biologique,* 1892, 671

28 George H. Bornside, 'Waldemar Haffkine's Cholera Vaccine and the Ferran-Haffkine Priority Dispute', *Journal of the History of Medicine and Allied Sciences,* 37 (4), October 1982, 399–422
29 Louis Pasteur to Jacques-Joseph Grancher, *Correspondence,* (ed R. Vallery-Radot), vol 4, 342–344
30 Ernest Hanbury Hankin, 'Remarks on Haffkine's Method of Protective Inoculation against Cholera', *British Medical Journal,* 2 (1654), 10 September 1892, 569–571
31 The Hamburg cholera outbreak of 1892–3 and its epidemiological, social and political consequences are analysed in the magisterial account by Richard J. J. Evans, *Death in Hamburg: Society and Politics in the Cholera Years* (Oxford, 1987)
32 Hanhart, 65, note 272
33 Louis Pasteur, 'Sur les maladies virulentes et en particulier sur le maladie appellée vulgairement choléra des poules', *Centre des Recherches de l'Academie des Sciences,* 90, 1880; Kendall A. Smith, op cit
34 Popovsky, 49–54; Hanhart, 72–73
35 David Arnold, 'Cholera and Colonialism in British India', *Past and Present,* 113, November 1986, 113
36 Britton Martin, 'Lord Dufferin and the Indian National Congress, 1885–1888', *Journal of British Studies,* November 1967, 68–96. Rightly seeking to correct the overly generous appraisal of Dufferin's liberalism, this possibly goes too far in the opposite direction, given that the Viceroy not only allowed but encouraged the formation of the Indian National Congress in 1885 (before dismissing it a few years later) and laid the basis of what would be the Provincial Councils Act of 1892 establishing local elections in India.
37 Ratan Lal Chakraborty, 'The Unpublished Part of the Dufferin Report, 1888' (principally on eastern Bengal and the country around Dhaka)
38 Queen Victoria's Journal, 13 July 1881, in Gerard Vallée (ed), *Florence Nightingale on Social Change in India: Collected Works of Florence Nightingale,* vol 10 (Waterloo, Ontario, 2007), 519
39 Samiksha Sehrawat, 'Feminising Empire: The Association of Medical Women and the Campaign to Found a Women's Medical Service', *Social Scientist,* 41, May–June 2013, 65–81. This argues strongly for the imperial convenience of the Fund's social conservatism, but underestimates the strong resistance to male medics in much poorer sections of the Indian population. It was not unreasonable for the grossly discrepant rates of attendance at hospital between the genders to be a legitimate concern of those wanting to encourage women, especially Indian women physicians. Also, Antoinette Burton, 'Contesting the Zenana: The Mission to Make "Lady Doctors for India" 1874–1885', *Journal of British Studies,* 35 (3), July 1996, 368–397; Maneesha Lal, 'The Politics of Gender in Colonial India: The Effect of the Lady Dufferin Fund, 1885–8', *Bulletin of the History of Medicine,* 68 (1), 1994; Mridula Ramanna, 'Women Physicians

as Vital Intermediaries in Colonial Bombay', *Economic and Political Weekly*, 43 (12), March 2008, 71–78

40 HA 325.03.327; Haffkine's cholera work

41 Ibid, March 1893

42 For Simpson's early advocacy of Haffkine's cholera vaccine, see W. J. R. Simpson, *Cholera in Calcutta in 1894 and Anti-Choleraic Inoculation* (Calcutta, 1895). Members of the Indian vaccinating team are identified by Haffkine in his *Protective Inoculation Against Cholera* (Calcutta, 1913), 39. As Mark Harrison points out – in *Public Health in British India: Anglo-Indian Preventive Medicine 1859–1914* (Cambridge, 1994), 213 ff – Simpson's zeal for radical improvements in Calcutta's sanitation, not to mention his bouts of political insensitivity and tactlessness, made him a controversial figure with both Indian and British opinion. But then – and to Haffkine's benefit – Simpson did not belong to the Indian Medical Service, so was regarded by many in the IMS as an intruder.

43 HA 325.01.105; Paris notes

44 Ibid

45 Haffkine referred to the first location of the inoculation campaign in the spring of 1894 as 'the suburbs' of Calcutta. HA 325.01.105; Paris notes

46 Kavita Misra, 'Productivity of Crises: Disease, Scientific Knowledge and State in India', *Economic and Political Weekly*, 35 (43–44), 21 October–3 November 2000, 3885–3896

47 Ibid, 3889

48 Isaacs, 289

49 Ibid, 299

50 Later, Professor of Hygiene and Tropical Medicine at Kings College Hospital, London and co-founder of the School of Tropical Medicine as well as ardent champion of Haffkine. W. J. Simpson, *Cholera in Calcutta in 1894 and Anti-Choleraic Inoculation* (Calcutta, 1895)

51 The article by W. Theobald (the former head of geological Survey of India), in *The Pall Mall Gazette* on 28 May 1894, was so concerned about the reaction of native troops as to write that the matter was 'of national importance' and that 'if a catastrophe is to be avoided . . . this Pasteurism is to be prohibited in India and no time lost in the matter'. Misra, 3892

52 Surgeon-Captain E. Harold Brown, *Anticholeraic Inoculation During an Outbreak of Cholera in the Darbhanga Jail* (Calcutta, 1896)

53 Surgeon-Major R. Macrae, 'Cholera and Preventive Inoculation in Gya Jail', *The Indian Medical Gazette*, September 1894, 29 (9), 335

54 HA 325.03.335; correspondence re: cholera

55 HA 325.01.105; Paris notes

56 Deepak Kumar, '"Colony" Under a Microscope: The Medical Works of W.M. Haffkine', *Science, Technology and Society*, 4 (2), 1999

57 Ibid

58 HA 325.01.104; list of personal expenses

CHAPTER SIX: The death of rats

1 Robert Peckham, in his important and suggestive article, 'Hong Kong Junk: Plague in the Economy of Chinese Things' (*Bulletin of the History of Medicine,* 90, Spring 2016, 32–60) suggests that the assertion that plague was 'brought' to Hong Kong says more about British fantasies and stereotypes about the opium-depraved Chinese than reality. But it's possible that those stereotypes, doubtless grotesque caricatures, do not preclude the actual habitation of rodents aboard shipping, much of which was still certainly carrying opium.

2 It may be significant that Yunnan and the border region with Laos and Viet Nam is also the region in which horseshoe bats have been discovered to host viruses whose genomic sequence is close to that of SARS-CoV-2.

3 Quoted in Carol Benedict, *Bubonic Plague in Nineteenth Century China* (Stanford, 1996)

4 Oxford University, 'News and Events', 26 July 2021, online

5 Elizabeth Gamillo, 'Chipmunks Test Positive for the Bubonic Plague in Several South Lake Tahoe Locations', *Smithsonian Magazine,* August 2021

6 The extraordinary history is analysed in gripping detail by David Atwill, 'Blinkered Visions: Islamic Identity, Hui Ethnicity and the Panthay Rebellion in South West China 1856–73', *Journal of Asian Studies,* 62 (4), November 2003, 1079–1108

7 There is a rich and learned literature on this. See especially, David Arnold, *Colonizing the Body: State Medicine and Epidemic Disease in Nineteenth-Century India* (Berkeley, Los Angeles and London, 1994); Mark Harrison, *Public Health in British India* (Cambridge, 1994); Bridie Andrews and Mary P. Sutphen, *Medicine and Colonial Identity* (London, 2015)

8 Alexander Rennie, 'Report on the Plague prevailing in Canton during the spring and summer of 1894', *China Imperial Maritime Customs Reports Medical Journal*, 48, 1895, 74

9 The speed at which aerosol-born pneumonic plague brings on organ failure has been, and still is, a challenge for antibiotics which require delivery at a rate of urgency that is often impossible, especially in rural areas of the world with limited road or air connections.

10 '*Mr Chadwick's Reports on the Sanitary Conditions of Hong Kong*' (Colonial Office, London, 1882), appendix 2, *Public Latrines*, 54–55

11 On social conditions and the unfolding of the plague, Myron Echenberg, *Plague Ports: The Global Impact of Bubonic Plague, 1894–1901* (New York, 2007), 16–46. For details on the living conditions of the Chinese poor, David Faure, 'The Common People in Hong Kong History: Their Livelihood and Aspirations Until the 1930s' in David Faure, *Colonialism and the Hong Kong Mentality* (Hong Kong, 2003)

12 Pui-Tak Lee, 'Colonialism versus Nationalism: The Plague of Hong Kong in 1894', *Journal of North East Asian History,* 10 (1), Summer 2013, 120

13 Peckham, op cit, 44
14 G. H. Choa, 'The Lowson Diary: A Record of the Early Phase of the Hong Kong Bubonic Plague, 1894', *Journal of the Hong Kong Branch of the Royal Asiatic Society*, 33, 1993, 129–145, 139; W. J. Simpson, *Report on the Causes and Continuance of Plague in Hong Kong with suggestions as to Remedial Measures* (London, 1903)
15 Ibid, 134
16 James R. Bartholomew, *The Formation of Science in Japan: Building a Research Tradition* (New Haven, 1989), 141
17 Tom Solomon, 'Hong Kong 1894: The Role of James A Lowson in the Controversial Discovery of the Plague Bacillus', *The Lancet*, 350, 5 July 1997, 59; W. I. Yule, 'A Scottish Doctor's Association with the Discovery of the Plague Bacillus', *Scottish Medical Journal*, 40, 1995, 184–6; J. A. Lowson, *The Epidemic of Bubonic Plague in Hong Kong 1894* (Hong Kong, 1895); Choa, op cit, 137
18 Aro Velmet, *Pasteur's Empire: Bacteriology and Politics in France, its Colonies and the World* (Oxford, 2020)
19 Antonis A. Kousoulis et al, 'Alexandre Yersin's Explorations 1892–4 in French Indochina Before his Discovery of the Plague Bacillus', *Acta Medico-Historica Adriatica*, 10 (2), 2012, 303–310
20 Quoted in Velmet, op cit
21 Robert Peckham, 'Matshed Laboratory, Colonial Cultures and Bacteriology', in Robert Peckham and David M Pomfret (eds), *Imperial Contagions: Medicine, Hygiene and Cultures of Planning in Asia* (Hong Kong, 2013), 129ff, makes a compelling argument about the radical significance of improvised 'mobile' laboratory work in distinction to metropolitan and imperial assumptions about institutionialised science. For details on Yersin's hut lab, Noel Bernard, *Yersin, pionnier, savant, explorateur* (1863–1943) (Paris, 1955), 91; Henri H. Molleret and Jacqueline Brosselet, *Alexandre Yersin ou le vainqueur de la peste* (Paris, 1983)
22 Alexandre Yersin, 'La Peste Bubonique a Hong Kong', *Annales de l'Institut Pasteur*, 1894, 662. A year later, it was followed by a second note co-signed with Calmette venturing the possibility of an anti-plague, serum-based vaccine, the one Yersin would try out in Bombay. On Yersin's inspired improvisation and discipline, Robert Peckham, 'Matshed Laboratory', op cit, 123–147

CHAPTER SEVEN: Calamity snaps

1 Beheroze Shroff, '"Goma is Going on": Sidis of Gujarat', *African Arts*, 46, 2013, 18–25
2 Ira Klein, 'Plague, Policy and Popular Unrest in British India', *Modern Asian Studies*, 22 (4), 1988, 737. This article stresses the importance of the expansion of the Indian grain trade as a major facilitator of rat-reservoir for the bacillus.

3 Cynthia Deshmukh, 'The Bombay Plague of 1896–7', *Proceedings of the Indian Historical Congress*, 49, 1988, 478–9
4 Mridula Ramanna, *Health Care in the Bombay Presidency, 1896–1930* (New Delhi, 2012), 15
5 Quoted in Arnold, *Colonising the Body*, 209. Risley, an eminent ethnographer as well as secretary of the Financial Department, added that 'the nature of the microbe was such that it was highly portable'. On the (largely non-) reception of germ theory into British colonial sanitary practice, Mary J. Sutphen, 'Not What But Where: Bubonic Plague and the Reception of Germ Theory in Hong Kong and Calcutta, 1894–1897', *Journal of the History of Medicine and Allied Sciences*, 52 (1), January 1997, 81–113
6 M. E. Couchman (under the Orders of DeCourcy Atkins) (sic), *Account of Plague Administration in the Presidency of Bombay from September 1896 to May 1897* (Bombay, 1897), Part 1, section 5, 11. On the nervousness of the official mind faced with plague, David Arnold, 'Disease, rumour and panic: India's Plague and Influenza Epidemics 1896–1919', in Robert Peckham (ed), *Empires of Panic: Epidemics and Colonial Anxieties* (Hong Kong, 2015)
7 Official assumptions that all infectious diseases arose from the 'filth' of the poor drove what was in effect an urban war on their dwelling places. Prashant Kidambi, 'An Infection of Locality: Plague, Pythogenesis and the Poor in Bombay, c. 1896–1905', *Urban History*, 31 (2), August 2004, 249–262
8 Adrien Proust, *La Defense de l'Europe contre la Peste et la Conference de Venise de 1897* (Paris, 1897), 24; Marie Miguet-Ollagnier, 'La "Recherche": Tombeau Adrien Proust?', *Bulletin d'Informations Proustiennes*, 1991, 102–103
9 Quoted in Arnold, *Colonizing the Body*, 214; idem, 'Touching the Body: Perspectives on the Indian Plague 1896–1900', in Ramajit Guha (ed), *Subaltern Studies: Volume V: Writing on South Asian History and Society* (Delhi and Oxford, 1987), 55–90
10 Deshmukh, op cit, 481
11 Natasha Sarkar, 'Plague Germs Can Penetrate the Celestial Dress but Plague Measures Cannot: Mapping Plague Narratives in British India, 1890–1925', Center for Historical Research, Ohio State University, October 2012, online
12 Ibid, 11
13 Couchman, *Account*, 11
14 Ibid, 13
15 *British Medical Journal*, 21 December 1895
16 Correspondence on cholera inoculations in Bengal 1896–7. HA 325.03. 332
17 Ibid
18 The entire crew, except one seaman aboard the ship *Majestic* docked in Calcutta, had been inoculated against cholera. The un-inoculated sailor was hospitalised and died from cholera.
19 HA 325.03.332; correspondence re: Bengal anti-cholera innoculation

20 Surgeon-Captain E. H. Brown, M.D. I.M.S., 'Anticholeraic inoculation during an outbreak of cholera at Darbhanga Jail', April 1896, HA 325.03.332; correspondence re: Bengal anti-cholera innoculation
21 Haffkine to Under-Secretary of India, 21 May 1906, HA 325.03.364.9; correspondence with the Under-Secretary of State for India re: Haffkine's return to India
22 The correspondence concerning the Sepoy Jajajit Mal case, including Hare's letters to Haffkine. HA.325.03.332; correspondence re: Bengal anti-cholera innoculation
23 Selman Waksman, *The Brilliant and Tragic Life of W. M. W. Haffine, Bacteriologist* (New Brunswick, New Jersey, 1964), 21
24 James Knighton Condon, *The Bombay Plague: Being a History of the Progress of the Plague in the Bombay Presidency from September* 1896 *to June 1899* (Bombay, 1900), 113
25 HA 325.03.345.6; instructions on use of Haffkine's prophylactic plague serum
26 Haffkine's own account of these experiments (along with all his submissions to the Indian government) can be found in the collection of documents, many concerning other Indian cities like Calcutta and Surat, assembled by R. Nathan I.C.S., *The Plague in India, 1896–7* (Simla, 1898)
27 Haffkine, 'Report on inoculations in the [Portugese] Goan district of Lower Damaun (Daman)', in Nathan, 38. Goa, which had a long, mixed experience in training physicians in persuading (or failing to persuade) its native population to accept smallpox vaccination rather than the older practices of inoculation, provided a pool of nurses and doctors for neighbouring British India as well. Cristiana Bastos's scrupulously researched and nuanced work – 'Borrowing, Adapting and Learning the Practices of Smallpox: Notes from Colonial Goa', *Bulletin of the History of Medicine*, 83, 2009, 146–162 – resists what she takes to be the over-simplified polarisation between colonial imposition of smallpox vaccination and native resistance or uniform preference for variolation. The open-minded experience of her major figure in the 1850s, Dr Eduardo Freitas de Almeida, has many echoes in the seriousness with which Haffkine would pay attention to native Indian attitudes and practices and his belief in the indispensability of Indian assistants and colleagues.
28 W. M. Haffkine, 'A Discourse on Preventive Inoculation', delivered to the Royal Society, 8 June 1899, *The Lancet,* 153 (3965), 24 June 1899, 669–667; Hawgood, 13; Barbara J. Hawgood, 'Waldemar Mordecai Haffkine: Prophylactic Vaccination Against Cholera and Bubonic Plague in British India', *Journal of Medical Biography,* vol 15, 2007, 9–19
29 Predictably, the very first automobile in Bombay was bought by Jamshedji Tata in 1898. Pneumatic tyres were available the next year.
30 A. Lustig and G. Galeotti, 'The Prophylactic and Curative Treatment of Plague', *British Medical Journal,* 26 January 1901, 206–208
31 Proust, *La Defense de l'Europe contre la Peste*, 372–449; Howard-Jones, *Scientific Background*, 78–80

32 Like Haffkine, Zabolotny, a fascinating figure in his own right, had participated in student politics at the Novorussiya University, had been imprisoned and remained a police suspect while pursuing micro-biological research.
33 Brigadier-General W. F. Gatacre, *Report on the Bubonic Plague in Bombay 1896–7* (Bombay, 1897), 76–78
34 The one major exception was Nasarwanji Choksy at Arthur Road who came to have more faith in Alessandro Lustig's serum and less in Haffkine's.
35 R. Nathan, *The Plague in India 1896–7* (Simla, 1898), vol 1, 35
36 Ibid
37 Gatacre, 214 ff
38 Sophia Jex-Blake's account, quoted in Catriona Blake, *The Charge of the Parasols: Womens' Entry into the Medical Profession* (London, 1990), 126
39 Ibid, 135
40 On the relationship of feminism and medical training, Antoinette Burton, *Burdens of History, British Feminists, Indian Women and Imperial Culture* (Chapel Hill, 1994); Ambalika Guha, 'The "Masculine Female": the Rise of Women Doctors in Colonial India 1870–1940', *Social Scientist*, 5 (6) May–June 2016, 49–64. On the first generation of women doctors, Samiksha Sehrawat, 'Feminising Empire: The Association of Medical Women in India and the Campaign to Found a Women's Medical Service', *Social Scientist*, 41, May–June 2013, 65–81
41 Sunil Pandya, *Medical Education in Western India: Grant Medical College and the Sir Jamsetjee Jejeebhoy's Hospital* (Newcastle-upon-Tyne, 2019), 329 ff
42 Walke is routinely described in the historical literature as Goan, but I can't find any evidence of this and her parental and sibling names suggest otherwise.
43 On the plague and inoculation in Dharwar district in 1898–9 (including the towns of Hubli and Gadag), HA 325.03.345.3; W. H. Haffkine, *Summarised Report of the Bombay Plague Research Laboratory, 1896–1902* (Bombay, 1903); Olive Renier, *Before the Bonfire* (Shipston-on-Stour, 1984)
44 Haffkine's celebration of Alice Corthorn's 'record' number of inoculations in a single day is in his speech to the Liverpool School of Tropical Medicine, 21 October 1907
45 Renier, op cit 18, 32
46 A. M. Corthorn and C. J. R. Milne, 'Plague in Monkeys and Squirrels', *Indian Medical Gazette*, March 1899, 34
47 HA 325.03.345.3. A full report on the experience of Dharwar district in 1898–9 was published in May 1901, with comparative figures on inoculated and un-inoculated populations as well as details of the procedures used.
48 A minority – but a very significant one – of Ismailis in Bombay were Bohra, believing in the self-concealment of the 21st Imam in the late 9th/early 10th century, thus rejecting the authority of the Aga Khanate.
49 Sultan Mahomed Shah, Aga Khan III, *The Memoirs of the Aga Khan: World Enough and Time* (New York, 1954)
50 Subhendu Mund, 'Colonialism and the Politics of Epidemiology: The Rise of Radical Nationalism in India', HAL archive ouvertes, hal-03350204;

Ian Catanach, 'Poona politicians and the plague', *South Asia Journal of South Asian Studies*, December 1984, 1–18

51 The classic account is S. Wolpert, *Tilak and Gokhale: Revolution and Reform in the Making of Modern India* (Berkeley and Los Angeles, 1961)

52 On Gokhale, Bal Ram Nanda, *Gokhale: The Indian Moderates and the British Raj* (Princeton, 2016)

53 Damodar Hari Chapekar, *Musings From Gallows: Autobiography of Damodar Hari Chapekar* (English translation) (Ajmer, 2021), 95

54 Alok Deshpande, 'Maharashtra to unlock history in prisons', *The Hindu*, 23 January 2021

55 An important account of the strikes of 1897–8 and their implications for structural change in labour-capital relations in the Bombay textile industry, Aditya Sarkar, 'The Tie that Snapped: Bubonic Plague and Mill Labour in Bombay 1896–8', *International Review of Social History*, 59 (2), June 2014, 181–214; see also Prashant Kindambi, 'Contestation and Conflict: Workers' Resistance and the "Labour Problem" in the Bombay Cotton Mills, *c.* 1898–1919', in Marcel van der Linden and Prabha Mohapatra (eds), *Labour Matters: Towards Global Histories: Studies in Honour of Sabyasachi Bhattacharya* (Delhi, 2009), 106

56 Sarkar, 188–95

57 HA 325.01.159; 'On Study' (speech)

58 James Knighton Condon, *The Bombay Plague, being a History of the Plague in the Bombay Presidency from September 1896 to June 1899* (Bombay, 1900); proceedings reprinted from *The Times of India*, 115–118

59 Hanhart, 134; Curzon Papers, Eur Fiii/158, 58c, 92d; Deepak Kumar, '"Colony" Under a Microscope: The Works of W. M. Haffkine', *Science, Technology and Society*, 1999, 4, 265

60 Nayana Goradia, *Lord Curzon: The Last of the British Moghuls* (Oxford, 1993), 150

61 Mridula Ramanna, *Health Care in the Bombay Presidency, 1896–1930* (New Delhi, 2012), 30

62 H. Bennett and W. B. Bannerman 'Inoculation of an entire community with Haffkine's Plague Vaccine', *Indian Medical Gazette*, June 1899

63 Bennett and Bannerman, op cit

64 HA 325.03.444.6; material related to staff problems

65 Hanhart, 121

66 HA 325.03.362; Haffkine's draft for articles and letter re: Malkowal Commission

CHAPTER EIGHT: Carbolic

1 [W. Haffkine] *Summarised Report of the Bombay Plague Research Laboratory 1896–1902* (Bombay, 1903), HA 325.03.426. Details of production, together with appeals for a greatly expanded staff and space are in Haffkine's 'Report on Manufacture', 1900, HA 325.03.345.4

2 HA 325.03.369.2; The Plague Research Laboratory
3 I am basing this later count on photographs of the staff at the PRL taken in 1901, HA 325.04.161.3
4 Hanhart, 122. Haffkine was (not unreasonably) much exercised over ensuring that his supply and maintenance of equipment could be ramped up to meet the sudden demand. HA 325.03.369–2
5 Myron Echenberg, *Plague Ports: The Global Urban Impact of Bubonic Plague, 1894–1901* (New York, 2010), 107–130
6 Anton Chekhov to A. C. Suvorin, 19 August 1899, *Complete Letters*, vol 8 (Moscow 1980), 242–243
7 Kavita Sivaramakrishnan, *Old Potions, New Bottles: Recasting Indigenous Medicine in Colonial Punjab, 1850–1945* (London, 2006); Sasha Tandon, *The Social History of Plague in Colonial Punjab* (New Delhi, 2015); Natasha Sarkar, 'Fleas, Faith and Politics: Anatomy of an Indian Epidemic, 1890–1925', PhD thesis, University of Singapore, 2011, online; Ian Catanach, 'Plague and the Indian Village 1896–1914' in Peter Robb (ed), *Rural India, Land, Power and Society Under British Rule* (London, 1983). W. Glen Liston, *The Causes and Prevention of Plague in India* (Bombay, 1908) supplies some details of indigenous medicine, seldom with much approval.
8 Ramanna, 12
9 Kumar, 263
10 Ira Klein, 'Plague, Policy and Popular Unrest in British India', *Modern Asian Studies*, 22 (4), 1988, 747
11 Waksman, 48
12 HA 325.03.426, *Summarised Report*, 16
13 Surgeon-General C. H. James, *Report on the Outbreak of Plague in Jullundur and Hoshiapur Districts in the Punjab 1897–98*, 1898, 133 ff
14 Ibid, 98
15 HA 325.01.162; speech at the Incorporated Liverpool School of Tropical Medicine
16 Ramanna, 12
17 Kumar, 255; Curzon to Hamilton, 5 November 1902; Curzon Papers, Mss Eur F, 111/201, 401
18 HA 325.03.348; Haffkine's evidence to the Malkowal Commission
19 HA 325.03.444.6; material related to staff problems
20 HA 325.03.444.6; material related to staff problems
21 HA 325.01.36; invitations to official ceremonies
22 Hamilton to Curzon, 4 December 1902, Kumar, 257
23 HA 325.03.364.9; Haffkine letter to the Under-Secretary for India, 21 May 1906
24 Ibid. Haffkine later noted, in his long letter to the Government of India during the battle for the truth, that the policy of segregation and disinfection had been acknowledged by Sandhurst's successor as Governor of Bombay, Lord Lexington, to have been a failure in arresting the plague and had been abandoned. This was not a universal view, especially among

the IMS, and almost certainly made more enemies for Haffkine.

25 HA 325.03.345.6; instructions on use of Haffkine's prophylactics
26 HA 325.03.362; Haffkine's draft for articles and letter re: Malkowal Commission
27 Letter to Godley HA 325.03.364.9; correspondence with the Under-Secretary of State for India re: Haffkine's return to India
28 Ibid
29 Ibid, Godley to Haffkine, 5 December 1906
30 HA 325.03.345.6; instructions on use of Haffkine's prophylactics
31 *British Medical Journal,* 1907, 277–278
32 Sir Ronald Ross, *Memoirs: With a Full Account of the Great Malaria Problem and its Solution* (London, 1923), 181
33 Ibid, 184
34 Ibid, 205
35 Ronald Ross, 'The Inoculation Accident at Mulkowal', *Nature,* 75, 21 March 1907, 486–487
36 Eli Chernin, 'Ross Defends Haffkine: The Aftermath of the Vaccine-Associated Mulkowal Disaster of 1902', *Journal of the History of Medicine and Allied Sciences,* 46 (2), April 1991, 207
37 Ibid, 214
38 *The Times,* 29 July 1907
39 Chernin, 207
40 HA 325.03.364.9, Haffkine to Charles Hobhouse, Under-Secretary of State for India, 12 November 1907
41 Chernin, 213
42 HA 325.01.35; election to the Calcutta Light Horse
43 HA 325. 01.107; diaries 1898–1901

CHAPTER NINE: Departures

1 On the tour of the USA, Hanhart, 179 ff
2 Haffkine, *A Plea for Orthodoxy* (New York, 1916), 13. Hanhart (185–469) gives an extensive, intensive and deeply illuminating reading of Haffkine's evolving Judaism, his growing dismay at reformed Judaism which he saw dominating Jewish life in the United States, and his complicated and uneasy relationship with Zionism.
3 Haffkine's tri-lingual journal of his visit to the USSR (Russian, French and English) is HA 325.01.139.1; an edited version was published in two parts as 'Une mission en Russie' in *Paix et Droit,* 1927. Also by the Russian writer Mark Popovsky, *The Story of Dr Haffkine* (Moscow, 1963), 132. The return to Odessa is based on conversations with Haffkine's grand-niece, Y. A. Havkina, and the son of Dr Yakov Bardakh, one of the founders of the Pasteur Institute in Odessa.
4 On Metchnikoff and Tolstoy, Anna A. Berman, 'Of Phagocytes and Men: Tolstoy's Response to Mechnikov and the Religious Purpose of Science',

Comparative Literature, 68 (3), 2016, 296–311; Stephen Lovell, 'Finitude at the Fin de Siècle: Il'ia Mechnikov and Lev Tolstoy on Death and Life', *The Russian Review,* 63 (2), April 2004, 296–316

5 Jean-Marc Cavaillon and Sandra Legout, 'Centenary of the death of Elie Metchnikoff: a visionary and an outstanding team leader', Institut Pasteur, *Microbes and Infections,* 18 (2016), 578

6 On Yaffe and the final years, Hanhart, 513–527

7 'Insanitary condition of the Jharia coalfields', June 1908, Bihar State Archives, Patna. Published by the Government of Bengal, Digitized Collection of Epidemic and Vaccination Papers

CHAPTER TEN: And in the end...

1 Stephanie Nolen, 'Cholera outbreaks surge worldwide as vaccine supply drains', *New York Times,* 31 October 2022. Global shortages of the vaccine have been exacerbated by the Indian subsidiary of the French pharmaceutical company, Sanofi, deciding to end production at the close of 2023, so that the sole dependable supplier is now the South Korean firm EuBiologics. The International Center for Diarrhoeal Research, which pioneered the modern version of the cholera vaccine, is based in Dhaka, Bangladesh, not far from Haffkine's original base in Kolkata.

2 On this, Tom Nichols, *The Death of Expertise: The Campaign Against Established Knowledge and Why it Matters* (Oxford, 2017)

3 If only this were mere hyperbole, but characterisations of vaccine science and scientists as agents of a Satanic conspiracy is commonplace in the demonising language of conspiracy theorists, especially among QAnon adherents.

4 Fox News Report, 23 August 2022

5 Andrew Bridgen, Conservative MP for North-West Leicestershire, was suspended from the party for saying that COVID vaccines were the worst crime against humanity 'since the Holocaust'.

6 Jamelle Bouie, 'Ron DeSantis likes his culture wars for a reason', *New York Times,* 31 January 2023

7 Annie Karni and Katie Thomas, 'Trump says he's taking hydroxychloroquine, prompting warning from health experts', *New York Times,* 18 May 2020

8 David R. Boulware et al, 'A Randomized Trial of Hydroxychloroquine as Postexposure Prophylactic for Covid-19', *New England Journal of Medicine,* 6 August 2020, 517–525. Laura Ingraham, in particular, and in the face of scientific conclusions, seemed to be obsessed with championing the drug's effectiveness.

9 Leon Caly et al, 'The FDA-approved drug Ivermectin inhibits the replications of SARS-CoV-2 in vitro', *Antiviral Research,* 178, June 2020

10 Christina Szalinski, 'Fringe doctors' groups promote Ivermectin despite lack of evidence', *Scientific American,* 29 September 2021

11 Newsmax Report, 28 November 2021
12 Carlson, 11 May 2021
13 Felicia Goodrum et al, 'Virology Under the Microscope – a Call for Rational Discourse', *Journal of Virology*, 97 (2), January 2023
14 David Cyranoski, 'Bat cave solves mystery of deadly SARS virus – and suggests new outbreak could occur', *Nature*, 1 December 2017
15 Russell D. Bicknell et al, 'Central Nervous System of a 310-million-year-old Horseshoe Crab: Expanding the Taphonomic Window for Nervous System Preservation', *Geology*, 49 (11), 2021, 1381–1385
16 F. B. Bangs, 'The Toxic Effect of a Marine Bacterium on *Limulus* and the Formation of Blood Clots', *Biological Bulletin*, 105, 361–362
17 See Deborah Cramer, *The Narrow Edge: A Tiny Bird, an Ancient Crab and an Epic Journey* (Yale, 2015); Jack Sargent, *Crab Wars: A Tale of Horseshoe Crabs, Bioterrorism and Human Health* (Boston, 2021)
18 Deborah Cramer, 'When the horseshoe crabs are gone, we'll be in trouble', *New York Times*, 19 February 2023, has recently drawn attention to the urgency of recombinant alternatives receiving approval by the non-profit US Pharmacopeia without the need for further, drawn-out testing. A precious ecosystem is breaking.

A SHORT GUIDE TO FURTHER READING

While most of this book is based on primary sources, a growing number of which can be found online, I have confined myself here to secondary-source literature published in English; for articles, works in other languages and primary sources, both printed and manuscript, see the chapter notes.

Environment, pandemics and history

The crucial connection between environmental change and the spread of infectious disease as an integral part of the history of colonialism was first canvassed by **David Arnold** in a book of essays well ahead of its time, ***The Problem of Nature: Environment, Culture and European Expansion*** (Oxford, 1996). The unnervingly prophetic and eloquent tour de force on modern zoonotic disease is **David Quammen, *Spillover: Animal Infections and the Next Human Pandemic*** (New York, 2012), matched by his book on the response to COVID-19, ***Breathless: The Scientific Race to Defeat a Deadly Virus*** (New York, 2021). **Peter Baldwin, *Fighting the First Wave: Why the Coronavirus Was Tackled So Differently Across the Globe*** (Cambridge, 2021) is an important survey of this crucial subject. **Nicholas A. Christakis, *Apollo's Arrow: The Profound and Enduring Impact of Coronavirus on the Way We Live*** (New York, Boston and London, 2020) by the Sterling

Professor of Social and Natural Science at Yale is magisterially sobering. **Mary Beth Pfeiffer, *Lyme: The First Epidemic of Climate Change*** (Washington and London, 2018) is the definitive account of the subject.

There is an enormous literature on the historical impact of infectious diseases. The book that first made me think about the subject (when I was an undergraduate in the 1960s) was the extraordinary ***Rats, Lice and History*** by the bacteriologist **Hans Zinsser**, gripping and sinister; the classic on the subject is **William H. McNeill, *Plagues and Peoples*** (New York, 1977); and germs are made to do a great deal of explanatory work in **Jared Diamond**'s epic of Darwinian historical sociology, ***Guns, Germs and Steel*** (New York, 1997). **Michael B. A. Oldstone, *Viruses, Plagues and History: Past, Present and Future*** (Oxford, 2020) comes at the subject from the point of view of a virologist. Three excellent recent additions to the synoptic literature are **Kyle Harper, *Plagues Upon the Earth: Disease and the Course of Human History*** (Princeton, 2021); **Frank M. Snowden, *Epidemics and Society from the Black Death to the Present*** (Yale, 2019) (a Yale Open Curriculum course); and **Joshua S. Loomis, *Epidemics: The Impact of Germs and their Power over Humanity*** (Westport, 2018).

The best account of the politics of epidemic response, still highly relevant, is **Arthur Allen, *Vaccine: The Controversial Story of Medicine's Greatest Lifesaver*** (New York and London, 2008).

Eighteenth-century smallpox and inoculation before Jenner

The classic, and still invaluable, work is **Genevieve Miller, *The Adoption of Inoculation for Smallpox in England and France*** (Philadelphia, 1957). On measuring the success of inoculation, **Peter Razzell, *The Conquest of Smallpox: The Impact of Inoculation on Smallpox Mortality in Eighteenth-Century Britain*** (London, 1977), and more recently the superb and rigorous work of **Andrea Rusnock, *Vital Accounts: Quantifying Health and Population in Eighteenth Century England and France*** (Cambridge, 2002). Rusnock has also helpfully edited and published ***The Correspondence of James Jurin,***

1684–1750: Physician and Secretary to the Royal Society (Amsterdam and Atlanta, 1996). All the key documents published by the Royal Society as its *Philosophical Transactions* are now, and invaluably, accessible online. Lady Mary Wortley Montagu's life is definitively canvassed in **Isobel Grundy,** ***Lady Mary Wortley Montagu: Comet of the Enlightenment*** (Oxford, 1999), though the popular biography by **Jo Willett,** ***The Pioneering Life of Mary Wortley Montagu, Scientist and Feminist*** (Yorkshire and Pennsylvania, 2021), is also excellent. For her conversion moment to inoculation, ***Lady Mary Wortley Montagu: The Turkish Embassy Letters*** (eds **Teresa Heffernan** and **Daniel O'Quinn**) (London and Peterborough, Ontario, 2013). The older biography by **Robert Halsband** and his edition of her correspondence is still very valuable. The story of the epidemic in Boston is well told in **Arthur W. Boylston, M.D.,** ***Defying Providence: Smallpox and the Forgotten Eighteenth-Century Medical Revolution*** (North Charleston, 2012). The Suttons and their extraordinary commercialisation of inoculation have been recently chronicled by **Gavin Weightman,** ***The Great Inoculator: The Untold Story of Daniel Sutton and his Medical Revolution*** (New Haven, 2020) and Thomas Dimsdale's extraordinary story in Russia is brilliantly told by **Lucy Ward,** ***The Empress and the English Doctor: How Catherine the Great Defied a Deadly Virus*** (London, 2022). **Alicia Grant,** ***Globalisation of Variolation: The Overlooked Origins of Immunity for Smallpox in the Eighteenth Century*** (London, 2019) has valuable contributions on Scandinavia, Russia and the Ottoman world. The disturbing role that smallpox (and deliberately engineered infections) played in colonial and revolutionary America is told in another modern classic, **Elizabeth Fenn,** ***Pox Americana: The Great Smallpox Epidemic of 1775–1782*** (New York, 2002).

States, societies, epidemics in Europe and the rise of public health in the nineteenth century

The definitive comparative account of governmental response to infectious diseases is **Peter Baldwin,** ***Contagion and the State in***

Europe, 1830–1930 (Cambridge, 1999); **Norman Howard Jones,** *The Scientific Background of the International Sanitary Conferences, 1851–1938* (Geneva, 1975), now digitally accessible, has a wealth of information on those pioneering but often politically frustrated international meetings dealing with cholera, yellow fever and bubonic plague. Many of the most illuminating studies are local, including the gripping tour de force of analytical narrative history, **Richard J. Evans,** *Death in Hamburg: Society and Politics in the Cholera Years* (London, 1987). Other important works are **Margaret Pelling,** *Cholera Fever and English Medicine, 1825–1918* (Oxford, 1965); **Michael Durey,** *The Return of the Plague: British Society and the Cholera of 1831–32* (London, 1979); **Francois Delaporte (tr. Arthur Goldhammer),** *Disease and Civilization: Cholera in Paris, 1832* (Cambridge, Mass., 1989). On the challenges to epidemic containment posed by the opening of the Suez Canal, **Valeska Huber,** *Channelling Mobilities: Migration and Globalisation in the Suez Canal Region and Beyond 1869–1914* (Cambridge, 2013).

Imperialism and epidemics, especially in India

That this is now a flourishing and prolific field of historical enquiry owes much to the pathbreaking research of two scholars, **David Arnold,** *Colonizing the Body: State Medicine and Epidemic Disease in Nineteenth-Century India* (Berkeley, Los Angeles and London, 1993) and **Mark Harrison,** *Public Health in British India: Anglo-Indian Preventive Medicine 1859–1914* (Cambridge, 1994). Anyone working in this field, including this author, is in their debt. They have also been responsible for many invaluable essay collections that have built an entire field including **David Arnold (ed),** *Imperial Medicine and Indigenous Societies* (Manchester, 1988); idem, with **Ramachandra Guha,** *Nature, Culture and Imperialism: Essays on the Environmental History of South Asia* (Oxford, 1995). **Myron Echenburg,** *Plague Ports: The Global Impact of Bubonic Plague 1894–1901* (New York and London, 2007) charts the global spread of the pandemic from Hong Kong and Bombay to Egypt, the Americas and Australia. More recently, the essays collected by **Robert Peckham,** especially his own introductions to ***Empires of Panic: Epidemics and Colonial Anxieties*** (Hong Kong, 2015) and with **David M. Pomfret,** ***Imperial Contagions: Medicine, Hygiene and Cultures of Planning in Asia*** (Hong Kong, 2013) have opened up stimulating new approaches through studies in collective social psychology and ethnography. Another bracingly fresh approach is taken by **Rohan Deb Roy,** ***Malarial Subjects: Empire, Medicine and Nonhumans in British India 1820–1909*** (Cambridge, 2017). On French colonial medicine, a brilliant study by **Aro Velmet,** ***Pasteur's Empire: Bacteriology and Politics in France, Its Colonies and the World*** (Oxford, 2020).

The microbiologists

The classic popular narrative by **Paul de Kruif,** ***Microbe Hunters*** (New York, 1926), though nearly a century old and inevitably on the 'triumphalist' side (and why not?), is still a wonderful read, as is **Luba Vikhanski,** ***Immunity: How Elie Metchnikoff Changed the Course of Modern Medicine*** (Chicago, 2016). Pasteur has been the subject of many biographies but **Gerald Geison,** ***The Private Science of Louis Pasteur*** (Princeton, 1995) was the first to take account of his private laboratory notebooks that seriously qualified the standard heroic story.

There are only two English-language biographies of Waldemar Haffkine, both brief, out of print and often unreliable: **Selman Waksman,** ***The Brilliant and Tragic Life of W. M. W. Haffkine, Bacteriologist*** (Chicago, 1964) and the Soviet writer **Mark Popovsky**'s ***The Story of Dr. Haffkine*** (Moscow, 1964). They have been replaced in French by the comprehensive monograph of **Joel Hanhart,** ***Waldemar Mordekhai Haffkine (1860–1930): Biographie Intellectuelle*** (Paris, 2016) which emphasises throughout Haffkine's Jewish identity and his evolving relationship with Judaism itself. Though our approaches differ somewhat, Dr Hanhart's monumental work has been exceptionally helpful in the writing of this book.

PICTURE CREDITS

p. 1: Anton Sorokin/Alamy
p. 11: Maurice Savage/Alamy
p. 20: Fabian von Poser/ImageBROKER/via Getty Images
p. 21: Chris Bjornberg/Science Photo Library
p. 23: Bridgeman Images
p. 58: Bridgeman Images
p. 70: Pictorial Press Ltd/Alamy
p. 79: © The Royal Society
p. 88: © The Royal Society
p. 105: Dartmouth College Electron Microscope Facility
p. 107: Courtesy of the author
p. 109: Apic/Hulton Archive/via Getty Images
p. 122: Bridgeman Images
p. 124: © Bibliothèque de l'Académie nationale de médecine. Bibliothèque de l'Académie nationale de médecine, Ms 1200 (2071) n° 35x0042
p. 139: © Conseil départemental d'Eure et Loir/Société des Amis de Marcel Proust
p. 142: Bibliothèque nationale de France
p. 143: The Bodleian Libraries, University of Oxford, MS. Minn 169 no. 4
p. 151: The Bodleian Libraries, University of Oxford, MS. Minn 202 no. 9

p. 158: Elie Metchnikoff. Photograph. Wellcome Collection
p. 164: B. Gottlieb. Haffkine Archive, National Library of Israel, W. M.Haffkine, ARC. Ms. Var 325 04 121
p. 172: A. A. Drozdovsky Collection/Odesa National Scientific Library
p. 173: Boissonnas. Haffkine Archive, National Library of Israel, W. M. Haffkine, ARC. Ms. Var. 325 04 122
p. 179: © Institut Pasteur/Archives Elie Metchnikoff – photo Eugène Pirou
p. 185: Reproduced by kind permission of Cambridge Antiquarian Society and the Syndics of Cambridge University Library (UA/CAS H62)
p. 191 left: Chris Hellier/Alamy
p. 191 right: © British Library Board. All Rights Reserved/ Bridgeman Images
p. 197: Haffkine Archive, National Library of Israel, Anticholera inoculation in Calcutta, ARC. Ms. Var. 325 04 179
p. 207: Haffkine Archive, National Library of Israel, Anticholera inoculations at Chaibassa, Bengal, ARC. Ms. Var. 325 04 165
p. 208: Lala Deen Dayal & Sons, The National Library of Israel, Waldemar Mordecai Wolff Haffkine archive, W. H. Haffkine, ARC. Ms. Var. 325 04 155
p. 209: Dr Gary Gaugler/Science Photo Library
p. 222: Hong Kong Museum of Medical Sciences Society
p. 224: Courtesy of Noonans
p. 226: By kind permission of Soldiers of Shropshire Museum
p. 230: Fukuzawa Memorial Center for Modern Japanese Studies, Keio University
p. 233: Courtesy of The Kitasato Institute
p. 235: Pascal Deloche/Godong/Universal Images Group/via Getty Images
p. 243: The Bombay plague epidemic of 1896–1897: work of the Bombay Plague Committee. Photographs attributed to Capt C. Moss, 1897. Wellcome Collection
p. 248: The Bombay plague epidemic of 1896–1897: work of

the Bombay Plague Committee. Photographs attributed to Capt C. Moss, 1897. Wellcome Collection

p. 249: The Bombay plague epidemic of 1896–1897: work of the Bombay Plague Committee. Photographs attributed to Capt C. Moss, 1897. Wellcome Collection

p. 250: The Bombay plague epidemic of 1896–1897: work of the Bombay Plague Committee. Photographs attributed to Capt C. Moss, 1897. Wellcome Collection

p. 251: The Bombay plague epidemic of 1896–1897: work of the Bombay Plague Committee. Photographs attributed to Capt C. Moss, 1897. Wellcome Collection

p. 252 both: The Bombay plague epidemic of 1896–1897: work of the Bombay Plague Committee. Photographs attributed to Capt C. Moss, 1897. Wellcome Collection

p. 253 both: The Bombay plague epidemic of 1896–1897: work of the Bombay Plague Committee. Photographs attributed to Capt C. Moss, 1897. Wellcome Collection

p. 255: The Bombay plague epidemic of 1896–1897: work of the Bombay Plague Committee. Photographs attributed to Capt C. Moss, 1897. Wellcome Collection

p. 272: Haffkine Archive, National Library of Israel, Operations against the plague, ARC. Ms. Var. 325 04 162

p. 274: Dr Saman Habib from her family album

p. 281 both: The Bombay plague epidemic of 1896–1897: work of the Bombay Plague Committee. Photographs attributed to Capt. C. Moss, 1897. Wellcome Collection

p. 285: The Bombay plague epidemic of 1896–1897: work of the Bombay Plague Committee. Photographs attributed to Capt. C. Moss, 1897. Wellcome Collection

p. 286 both: The Bombay plague epidemic of 1896–1897: work of the Bombay Plague Committee. Photographs attributed to Capt. C. Moss, 1897. Wellcome Collection

p. 288: © National Portrait Gallery, London

p. 291: Historic Collection/Alamy

p. 294: By permission of the British Library (X.809/65336)

p. 298 left: Reproduced with permission from the Sassoon Family Album, *Ashley Park*
p. 298 right: Mary Evans Picture Library
p. 300: Haffkine Archive, National Library of Israel, W. M. Haffkine, ARC. Ms. Var. 325 04 161
p. 318 top: Dr I. M. Gibson, Haffkine Archive, National Library of Israel, The Punjab Inoculation Campaign, ARC. Ms. Var. 325 04 236
p. 318 bottom: Dr Maitland Gibson, The National Library of Israel, Waldemar Mordecai Wolff Haffkine archive, The Punjab Inoculation Campaign, ARC. Ms. Var. 325 04 236
p. 320: Dr Maitland Gibson, Haffkine Archive, National Library of Israel, The Punjab Inoculation Campaign, ARC. Ms. Var. 325 04 236
p. 339: Mary Evans Picture Library
p. 347: A dissection of malariated mosquito, R. Ross. Wellcome Collection. Attribution 4.0 International (CC BY 4.0)
p. 357: The National Library of Israel, Waldemar Mordecai Wolff Haffkine archive, ARC. Ms. Var. 325 01 20
p. 377: Science History Images/Alamy
p. 385: Dr Waldemar Mordecai Haffkine Museum
p. 388: Haffkine Archive, National Library of Israel, Anticholera inoculations at Jharia, ARC. Ms. Var. 325 04 167
p. 401: Benjamin Jesty. Oil painting by M. W. Sharp, 1805. Wellcome Collection
p. 402: John Cancalosi/Alamy
p. 409: Andrew J. Martinez/Science Photo Library

INDEX

Page references in *italics* indicate images.